FUNDAMENTAL INTERACTIONS AND THE NUCLEUS

FUNDAMENTAL INTERACTIONS AND THE NUCLEUS

R. J. BLIN-STOYLE

Professor of Theoretical Physics, University of Sussex

1973

NORTH-HOLLAND PUBLISHING COMPANY, AMSTERDAM • LONDON
AMERICAN ELSEVIER PUBLISHING COMPANY, INC. – NEW YORK

Library of Congress Catalog Card Number: 72-93086
North-Holland ISBN: 0 7204 0268 9
American Elsevier ISBN: 0444 104046 1

Publishers:

NORTH-HOLLAND PUBLISHING COMPANY – AMSTERDAM
NORTH-HOLLAND PUBLISHING COMPANY LTD. – LONDON

Sole distributors for the U.S.A. and Canada

AMERICAN ELSEVIER PUBLISHING COMPANY, INC,
52 VANDERBILT AVENUE
NEW YORK, N.Y. 10017

PRINTED IN THE NETHERLANDS

To Peggy

PREFACE

Many studies in the area of what is conventionally known as low energy nuclear physics have over the years thrown light on different aspects of the fundamental interactions of elementary particles. The object of this book is to provide a fairly detailed, up-to-date account of these studies for the benefit of theoretical and experimental nuclear and elementary particle research physicists at all levels. It is hoped that the experimental nuclear physicist will find some incentive to carry out further experiments aimed at unravelling one or other aspect of the fundamental interactions, that the theoretical nuclear physicist will respond to the many issues and problems raised throughout the book which require his skills and that the elementary particle physicist will learn about the ways in which low energy nuclear physics processes can reveal significant information about the elementary particle interactions.

The book is pitched at a level which should enable a physicist who has been through, say, a first year postgraduate course in basic nuclear and elementary particle physics to understand virtually all sections. Nevertheless, because of its broad, and to some extent unconventional coverage, it has been thought important to give extensive references to original papers and reviews so that the reader can readily follow up work in any area which particularly interests him. Inevitably, also, the nuclear physics topics dealt with are selective, the continual criterion being that they can help with our understanding of the fundamental interactions.

The book clearly reflects the spread and aim of my own research interests and no apology is made for this. It also reflects the enthusiasm and interest in this broad field generated in me throughout the years by many teachers and colleagues and, in particular, I would like to acknowledge with pleasure the influence of Professor Sir Rudolf Peierls, Professor H. Primakoff, Professor V.F. Weisskopf and Professor D.H. Wilkinson. Special thanks are due to Professor W.E. Burcham for his continual encouragement and his kindness in reading the full manuscript and making many suggestions for its improvement. I am also indebted to Dr. D. Bailin and Dr. W.D. Hamilton for commenting on parts of the manuscript. Finally I wish to express my deep gratitude to Mrs. Sue Newman who, knowing nothing about nuclear physics, translated a highly illegible manuscript into a perfect typescript with remarkable speed.

R.J. Blin-Stoyle

CONTENTS

CHAPTER I

FUNDAMENTAL INTERACTIONS AND THE NUCLEUS

I-1. Introduction

The history of our developing understanding of the fundamental interactions of elementary particles is part and parcel of the history of nuclear physics. Early studies of the properties of nuclei led to the deduction that they were structures consisting of neutrons and protons interacting through a complicated strong and short-range nuclear force. The force was clearly spin dependent and it also emerged that it had essentially the same form and strength between two neutrons, a neutron and a proton or two protons when the two particles concerned were in the same quantum state. Hence arose the concept of isobaric spin which, of course, is now used to classify all strongly interacting elementary particles and which is conserved by the strong interactions.

Before the nucleus had even been identified as such the phenomena of α, β and γ-radioactivity were discovered. Subsequent investigations of the β-decay process established the existence of the neutrino and that the interaction responsible for β-decay was extremely weak compared with that responsible for nuclear forces. Later on it was also possible by precise measurements of β-decay phenomena to determine the detailed form of the β-decay interaction and, in particular, to establish that it does not conserve parity.

Similarly, studies of γ-decay processes and the electromagnetic properties of nuclei in general (e.g. nuclear magnetic moments) led to increasing understanding of the electromagnetic interactions of the neutron and proton.

Thus many of the basic and key factors in elementary particle physics have been determined from theoretical and experimental investigations of nuclear properties.

However, with the development of particle accelerators of higher and higher energies and sensitive particle detectors, the main stream of progress in elementary particle physics during the last two or three decades has not generally involved the nucleus. Attention has been devoted mainly to the creation of new particles and resonances and the study of their interactions and properties in the simplest of situations. Correspondingly in the field of low energy nuclear physics concern has been primarily with the study of the many-body problem that is presented by the nucleus. Highly sophisticated models and methods have been developed to account for the detailed properties of nuclear energy levels, transition probabilities, reaction phenomena and so forth and the nucleus has proved an excellent testing ground for, and generator of, many-body theories.

Nevertheless, although elementary particle physics and low energy nuclear physics are to a large extent now going their separate ways, nuclear studies can still provide significant and important information about some aspects of the fundamental interactions of elementary particle physics and it is with these things that this book is concerned. We now very briefly review current thinking about the fundamental interactions and then go on to consider why and how nuclear studies can throw light on them.

I-2. The fundamental interactions

To summarise the current status of elementary particle physics in a few paragraphs is an impertinence and, of course, an impossibility. The object of this section is, however, not quite this but rather, assuming that the reader has more than a nodding acquaintance with elementary particle physics, to introduce a few aspects of the fundamental interactions which need to be referred to later on and to establish some notation. We start from the basic assumption that most of the main features of elementary particle physics can be understood broadly in terms of the operation of three basic interactions – strong, electromagnetic and weak. In addition we recognise that it is sometimes necessary to pre-suppose other interactions (e.g. medium-strong, superweak) in order to account for some observed phenomena. We consider now some properties of these different interactions.

I-3. The strong interaction

The strong interactions are responsible for the main framework of elementary particle physics and our understanding of them has moved forward considerably during the last decade or so through theoretical studies of the analytic properties of scattering amplitudes (see e.g. Chew (1966); Eden *et al.* (1966)) and through the exploitation of the different symmetries attributed to these interactions. Early work in this latter respect had established that they were invariant under isospin transformations (SU2 symmetry) and that strongly interacting particles could be classified by charge (Q), 3-component of isospin (T_3), baryon number (B) and strangeness (S) and that these quantities were simply related as follows (Gell-Mann (1953); Nakano and Nishijima (1953))

$$\begin{aligned} Q &= T_3 + \tfrac{1}{2}(B+S) \\ &= T_3 + \tfrac{1}{2}Y \end{aligned} \tag{I-1}$$

where Y signifies hypercharge. A further major step forward was the recognition that the strong interactions approximately satisfy even higher symmetries, in particular SU3 (Goldberg and Ne'eman (1963)), and this led to many clarifications and a great increase in understanding (see e.g. Gell-Mann and Ne'eman (1964); Carruthers (1966)). The concept of SU3 symmetry also brought in its train the suggestion (Gell-Mann (1964a); Zweig (1964) that the basic representation (a triplet with $B = \frac{1}{3}$, splitting into a non-strange isodoublet ($T = \frac{1}{2}, S = 0$) and a strange isosinglet ($T = 0, S = -1$) with $Q = (\frac{2}{3}, -\frac{1}{3}, -\frac{1}{3})$) might correspond to a set of actual particles (quarks) with the above quantum numbers. So far such particles have not been seen.

Even so, the earlier Lagrangian formulation of strong interactions based on Yukawa type interactions has continued to play an important role in theories of strong interaction processes in spite of the difficulty of using perturbation theory. In this description the pion-nucleon interaction, for example, is represented by an effective Lagrangian $L_{\pi\mathrm{NN}}$ having the form *

$$L_{\pi\mathrm{NN}} = \mathrm{i}\, G_{\pi\mathrm{NN}} \bar{N} \gamma_5 \boldsymbol{\tau} \cdot \boldsymbol{\pi} \mathrm{N} \tag{I-2}$$

where $G_{\pi\mathrm{NN}}$ is the rationalised, renormalised pion-nucleon coupling constant ($G^2_{\pi\mathrm{NN}}/4\pi \approx 14.6$) and N and π are renormalised field operators for the nucleon and pion respectively. This Lagrangian is a scalar in isospin space and therefore conserves isospin. It also implies certain relationships between the specific proton-neutron-pion coupling constants, namely:

$$G_{\pi^0\mathrm{pp}} = -G_{\pi^0\mathrm{nn}} = 2^{-\frac{1}{2}} G_{\pi^+\mathrm{pn}} = 2^{-\frac{1}{2}} G_{\pi^-\mathrm{np}} \tag{I-3}$$

and reflects SU2 symmetry.

The recognition that the elementary particles could be classified into

* In general, but not invariably, units will be used for which $\hbar = c = 1$.

octets and decuplets according to an SU3 symmetry scheme led to an immediate generalisation of the simple Lagrangian given in eq. (I-2) of the form

$$L_{\mathrm{MBB}} = 2i\, G_{\mathrm{MBB}}\, \bar{\mathrm{B}} \gamma_5 [\alpha D_i + (1-\alpha) F_i] \mathrm{M}_i \mathrm{B} \tag{I-4}$$

where B and M refer to the baryon octet (p, n, Λ^{o}, $\Sigma^+ \Sigma^{\mathrm{o}} \Sigma^-$, Ξ^{o}, Ξ^-) and pseudoscalar meson octet (π^+, π^-, π^{o}, K^+, K^{o}, $\bar{\mathrm{K}}^{\mathrm{o}}$, K^-, η^{o}) respectively. D_i and F_i ($i = 1, \ldots, 8$) are 8 × 8 matrices (analogous to the τ_i) which arise in the SU3 scheme of things and α is a parameter measuring the relative contributions of the two types of coupling (symmetric and antisymmetric respectively). The above expression for L_{MBB} clearly implies further relations of the type given in eq. (I-3) between baryon-pseudoscalar meson coupling constants. Similar Lagrangians can also be constructed to describe interactions between other baryon-meson multiplets. As far as the work of this book is concerned the effects of strong interactions are almost invariably represented through interactions of the type given in eqs. (I-2) and (I-4). It should be remembered, in addition, that SU3 symmetry is not exact and that a medium strong SU3 symmetry breaking interaction has to be introduced in order to account more exactly for the experimental data (e.g. mass splitting *between* the different SU2 multiplets within an SU3 multiplet *).

Of particular interest for our purposes is the fact that using Lagrangians of the above form it has been found possible to calculate forms for the internucleon potential in terms of the exchange of different bosons which agree with experimental data on nucleon-nucleon scattering (see Signell (1969) for a review). This work generalises and up dates the original Yukawa approach to nuclear forces and will be touched upon in Chapters VIII, XI and XII.

I-4. The electromagnetic interaction

The considerable success of quantum electrodynamics in accounting for the properties of the electron-photon interaction is well known and the methods developed in this respect have been taken over to deal with the electromagnetic interactions of all elementary particles. However, in the case of hadrons, the strong interactions have important modifying effects and must clearly be taken into account.

The general electromagnetic Lagrangian density can be written in the

* This is analogous to the well-known mass splitting due to the electromagnetic interaction *within* an isospin (SU2) multiplet.

form

$$\mathcal{L}_{\text{e.m.}}(x) = e\mathcal{J}_\lambda^{(\text{e.m.})}(x)A_\lambda(x) \tag{I-5}$$

where $\mathcal{J}_\lambda^{(\text{e.m.})}$ is the total electromagnetic current and A_λ is the four-vector potential describing the electromagnetic field. $\mathcal{J}_\lambda^{(\text{e.m.})}$ is a conserved current and therefore satisfies the equation

$$\partial_\lambda \mathcal{J}_\lambda^{(\text{e.m.})} = 0\,. \tag{I-6}$$

Further $\mathcal{J}_\lambda^{(\text{e.m.})}$ can be decomposed into hadronic and leptonic parts each separately conserved, thus

$$\mathcal{J}_\lambda^{(\text{e.m.})} = J_\lambda + \ell_\lambda^{(\text{e.m.})}\,. \tag{I-7}$$

As far as the leptonic part is concerned there are contributions from both electrons and muons and the electron part, for example, has the explicit form

$$\ell_\lambda^{(\text{e.m.})} = \tfrac{1}{2}\mathrm{i}\,[\bar{\psi}_\mathrm{e}(x)\gamma_\lambda, \psi_\mathrm{e}(x)] \tag{I-8}$$

where ψ_e is a Dirac field function for an electron.

In the case of J_λ it is generally assumed that it consists of isoscalar and isovector parts and that its form can, in principle, be written down within the framework of Lagrangian field theory. However, because of strong interaction effects this is, in general, not particularly useful. Of more importance for our purposes is the use of invariance arguments to enable the matrix element of J_λ between hadronic states, for example, to be written down in terms of form factors (see App. F for details). Thus, between proton states, we have

$$\langle p'\,|J_\lambda(0)|\,p\rangle = \mathrm{i}(\bar{u}_{p'}\,|F_1^{(p)}(k^2)\gamma_\lambda - F_2^{(p)}(k^2)\,\sigma_{\lambda\nu}k_\nu\,|u_p) \tag{I-9}$$

where $u_{p'}$ and u_p are spinor functions for protons of four-momenta p' and p respectively. $F_{1,2}^{(\mathrm{p})}(k^2)$ are proton form factors which are functions of k^2 where k $(= p' - p)$ is the four-momentum transfer. In the limit of zero-momentum transfer $F_1^{(p)}$ and $F_2^{(p)}$ have the following values

$$\begin{aligned} F_1^{(p)}(0) &= 1 \\ F_2^{(p)}(0) &= \frac{\kappa_\mathrm{p}}{2M} \end{aligned} \tag{I-10}$$

where κ_p ($= \mu_\mathrm{p} - 1$) is the proton anomalous magnetic moment.

An exactly equivalent expression to eq. (I-9) can be written down for neutron states, namely

$$\langle n' | J_\lambda(0) | n\rangle = \mathrm{i}(\bar{u}_{n'} | F_1^{(n)}(k^2)\gamma_\lambda - F_2^{(n)}(k^2)\sigma_{\lambda\nu}k_\nu | u_n) \qquad \text{(I-11)}$$

where

$$F_1^{(\mathrm{n})}(0) = 0$$

$$F_2^{(\mathrm{n})}(0) = \frac{\kappa_\mathrm{n}}{2M} \qquad \text{(I-12)}$$

and κ_n is the neutron anomalous magnetic moment.

Further, eqs. (I-9) and (I-11) can be gathered together into a single equation using an isospin notation, thus:

$$\langle N' | J_\lambda(0) | N\rangle = \mathrm{i}(\bar{u}_{N'} | (F_1^{(\mathrm{s})}(k^2) + F_1^{(\mathrm{v})}(k^2)\tau_3)\gamma_\lambda$$
$$- (F_2^{(\mathrm{s})}(k^2) + F_2^{(\mathrm{v})}(k^2)\tau_3)\sigma_{\lambda\nu}k_\nu | u_N) \qquad \text{(I-13)}$$

where the isoscalar and isovector form factors are defined as follows

$$F_1^{(\mathrm{s})} = \tfrac{1}{2}(F_1^{(\mathrm{p})} + F_1^{(\mathrm{n})})\ ; \qquad F_2^{(\mathrm{s})} = \tfrac{1}{2}(F_2^{(\mathrm{p})} + F_2^{(\mathrm{n})})$$
$$F_1^{(\mathrm{v})} = \tfrac{1}{2}(F_1^{(\mathrm{p})} - F_1^{(\mathrm{n})})\ ; \qquad F_2^{(\mathrm{v})} = \tfrac{1}{2}(F_2^{(\mathrm{p})} - F_2^{(\mathrm{n})})\,. \qquad \text{(I-14)}$$

Explicitly, for $k^2 \to 0$, we have

$$F_1^{(\mathrm{s})}(0) = \tfrac{1}{2}\,, \qquad F_2^{(\mathrm{s})}(0) = \frac{\kappa_\mathrm{p} + \kappa_\mathrm{n}}{4M}$$
$$F_1^{(\mathrm{v})}(0) = \tfrac{1}{2}\,, \qquad F_2^{(\mathrm{v})}(0) = \frac{\kappa_\mathrm{p} - \kappa_\mathrm{n}}{4M}\,. \qquad \text{(I-15)}$$

The above expressions will be used at various points in this book (e.g. Chaps. III and XI) and clearly could be generalised for other baryons.

I-5. The weak interaction

The weak interaction is responsible for pure leptonic processes such as μ-decay ($\mu^- \to e^- + \tilde{\nu}_e + \nu_\mu$), strangeness conserving semi-leptonic decays (e.g. nuclear β-decay, $\pi^- \to \mu^- + \tilde{\nu}_\mu$), strangeness changing semi-leptonic decays (e.g. $\Lambda^0 \to p + e^- + \tilde{\nu}_e$, $K^+ \to \mu^+ + \nu_\mu$) and non-leptonic decays (e.g. $\Lambda^0 \to p + \pi^-$, $K^+ \to \pi^+ + \pi^0$). All such phenomena have been very successfully described in terms of a weak interaction theory involving self-interacting vector currents (see Bailin (1972) for a general review).

Basically the weak interaction Hamiltonian is taken to have the form

$$\mathcal{H}_W = \frac{-G}{2\sqrt{2}}\{J_\lambda, J_\lambda^*\}_+ \equiv \frac{-G}{2\sqrt{2}}[J_\lambda J_\lambda^* + J_\lambda^* J_\lambda] \tag{I-16}$$

where the vector current J_λ can be decomposed as follows:

$$J_\lambda = \cos\theta\, \mathcal{G}_\lambda + \sin\theta\, \mathcal{S}_\lambda + \ell_\lambda\,. \tag{I-17}$$

$\mathcal{G}_\lambda$ is a hadronic current conserving strangeness and can be further decomposed into polar vector and axial vector parts, thus

$$\mathcal{G}_\lambda = \mathcal{V}_\lambda + \mathcal{A}_\lambda\,. \tag{I-18}$$

$\mathcal{S}_\lambda$ is a strangeness changing hadronic current also comprising polar and axial vector parts, thus

$$\mathcal{S}_\lambda = \mathcal{S}_\lambda^{(V)} + \mathcal{S}_\lambda^{(A)}\,. \tag{I-19}$$

ℓ_λ is a leptonic current and is taken to have the explicit form

$$\begin{aligned}\ell_\lambda &= \ell_\lambda^{(e)} + \ell_\lambda^{(\mu)} \\ &= -i\bar{\psi}_{\nu_e}\gamma_\lambda(1+\gamma_5)\psi_e - i\bar{\psi}_{\nu_\mu}\gamma_\lambda(1+\gamma_5)\psi_\mu\end{aligned} \tag{I-20}$$

where ψ_ℓ is a field operator for the lepton indicated. Such a form ensures the additive conservation of electron-number (N_e) and muon-number (N_μ). Here it is conventional to allocate lepton-numbers as follows: (a) e^- and ν_e have $N_e = +1, N_\mu = 0$; (b) e^+ and $\tilde{\nu}_e$ have $N_e = -1, N_\mu = 0$; (c) μ^- and ν_μ have $N_e = 0, N_\mu = 1$; (d) μ^+ and $\tilde{\nu}_\mu$ have $N_e = 0, N_\mu = -1$. Finally θ is the Cabibbo angle (Cabibbo (1963)) whose value (≈ 0.2) is chosen to fit the experimental data *.

It is further generally accepted that $\mathcal{V}_\lambda$, $\mathcal{V}^*_\lambda$, $\mathcal{S}^{(V)}_\lambda$ and $\mathcal{S}^{(V)*}_\lambda$ are the charged members of an SU3 octet of *polar* vector currents $\mathcal{F}^{(i)}_\lambda$ (i = 1, ..., 8), thus

$$\begin{aligned}
\mathcal{V}_\lambda &= \mathcal{F}^{(1)}_\lambda + \mathrm{i}\,\mathcal{F}^{(2)}_\lambda \\
\mathcal{V}^*_\lambda &= \mathcal{F}^{(1)}_\lambda - \mathrm{i}\,\mathcal{F}^{(2)}_\lambda \\
\mathcal{S}^{(V)}_\lambda &= \mathcal{F}^{(4)}_\lambda + \mathrm{i}\,\mathcal{F}^{(5)}_\lambda \\
\mathcal{S}^{(V)*}_\lambda &= \mathcal{F}^{(4)}_\lambda - \mathrm{i}\,\mathcal{F}^{(5)}_\lambda \,.
\end{aligned} \tag{I-21}$$

The hadronic electromagnetic current, J_λ is also taken to be composed of neutral members of this same octet, thus

$$J_\lambda = \mathcal{F}^{(3)}_\lambda + \frac{1}{\sqrt{3}}\mathcal{F}^{(8)}_\lambda \,. \tag{I-22}$$

Here $\mathcal{F}^{(8)}_\lambda$ is the isoscalar part and $\mathcal{F}^{(3)}_\lambda$ the isovector part of J_λ. The latter, together with $\mathcal{F}^{(1)}_\lambda$ and $\mathcal{F}^{(2)}_\lambda$ form an isotriplet – the isospin current.

Further, the charge operator Q, the isospin operator T_k (k = 1, 2, 3) and the hypercharge operator Y are given by

$$\begin{aligned}
Q &= -\mathrm{i}\int J_4(\boldsymbol{x}, t)\, \mathrm{d}^3x \\
T_k &= -\mathrm{i}\int \mathcal{F}^{(k)}_4(\boldsymbol{x}, t)\, \mathrm{d}^3x \\
Y &= \frac{-2i}{\sqrt{3}}\int \mathcal{F}^{(8)}_4(\boldsymbol{x}, t)\, \mathrm{d}^3x
\end{aligned} \tag{I-23}$$

so that integrating the 4th component of eq. (I-22) over d^3x gives the Gell-Mann–Nishijima relation (eq. (I-1)) namely

$$Q = T_3 + \tfrac{1}{2}Y \,. \tag{I-24}$$

* Attempts to "explain" θ are summarized, for example, by Bailin (1972).

An additional consequence of the identification given in eqs. (I-21) and (I-22) is that to the extent that isospin is conserved

$$\partial_\lambda \mathcal{V}_\lambda = \partial_\lambda \mathcal{V}^*_\lambda = 0 \tag{I-25}$$

and in the exact SU3 limit

$$\partial_\lambda \mathcal{F}^{(i)}_\lambda = 0 \,. \tag{I-26}$$

This is, of course, the well established Conserved Vector Current hypothesis (CVC theory) (Feynman and Gell-Mann (1958); Gershtein and Zeld'ovitch (1958)) and the extent to which it is supported by nuclear data is discussed in detail in Chapter III.

The axial vector currents $\mathcal{A}_\lambda$, $\mathcal{A}^*_\lambda$, $\mathcal{S}^{(A)}_\lambda$ and $\mathcal{S}^{(A)*}_\lambda$ are also taken to be the charged members of an SU3 octet of *axial* vector currents $\mathcal{F}^{5(i)}_\lambda$ ($i = 1, ..., 8$), thus *

$$\begin{aligned} \mathcal{A}_\lambda &= \mathcal{F}^{5(1)}_\lambda + \mathrm{i}\,\mathcal{F}^{5(2)}_\lambda \\ \mathcal{A}^*_\lambda &= \mathcal{F}^{5(1)}_\lambda - \mathrm{i}\,\mathcal{F}^{5(2)}_\lambda \\ \mathcal{S}^{(A)}_\lambda &= \mathcal{F}^{5(4)}_\lambda + \mathrm{i}\,\mathcal{F}^{5(5)}_\lambda \\ \mathcal{S}^{(A)*}_\lambda &= \mathcal{F}^{5(4)}_\lambda - \mathrm{i}\,\mathcal{F}^{5(5)}_\lambda \,. \end{aligned} \tag{I-27}$$

It is further assumed (Gell-Mann (1962)), and this is one of the essential elements of current algebra (see Adler and Dashen (1968) for a general account of current algebra), that the following equal time commutation relations are satisfied

$$\begin{aligned} [F^{(i)}(t), F^{(j)}(t)] &= \mathrm{i}\, f^{ijk} F^{(k)}(t) \\ [F^{5(i)}(t), F^{5(j)}(t)] &= \mathrm{i}\, f^{ijk} F^{(k)}(t) \\ [F^{(i)}(t), F^{5(j)}(t)] &= \mathrm{i}\, f^{ijk} F^{5(k)}(t) \end{aligned} \tag{I-28}$$

* The superfix 5 symbolises the axial vector nature of $\mathcal{F}^{5(i)}_\lambda$ and the relation between $\mathcal{F}^{(i)}_\lambda$ and $\mathcal{F}^{5(i)}_\lambda$ is analogous to that between the Dirac operators γ_λ and $\gamma_5\gamma_\lambda$.

where * the vector and axial vector "charges" are defined as follows

$$F^{(i)}(t) = -\mathrm{i} \int \mathcal{F}_4^{(i)}(x, t)\, \mathrm{d}^3 x$$
$$F^{5(i)}(t) = -\mathrm{i} \int \mathcal{F}_4^{5(i)}(x, t)\, \mathrm{d}^3 x \tag{I-29}$$

and the f^{ijk} are the structure constants for the group SU3. In the exact SU3 limit the $F^{(i)}$ and $F^{5(i)}$ are time independent. Breaking of SU3 by the usual strong interaction leaves the $i = 1, 2, 3$ and 8 components of $F^{(i)}$ time independent, and breaking by the electromagnetic interaction leaves only the $i = 3$ and 8 components time independent. In addition, more restrictive (i.e. less generally valid), current commutation relations have been postulated and used. In the most extreme form these take the form of commutators between members of the two current octets $\mathcal{F}_\lambda^{(i)}$ and $\mathcal{F}_\lambda^{5(i)}$ and, for example, contain undetermined "Schwinger" terms (Schwinger (1959)). Once integrated forms of these commutation relations are less ambiguous and do not involve these terms.

The first two commutators are exact in the SU3 limit. However, SU3 is not an exact symmetry and they together with the third commutator are assumptions, justified to some extent by model field theories, but mainly by the success of the resulting physical predictions (see e.g. Marshak *et al.* (1968) and Bailin (1972) for some discussion of these aspects of current algebra).

A final important assumption about the $\mathcal{F}^{5(i)}$ and, in particular about $\mathcal{A}_\lambda$ is that it is "partially conserved". It is clear that it cannot be exactly conserved since this would forbid the $\pi_{\ell 2}$ decays (Taylor (1958)). However, Gell-Mann and Levy (1960) proposed that the divergence of $\mathcal{A}_\lambda$ is proportional to the pion field – the Partially Conserved Axial Vector Current Hypothesis (PCAC theory). This can be written formally as follows

$$\partial_\lambda \mathcal{A}_\lambda = c\phi_{\pi^-} \tag{I-30}$$

or, more generally

$$\partial_\lambda \mathcal{F}_\lambda^{5(i)} = cM^{(i)} \tag{I-31}$$

where c is a constant (see Chapter IV), the field operator ϕ_{π^-} annihilates a π^-

* Note that $F^{(1,2,3)}(t) = T_{1,2,3}$ where $\boldsymbol{T}$ is the isospin operator.

and the $M^{(i)}$ are field operators for the octet of pseudoscalar mesons. Some consequences of eq. (I-30) are discussed in Chapter IV.

Turning now to the basic weak interaction Hamiltonian (eq. (I-16)) it is clear that it contains parts which can account, in principle at least, for leptonic, semi-leptonic and non-leptonic processes. We now consider these briefly.

I-5.1. *Pure leptonic interactions*

The part of $\mathcal{H}_W$ involving only leptons has the form

$$\mathcal{H}_W^{(\ell)} = \frac{-G}{2\sqrt{2}}\{\ell_\lambda, \ell_\lambda^*\}_+ . \tag{I-32}$$

Of the different possible processes predicted by this interaction, only μ-decay (e.g. $\mu^- \to e^- + \tilde{\nu}_e + \nu_\mu$) has been studied in the laboratory and will be discussed here. In terms of the lepton field operators, the μ-decay Hamiltonian has the usual V–A four-fermion interaction form (substitute eq. (I-20) into eq. (I-32)),

$$\mathcal{H}_{\mu\text{-decay}} = \frac{-G}{\sqrt{2}}[(\overline{\psi}_{\nu_\mu}\gamma_\lambda(1+\gamma_5)\psi_\mu)(\overline{\psi}_e\gamma_\lambda(1+\gamma_5)\psi_{\nu_e}) + \text{h.c.}] . \tag{I-33}$$

An experimental study of μ-decay can thus enable a value for the universal weak coupling constant G to be determined. Unfortunately there is one complicating factor to be taken into account, namely the effect of electromagnetic radiative corrections (see §III-2.3 for a discussion of the more complicated, but equivalent corrections in β-decay). Since it is presumed that the quantum electrodynamics of electrons and muons are relatively well understood this should be fairly straightforward (see Kinoshita and Sirlin (1959)). However, there are some technical divergence problems and a little uncertainty remains (Matsson (1969); Roos and Sirlin (1971)). Using the most recent values of the fundamental constants (Taylor *et al.* (1969)) and Matsson's (1969) estimate of the radiative corrections, Blin-Stoyle and Freeman (1970) obtain

$$G = G_\mu = (1.4343 \pm 0.0007) \times 10^{-49} \text{ erg. cm}^3 . \tag{I-34}$$

On the other hand, using the work of Roos and Sirlin (1971) gives

$$G = G_\mu = (1.4354 \pm 0.0003) \times 10^{-49} \text{ erg. cm}^3 . \tag{I-35}$$

and an uncertainty of the order 0.1% exists. Clearly, though, $G \approx 1.4 \times 10^{-49}$ erg. cm^3 or, dimensionlessly, $G \approx 10^{-5}/M^2$ where M is the nucleon mass.

All other measurements on μ-decay (e.g. of energy spectrum, angular correlations etc.) are consistent with the V-A Hamiltonian given in eq. (I-33) (see Bailin (1972)) but, nevertheless, as pointed out by Jarlskog (1966), Fryberger (1968) and Derenzo (1969) the experimental data also allow substantial deviation from this form.

I-5.2. *Semi-leptonic interactions*

Returning to eq. (I-16) for the total weak interaction, two separate parts can be identified as responsible for semi-leptonic processes. These are

$$\mathcal{H}_{\mathrm{W}}^{(\mathrm{s.}\ell.:\Delta S=0)} = -\frac{G}{\sqrt{2}}\cos\theta\,[\mathcal{G}_\lambda \ell_\lambda^* + \mathrm{h.c.}] \tag{I-36}$$

$$\mathcal{H}_{\mathrm{W}}^{(\mathrm{s.}\ell.:\Delta S=\pm 1)} = -\frac{G}{\sqrt{2}}\sin\theta\,[\mathcal{S}_\lambda \ell_\lambda^* + \mathrm{h.c.}]\ . \tag{I-37}$$

$\mathcal{H}_{\mathrm{W}}^{(\mathrm{s.}\ell.:\Delta S=0)}$ is responsible for strangeness conserving semi-leptonic processes (e.g. nuclear β-decay, μ-capture, $\pi_{\ell 2}$, π_{e3}, $\Sigma \to \Lambda e\nu$ etc.) and will be studied extensively in later chapters. For example, the β-decay Hamiltonian can be written

$$\mathcal{H}_\beta = -\frac{G_\beta}{\sqrt{2}}\,[\mathcal{G}_\lambda \ell_\lambda^{(e)*} + \mathrm{h.c.}] \tag{I-38}$$

where G_β ($= G\cos\theta$) is the β-decay coupling constant and $\ell_\lambda^{(e)}$ is the "electron" part of the lepton current (see eq. I-20). A similar expression can be written down for the μ-capture Hamiltonian in which $\ell_\lambda^{(e)}$ is replaced by $\ell_\lambda^{(\mu)}$.

$\mathcal{H}_W^{(\mathrm{s.}\ell.:\Delta S=\pm 1)}$ is the part of $\mathcal{H}_W$ responsible for strangeness changing leptonic decays (e.g. $K_{\ell 2}$, $K_{\ell 3}$, $\Lambda \to Ne\nu$, $\Sigma \to Ne\nu$). Apart from the K_{e3} decay we shall not refer further to this type of decay. Suffice it to say here that as with the $\Delta S = 0$ part of $\mathcal{H}_W$, the experimental data is in excellent agreement with theory and, in particular with those aspects of the Hamiltonian deriving from the SU3 and current conservation assumptions (see e.g. Bailin (1972)).

I-5.3. *Non-leptonic interactions*

As in the previous section the non-leptonic interaction part of $\mathcal{H}_{\mathrm{W}}$ can be split into two parts according to whether strangeness is conserved or not.

Thus

$$\mathcal{H}_W^{(\mathrm{n.\ell.}:\Delta S=0)} = -\frac{G}{2\sqrt{2}}[\cos^2\theta\,\{\mathcal{J}_\lambda, \mathcal{J}_\lambda^*\}_+ + \sin^2\theta\,\{\mathcal{S}_\lambda, \mathcal{S}_\lambda^*\}_+] \tag{I-39}$$

$$\mathcal{H}_W^{(\mathrm{n.\ell.}:\Delta S=\pm 1)} = -\frac{G}{2\sqrt{2}}\sin\theta\,\cos\theta\,[\{\mathcal{J}_\lambda, \mathcal{S}_\lambda^*\}_+ + \{\mathcal{S}_\lambda, \mathcal{J}_\lambda^*\}_+]\,. \tag{I-40}$$

Effects generated by the $\Delta S = 0$ part, since it conserves strangeness, are virtually impossible to observe in elementary particle processes *per se* since they form a small ($\lesssim 10^{-6}$) background in what are overwhelmingly strong interaction processes. However, as is discussed in detail in Chaps. VIII and IX, the effects of $\mathcal{H}_W^{(\mathrm{n.\ell.}:\Delta S=0)}$ can be studied in nuclear phenomena where this part of $\mathcal{H}_W$ manifests itself as a weak parity violating internucleon potential.

The $\Delta S = \pm 1$ part is responsible for the mass of weak non-leptonic decay processes of elementary particles (e.g. $\Lambda \to N\pi$, $\Sigma \to N\pi$, $\Xi \to \Sigma\pi$, $K \to 2\pi$, 3π) and its success in this respect is well reviewed by Bailin (1972). Essentially the experimental data is consistent with the form of Hamiltonian given in eq. (I-40) although, because all the particles involved are strongly interacting there is much more theoretical uncertainty.

I-5.4. *The intermediate vector boson hypothesis*

The weak interaction theory just briefly discussed is based on the concept of self-interacting currents. Yukawa (1935) originally, and Schwinger (1957) more recently, suggested that the interaction might be propagated by a charged intermediate vector boson W of mass M_W. In this case the fundamental weak interaction is analogous to the electromagnetic interaction (see eq. (I-5)) and has the form

$$\mathcal{H}_W^{(\mathrm{IVB})} = f\,J_\lambda W_\lambda + \mathrm{h.c.} \tag{I-41}$$

where f is an appropriate coupling constant and W_λ describes a charged boson particle having unit spin. Eliminating the boson field to lowest order an effective weak interaction involving only the current J_λ is obtained and if M_W is allowed to approach infinity keeping f/M_W fixed then this Hamiltonian takes exactly the form given in eq. (I-16) with

$$\frac{f^2}{M_W^2} = \frac{G}{\sqrt{2}}. \tag{I-42}$$

There is currently no experimental evidence for the existence of the intermediate vector boson, although if it does exist, its mass is certainly greater than the kaon mass (to prevent the fast decay $K^+ \to W^+ + \gamma$). In any case a value of the mass $\lesssim M$ (the nucleon mass) would lead to detectable modifications of, for example, the μ-decay spectrum (Shaffer (1962); Bailin (1964, 1965); Derenzo (1969)). Further, the non-observation of the W in high energy neutrino-nucleus experiments (Block *et al.* (1966); Bernardini *et al.* (1966)) implies $M_W \gtrsim 2M$ and data on muon pairs (Lederman and Pope (1971)) imply $M_W \gtrsim 4.8\,M$. Theoretical arguments have also been given (e.g. Schwinger (1957); Salam and Ward (1959); Lee (1971); Weinberg (1971)) which attempt to unify electromagnetic and weak interaction phenomena and suggest that f and e are related. Lee, for example, obtained $f = (2\sqrt{2})^{-1}e$. Using eq. (I-42) and the known value of G (eqs. (I-34) and (I-35)) then gives $M_W = 37.29$ GeV $\approx 40M$. There is some slight evidence (see §III-2.6) which suggests that M_W might be in this region.

I-5.5. *Higher order effects of the weak interaction*

In all the foregoing discussions it has been assumed that because of the low value of G the weak interaction need only be treated in lowest order, and certainly this assumption seems to work for the processes considered. However, as was first pointed out by Heisenberg (1936) at high energies (100's of GeV) it becomes necessary to include higher order terms to avoid violating the unitarity principle. Further, with, for example, the four fermion interaction, higher order terms diverge and formal renormalisation procedures cannot be applied. Unfortunately, including the vector boson in the interaction, although it helps with the problem, does not solve it. There is here an outstanding problem (see Bailin (1972) for a review) but it has not inhibited the continual treatment of weak interaction processes in lowest order with a blind eye being turned towards this fundamental uncertainty.

I-6. *CP* violation *

It is now a well established part of elementary particle physics that in neutral kaon decay *CP*-invariance is not maintained. In brief, because weak interactions allow $K^0 \leftrightarrow \bar{K}^0$ transitions (in second order) only linear combinations of these states have definite lifetimes. These are conventionally

* *C*, *P* and *T* signify the charge conjugation, parity and time reversal operations respectively (see Apps. B, C and D).

referred to as K_S^o and K_L^o. If there were no *CP* violation K_S^o and K_L^o would be eigenstates of *CP* with eigenvalues + 1 and − 1 respectively *. This would then carry with it the further consequence that only the K_S^o can decay into two pions and the decay of the K_L^o into two pions would be completely forbidden. The experimental observation of this latter decay by Christenson *et al.* (1964) thus clearly established that *CP*-violation was taking place. The magnitude of the violation is indicated by the following experimental numbers

$$|\eta^{+-}| = \left|\frac{M(K_L^o \to \pi^+\pi^-)}{M(K_S^o \to \pi^+\pi^-)}\right| = (1.96 \pm 0.03) \times 10^{-3} \text{ (Particle Data Group (1972))}$$

$$|\eta^{oo}| = \left|\frac{M(K_L^o \to \pi^o\pi^o)}{M(K_S^o \to \pi^o\pi^o)}\right| = (2.09 \pm 0.12) \times 10^{-3} \text{ (Particle Data Group (1972))}$$

where $M(K \to \pi\pi)$ is the probability amplitude for the process indicated. Less precise information is available about the phases of η^{+-} and η^{oo}. In addition there is also evidence of *CP* violation from the $(K_L^o)_{e3}$ decays. Here it is useful to mention the so-called *CPT* theorem (Lüders (1954, 1957); Pauli (1955); Schwinger (1953, 1957)) which states that for all the usual forms of elementary particle interactions (in particular those described by a Lorentz invariant local Lagrangian theory) there is invariance under the combined *CPT* operation. Accepting this theorem it then follows that *CP*-violation implies *T*-violation.

To account for *CP*-violation various microscopic theories have been proposed none of which is absolutely ruled out and which can be summarised very briefly as follows (see e.g. Marshak *et al.* (1968); Bailin (1972) for more detailed accounts). Marshak *et al.* (1968) classify these theories according to the hypercharge selection rules satisfied by the *CP*-violating part of the interaction Hamiltonian (denoted by $\mathcal{H}_-$) namely $|\Delta Y| = 0$, 1 or 2. We consider the theories in turn.

* We would then also have $|K_S^o\rangle = |K_1^o\rangle = \frac{1}{\sqrt{2}}[|K^o\rangle - |\bar{K}^o\rangle]$ and $|K_L^o\rangle = |K_2^o\rangle = \frac{1}{\sqrt{2}}[|K^o\rangle + |\bar{K}^o\rangle]$. In general $|K_S^o\rangle = \frac{1}{\sqrt{|p|^2 + |q|^2}}[p|K^o\rangle - q|\bar{K}^o\rangle]$ and $|K_L^o\rangle = \frac{1}{\sqrt{|p|^2 + |q|^2}}[p|K^o\rangle + q|\bar{K}^o\rangle]$ where $\epsilon = (p-q)/(p+q)$ measures the magnitude of *CP*-violation.

I-6.1. $|\Delta Y| = 0$ *theories*

Here $\mathcal{H}_-$ is assumed to satisfy $\Delta Y = 0$, to conserve P and to violate C and T. CP violation then arises from interference between $\mathcal{H}_-$ and the usual weak interaction $\mathcal{H}_W$. The experimental values of η^{+-} and η^{oo} indicate that in this situation the dimensionless coupling constant, F, for $\mathcal{H}_-$ must be of the order 10^{-3}. $\mathcal{H}_-$ could therefore be regarded as a milli-strong interaction (e.g. Prentki and Veltman (1965); Cabibbo (1965); Lee and Wolfenstein (1965)) or, much more attractively (since $\alpha/2\pi \approx 10^{-3}$) as being connected with the electromagnetic interaction (e.g. Bernstein *et al.* (1965); Barshay (1965); Lee (1965a); Arzubov and Fillipov (1968)). This latter possibility is discussed in some detail in §X-3.

I-6.2 $|\Delta Y| = 1$ *theories*

In this case $\mathcal{H}_-$ has the same Y quantum number as parts of the usual weak interaction $\mathcal{H}_W$ and to account for the values of η^{+-} and η^{oo} its dimensionless coupling constant F must satisfy $F/GM^2 \approx 10^{-3}$. Such a theory is therefore referred to as milli-weak and implies CP-violating effects in weak processes in general. Various formulations have been attempted (Cabibbo (1964, 1965); Glashow (1965); Zachariasen and Zweig (1965); Nishijima and Schwank (1966, 1967a, b); Oakes (1968); Das (1968); Maiani (1968); Okubo (1968a)) and of particular interest for our purposes, since they have nuclear physics implications, are those which introduce phase angles between the V and A parts of the currents $\mathcal{J}_\lambda$ and $\mathcal{S}_\lambda$ in $\mathcal{H}_W$ (Glashow (1965); Oakes (1968); Das (1968)). The implications of this type of theory for nuclear β-decay are discussed in Chapt. V.

A more exotic approach was initiated by Nishijima and Swank (1966, 1967a, b) (see also Okubo (1968b)) which abandons the current-current theory of weak interactions and develops a new theory based on a fundamental CP-violating Hamiltonian which only contributes to real physical processes in 2nd (CP invariant), 3rd (CP violating) and higher orders. Its basic requirement therefore is that $F^2 \sim GM^2$ (in order to obtain the usual weak interaction strength from 2nd order terms – see §I-5.1) and $F^3 \approx 10^{-3}\,G$ to account for CP violation. There are problems with this type of theory, however, and it will not be considered further.

I-6.3. $|\Delta Y| = 2$ *theories*

Since $\mathcal{H}_-$ is a $|\Delta Y| = 2$ operator there is no contribution to the $|\Delta Y| = 1$ hadronic weak decays and CP-violation in the $K^o_L \to 2\pi$ process arises due to the CP-violating coupling of the K^o and $\bar{K}^o$ components of K^o_L through $\mathcal{H}_-$. Since $\mathcal{H}_W$ only connects K^o and $\bar{K}^o$ in second order it follows that a theory of this kind (Wolfenstein (1964); Lee and Wolfenstein (1965)) requires a strength F for $\mathcal{H}_-$ such that

$$\frac{F}{(GM^2)^2} \approx 10^{-3}$$

i.e. $F \approx 10^{-8} GM^2$, since $GM^2 \approx 10^{-5}$. It is therefore referred to as a superweak theory. Among its predictions are that $\eta^{+-} = \eta^{00}$ which is not inconsistent with present data. Since F is so small it further follows that the effect of $\mathcal{H}_-$ can only be observed within the neutral kaon complex and that, in particular, its observation in nuclear physics processes is ruled out.

I-7. Nuclear studies of the fundamental interactions

As has been pointed out earlier in this chapter, some of the major advances in our understanding of the fundamental interactions have derived from low-energy nuclear studies. It is the contention of this book that this is a continuing situation and that from the viewpoint of elementary particle physics there is still considerable profit to be derived from such studies.

There are various reasons which enable low energy phenomena in the nucleus to reflect important features of the fundamental interaction. Firstly, the component neutrons and protons of a nucleus experience all the interactions (strong, electromagnetic and weak) of elementary particle physics and so many properties of these interactions show up, albeit in a small way, in nuclear phenomena. Secondly, experimental techniques in low energy nuclear physics have been developed to such a level that extremely accurate measurements can be made. Such measurements then allow very small but highly significant effects to be detected (e.g. parity violation in gamma-decay processes). Thirdly, there is a very large number of nuclear species and states which can be investigated so that processes can be chosen which are particularly sensitive to the aspect of the interaction being studied (e.g. the existence of $0^+ \rightarrow 1^+$ allowed β-decay transitions clearly establishes the presence of the Gamow–Teller interaction). Fourthly, since the nucleus is a many body system, effects which do not appear in a simple two-body situation (e.g. 3-body forces) can be investigated. Fifthly, situations can be chosen in which particular effects are strictly forbidden (e.g. single photon γ-radiation in $0 \rightarrow 0$ transitions).

There is no doubt that some of the most exciting and fruitful contributions of nuclear physics to our understanding of elementary particle interactions have been in the area of the weak interaction. This started with the unravelling of the beta-decay interaction, the establishment of its $V-A$ character and of two-component neutrino theory. This was followed by studies of the nature of the polar and axial vector currents (CVC and PCAC theory) in the context of the semi-leptonic interaction specified in eq. (I-36). It is with matters such as this and the consequent ramifications and possibilities (e.g. exchange phenomena, charge dependent effects, more complex currents, time reversal violation, the nature of the lepton-current, double β-decay etc.) that the next few chapters (II–VI) are concerned. Finally, in the area of semi-leptonic processes Chapt. VII is devoted to μ-capture which although requiring relatively high energy accelerators to generate the muons is much concerned with problems in nuclear structure and throws particular light on some details of the polar and axial vector currents.

Still in the field of weak interaction physics the next two chapters (VIII and IX) study the manifestation of the strangeness conserving non-leptonic Hamiltonian $\mathcal{H}_{\mathrm{W}}^{(\mathrm{n.\ell.}:\Delta S=0)}$ given in eq. (I-39) in nuclear processes – particularly α-decay and electromagnetic phenomena. Here the effects considered are generated by the parity violating internucleon potential ($V_{\mathrm{p.v.}}$) stemming from $\mathcal{H}_{\mathrm{W}}^{(\mathrm{n.\ell.}:\Delta S=0)}$. Although these effects are small, they can throw considerable light both on the form of $\mathcal{H}_{\mathrm{W}}^{(\mathrm{n.\ell.}:\Delta S=0)}$ and on the strong interaction assumptions made in determining $V_{\mathrm{p.v.}}$ from this Hamiltonian. Indeed, at the present time, it seems very unlikely that studies other than those in the low energy nuclear physics context could be carried out with sufficient accuracy to help with investigations of this aspect of $\mathcal{H}_{\mathrm{W}}$.

Chapter X deals with the manifestation of T-violating effects in electromagnetic and reaction phenomena in low energy nuclear physics. Of course, as pointed out in §I-6.3 if CP (or T)-violation stems from a superweak interaction then there is no chance of observing CP-violating effects other than in the $\mathrm{K}^{\mathrm{o}} \leftrightarrow \overline{\mathrm{K}}^{\mathrm{o}}$ system. However, although this is a popular theory, it is by no means firmly established and it is very important that searches for T-violating effects in nuclear phenomena having a strength $\sim 10^{-3} - 10^{-4}$ should be carried out.

Turning now to the electromagnetic interaction, its basic fundamental form is believed to be well understood. However, the way it shows itself in conjunction with strong interactions (e.g. in the charge dependence of nuclear forces, in electromagnetic exchange phenomena, nucleon form factors) is a subject about which there are still many uncertainties. Aspects such as these are the concern of Chapt. XI which also investigates the extent to which nuclear studies could be sensitive to possible deviations of the electromagnetic current J_λ from its commonly assumed form. Here the interest is particularly in whether or not J_λ contains an isotensor component as well as the usual isoscalar and isovector parts (see eq. (I-22)).

Finally there are the strong interactions. Here the uncertainties of the many-body problem have made it difficult in the past to obtain much more than qualitative information from nuclear physics about restricted parts of this interaction (i.e. essentially those parts not involving strange particles) and major developments in this area have derived from conventional elementary particle physics. However, it now seems that nuclear studies can throw more light on some aspects of the internucleon interaction which cannot be obtained through the analysis of nucleon-nucleon scattering experiments. In Chapt. XII such issues as the shape and off-energy-shell behaviour of the two-nucleon potential are discussed together with the evidence available about 3-body forces.

Some of the issues discussed in these different chapters are of considerable importance in heightening our understanding of the fundamental interactions. Others are more peripheral, but taken together, and quite apart from the theoretical interest of the nucleus as a unique many-body system, they justify much more theoretical and experimental research in the field of low energy nuclear physics.

CHAPTER II

NUCLEAR β-DECAY

II-1. The physics of nuclear beta-decay

Nuclear β-decay is perhaps the most fully investigated manifestation of the fundamental weak interaction and it has been comprehensively treated in a number of review papers and textbooks (e.g. Morita (1963); Blin-Stoyle and Nair (1966); Schopper (1966); Wu and Moszkowski (1966); Konopinski (1966)). In this chapter it will be assumed that the reader is familiar with the standard theory of β-decay although for convenience and to establish a notation its major features are described in this section.

The three basic processes to be considered are

$$(A, Z) \to (A, Z+1) + e^- + \tilde{\nu}_e$$

$$(A, Z) \to (A, Z-1) + e^+ + \nu_e$$

$$(A, Z) + e_K^- \to (A, Z-1) + \nu_e$$

where (A, Z) signifies a nucleus of mass number A and charge number Z and where the three processes correspond to β^- decay, β^+ decay and K capture * respectively. For many years it has been well established that the main characteristics of these processes can be satisfactorily accounted for in terms of a phenomenological interaction Hamiltonian H_β, having the essential form

$$H_\beta = \sum_i \left[\left(\frac{G_V}{\sqrt{2}} \gamma_\lambda^{(i)} - \frac{G_A}{\sqrt{2}} \gamma_\lambda^{(i)} \gamma_5^{(i)} \right) t_+^{(i)} \right] [\bar{\phi}_e(r_i) \gamma_\lambda (1 + \gamma_5) \phi_{\nu_e}(r_i)] + \text{h.c.} \tag{II-1}$$

where $t_+^{(i)}$ is the + component of the isobaric spin operator for the ith nucleon and converts a neutron state into a proton state ($t_+ |n\rangle = |p\rangle$), $\gamma_\lambda^{(i)}$ and $\gamma_5^{(i)}$ are

* Including L, M, N, ... and other possible forms of electron capture.

the usual Dirac operators for the ith nucleon whilst γ_λ and γ_5 are Dirac operators for the leptons. G_{V} and G_{A} are the polar vector (Fermi) and axial vector (Gamow–Teller) coupling constants and are taken to be real. They have the following approximate values:

$$G_{\text{V}} \approx 1.4 \times 10^{-49} \text{ erg. cm}^3$$
$$G_{\text{A}}/G_{\text{V}} \approx -1.24\,. \tag{II-2}$$

$\phi_{\text{e}}(r_i)$ is the Dirac wave function for the emitted electron moving in the Coulomb field of the nucleus and the surrounding atomic electrons evaluated at the position of the ith nucleon. $\phi_{\nu_{\text{e}}}(r_i)$ is similarly the wave function for an absorbed neutrino.

The implications of the interaction (eq. II-1) are that the two-component theory of the neutrino is valid and that only vector (V) and axial vector (A) interactions contribute to β-decay (i.e. there are no scalar (S), tensor (T) or pseudoscalar (P) terms present in the interaction). This latter implication as we shall see, is a slight approximation but certainly the main features of energy spectra, angular correlations, polarizations etc. are well accounted for by H_β.

With such a theory, all experimental quantities are calculated in terms of the matrix element M_β of H_β taken between appropriate initial and final nuclear states (say $|\text{i}\rangle$ and $|\text{f}\rangle$). Thus, for β^- decay

$$M_\beta = \langle \text{f}|H_\beta|\text{i}\rangle$$
$$= M_{\text{V}} + M_{\text{A}} \tag{II-3}$$

where

$$M_{\text{V}} = \frac{G_V}{\sqrt{2}} \langle \text{f}| \sum_i t_+^{(i)} \gamma_\lambda^{(i)} [\overline{\phi}_{\text{e}}(r_i)\gamma_\lambda(1+\gamma_5)\phi_{\nu_{\text{e}}}(r_i)]\,|\text{i}\rangle$$

and

$$M_{\text{A}} = -\frac{G_A}{\sqrt{2}} \langle \text{f}| \sum_i t_+^{(i)} \gamma_\lambda^{(i)} \gamma_5^{(i)} [\overline{\phi}_{\text{e}}(r_i)\gamma_\lambda(1+\gamma_5)\phi_{\nu_{\text{e}}}(r_i)]\,|\text{i}\rangle\,.$$

Similar expressions can be written down for β^+ decay and K capture.

Now it is well known that M_{V} and M_{A} can be expanded into a series of terms with decreasing order of magnitude and characterised by related nuclear parity and angular momentum selection rules. These matrix elements correspond to the classification of β-transitions as "allowed", "first forbidden", "second forbidden" etc. The expansion derives from two factors. Firstly, some Dirac operations (e.g. γ_4 and $\gamma_k \gamma_5$ ($k = 1, 2, 3$)) have matrix elements of the order unity whilst others (e.g. γ_k and $\gamma_4 \gamma_5$) have matrix elements of the order v/c when taken between nuclear states, where v is a typical nucleon speed in the nucleus. Secondly, the lepton wave functions can be expanded in powers of kr_i where k is the lepton momentum and r_i, of course, has the order of magnitude $r_i \approx R$ where R is the nuclear radius.

Since for most β-decays v/c and kR are small (of the order a few tenths) it is generally only necessary to consider the largest terms in the matrix element expansion which can contribute to the transition. In an allowed transition for example, if the further (good) approximation is made of neglecting the variation of the lepton functions over the nuclear volume and treating these functions as constants then only two matrix elements are relevant. They are conventionally defined as

$$M_V = \langle \mathrm{f} | \sum_i t_\pm^{(i)} | \mathrm{i} \rangle \equiv \langle t_\pm \rangle \left(= \int 1\right) \tag{II-4}$$

$$M_A = \langle \mathrm{f} | \sum_i \boldsymbol{\sigma}^{(i)} t_\pm^{(i)} | \mathrm{i} \rangle \equiv \langle \boldsymbol{\sigma} t_\pm \rangle \left(= \int \boldsymbol{\sigma}\right) \tag{II-5}$$

and correspond to the lepton pair having total angular momenta 0 or $\hbar$ respectively. In eq. (II-5) $\boldsymbol{\sigma}^{(i)}$ is the 2×2 Pauli spin operator for the ith nucleon and constant terms deriving from the lepton functions and other sources are omitted. To obtain the above expressions for M_V and M_A the non-relativistic approximations $\gamma_4 \to 1$ and $-\mathrm{i}\gamma_k \gamma_5 \to \sigma_k$ have been made. The bracketed terms in eqs. (II-4) and (II-5) signify another frequently used notation for M_V and M_A.

For a first forbidden transition the situation becomes much more complicated since v/r and kr terms now have to be included. The following matrix elements are relevant

$$\text{(i)} \quad \langle \mathrm{f} | \sum_i t_\pm^{(i)} \boldsymbol{\alpha}^{(i)} | \mathrm{i} \rangle \left(= \int \boldsymbol{\alpha}\right) \qquad \text{(iv)} \quad \langle \mathrm{f} | \sum_i t_\pm^{(i)} \boldsymbol{\sigma}^{(i)} \times \boldsymbol{r}^{(i)} | \mathrm{i} \rangle \left(= \int \boldsymbol{\sigma} \times \boldsymbol{r}\right)$$

$$\text{(ii)} \quad \langle \mathrm{f} | \sum_i t_\pm^{(i)} \boldsymbol{r}^{(i)} | \mathrm{i} \rangle \left(= \int \boldsymbol{r}\right) \qquad \text{(v)} \quad \langle \mathrm{f} | \sum_i t_\pm^{(i)} \boldsymbol{\sigma}^{(i)} \boldsymbol{r}^{(i)} | \mathrm{i} \rangle \left(= \int B_{ij}\right)$$

$$\text{(iii)} \quad \langle \mathrm{f} | \sum_i t_\pm^{(i)} \boldsymbol{\sigma}^{(i)} \cdot \boldsymbol{r}^{(i)} | \mathrm{i} \rangle \left(= \int \boldsymbol{\sigma} \cdot \boldsymbol{r}\right) \quad \text{(vi)} \quad \langle \mathrm{f} | \sum_i t_\pm^{(i)} \gamma_5^{(i)} | \mathrm{i} \rangle \left(= \int \gamma_5\right). \tag{II-6}$$

The matrix elements (i) and (ii) stem from the polar vector part of H_β and the matrix elements (iii) – (vi) stem from the axial vector part. Again the bracketed terms signify an alternative notation which has been used by various authors – unfortunately with varying sign conventions. This latter point is discussed in detail by Wu and Moszkowski (1966; p. 91). Going to second and higher forbidden transitions even more matrix elements apply and the situation becomes exceedingly complex.

It is in terms of these matrix elements that any analysis of experimental β-decay results is made and considerable success has been achieved in giving an overall interpretation of the phenomena of β-decay in terms of them. This is particularly so for allowed transitions where there are only two matrix elements to consider and where, because of the simplicity of the operators arising, it has been possible to obtain accurate theoretical estimates of the matrix elements in some cases.

This then is phenomenological beta-decay theory at its simplest. We now go on to study the way in which this phenomenology is related to the basic theory of weak interactions set out in Chapt. I and to show how investigations within the field of beta-decay can give information about the detailed nature of the weak interaction.

II-2. **The beta-decay interaction**

In §I-5.2 it was shown that the part of the assumed fundamental weak interaction Hamiltonian density responsible for nuclear beta-decay has the form

$$\mathcal{H}_\beta(x) = -\frac{G_\beta}{\sqrt{2}}[\mathcal{J}_\lambda(x)\ell_\lambda^{(e)*}(x) + \text{h.c.}] \tag{II-7}$$

where $\mathcal{J}_\lambda$ is the hadron current consisting of a polar vector part $\mathcal{V}_\lambda$ and an axial vector part $\mathcal{A}_\lambda$, thus

$$\mathcal{J}_\lambda = \mathcal{V}_\lambda + \mathcal{A}_\lambda\,. \tag{II-8}$$

$\ell_\lambda^{(e)}$ is the electron part of the lepton current which is assumed to have the explicit form

$$\ell_\lambda^{(e)} = -\mathrm{i}\,\bar{\psi}_{\nu_e}(x)\,\gamma_\lambda(1+\gamma_5)\,\psi_e(x) \tag{II-9}$$

where ψ_ℓ is the field operator for the lepton ℓ (= ν_e, e); G_β is the beta-decay coupling constant which is related to the basic weak interaction coupling constant G and the Cabibbo angle θ by $G_\beta = G\cos\theta$.

Consider now for simplicity the matrix element $\mathcal{M}$ for the symmetrical process $n + \nu_e \to p + e^-$; it is given by

$$\mathcal{M} = \langle \mathrm{pe}^- | \int \mathcal{H}_\beta(x)\, \mathrm{d}^4x\, | \mathrm{n}\nu_e \rangle \tag{II-10}$$

where $|\mathrm{p}\rangle$, $|\mathrm{n}\rangle$, $|\mathrm{e}^-\rangle$, $|\nu_e\rangle$ are state vectors for the real physical particles indicated with four momenta p, n, e^- and ν_e respectively. Because of translational invariance $\mathcal{H}_\beta(x)$ can be written

$$\mathcal{H}_\beta(x) = \exp(-\mathrm{i}P\cdot x)\,\mathcal{H}_\beta(0)\exp(\mathrm{i}P\cdot x) \tag{II-11}$$

where the four vector P is the total energy-momentum operator. Substituting this expression into eq. (II-10) gives

$$\begin{aligned}\mathcal{M} &= \int \exp[-\mathrm{i}(p - n + e^- - \nu_e)\cdot x]\, \mathrm{d}^4x\, \langle \mathrm{pe}^- | \mathcal{H}_\beta(0) | \mathrm{n}\nu_e\rangle \\ &= (2\pi)^4 \delta(p - n + e^- - \nu_e) \langle \mathrm{pe}^- | \mathcal{H}_\beta(0) | \mathrm{n}\nu_e \rangle\,.\end{aligned} \tag{II-12}$$

Using eq. (II-7) for $\mathcal{H}_\beta$ this becomes:

$$\mathcal{M} = -\frac{G_\beta}{\sqrt{2}}(2\pi)^4\delta(p - n + e^- - \nu_e)\,\langle \mathrm{p}|\mathcal{J}_\lambda(0)|\mathrm{n}\rangle\,\langle \mathrm{e}^-|\ell_\lambda^{(e)*}(0)|\nu_e\rangle \tag{II-13}$$

there being no contribution for this process from the $\mathcal{J}_\lambda^* \ell_\lambda^{(e)}$ term in $\mathcal{H}_\beta$. Neglecting electromagnetic interactions for the moment, it follows from eq. (II-9) and Apps. C and D that apart from a normalisation factor the lepton matrix element can be written

$$\langle \mathrm{e}^-|\ell_\lambda^*|\nu_e\rangle = -\mathrm{i}(\bar{u}_{e^-}|\gamma_\lambda(1+\gamma_5)|u_{\nu_e}) \tag{II-14}$$

where u_{e^-} and u_{ν_e} are positive energy spinors for an electron of four momentum e^- and a neutrino of momentum ν_e.

In the case of $\langle \mathrm{p}|\mathcal{J}_\lambda(0)|\mathrm{n}\rangle$ the situation is complicated by the presence of strong interactions and this matrix element cannot be written down explicitly. However, because of the covariant structure of $\langle \mathrm{p}|\mathcal{J}_\lambda(0)|\mathrm{n}\rangle$ it is possible to write down its general form in a fairly straightforward fashion. This is done in the next section.

II-2.1. *The matrix element* $\langle \mathrm{p} | \mathcal{J}_\lambda(0) | \mathrm{n} \rangle$

It is convenient to consider the polar and axial vector parts of $\mathcal{J}_\lambda(0)$ separately. We therefore write

$$\langle \mathrm{p} | \mathcal{J}_\lambda(0) | \mathrm{n} \rangle = \langle \mathrm{p} | \mathcal{V}_\lambda(0) | \mathrm{n} \rangle + \langle \mathrm{p} | \mathcal{A}_\lambda(0) | \mathrm{n} \rangle . \tag{II-15}$$

Now in App. F it is shown how expressions for the above two matrix elements can be obtained in terms of different form factors. The following results are obtained

$$\langle \mathrm{p} | \mathcal{V}_\lambda(0) | \mathrm{n} \rangle = \mathrm{i}(\bar{u}_\mathrm{p} | f_\mathrm{V} \gamma_\lambda + f_\mathrm{W} \sigma_{\lambda\nu} k_\nu + \mathrm{i} f_\mathrm{S} k_\lambda | u_\mathrm{n}) \tag{II-16}$$

$$\langle \mathrm{p} | \mathcal{A}_\lambda(0) | \mathrm{n} \rangle = \mathrm{i}(\bar{u}_\mathrm{p} | -f_\mathrm{A} \gamma_\lambda \gamma_5 + \mathrm{i} f_\mathrm{P} \gamma_5 k_\lambda + f_\mathrm{T} \sigma_{\lambda\nu} k_\nu \gamma_5 | u_\mathrm{n}) \tag{II-17}$$

where u_p and u_n are Dirac spinors for the proton and neutron with four momenta p and n respectively. The form factors f_i (i = V, W, S, A, P, T) are in general complex functions of k^2 where $k = p - n$. However, it will emerge in §V-2 that if time-reversal invariance holds then the f_i's are real functions of k^2. For the present we shall take this to be the case.

Considering the form factors in turn, it is clear that f_V and f_A define the strength of the usual polar vector and axial vector couplings (see eq. (II-1) – hence the notation. The term f_W is analogous to the anomalous magnetic moment term arising in the coupling between nucleons and the electromagnetic field (see §XI-2) and is, by convention, referred to as weak magnetism. The term in f_S has the form of an *induced scalar* interaction as can easily be shown by considering its contribution to the total β-decay matrix element $\mathcal{M}$ given in eq. (II-13). This contribution has the form

$$-\frac{G_\beta}{\sqrt{2}}(2\pi)^4 \delta(p - n + e^- - \nu_\mathrm{e})(\bar{u}_\mathrm{p} | \mathrm{i} f_\mathrm{S} k_\lambda | u_\mathrm{n})(\bar{u}_\mathrm{e} | \gamma_\lambda(1 + \gamma_5) | u_\nu) . \tag{II-18}$$

Because of the δ-function k_λ $(= (p - n)_\lambda)$ can be replaced by $(\nu_\mathrm{e} - e^-)_\lambda$ and inserted into the lepton bracket. Further, by the Dirac equation

$$\begin{aligned} &\bar{u}_\mathrm{e} \gamma_\lambda e_\lambda^- = \mathrm{i} m_\mathrm{e} \bar{u}_\mathrm{e} \\ &\nu_{\mathrm{e}\lambda} \gamma_\lambda u_\nu = 0 \end{aligned} \tag{II-19}$$

so that using these results the above expression reduces to

$$-\frac{G_\beta}{\sqrt{2}}(2\pi)^4\,\delta\,(p-n+e^- -\nu_e)(\bar{u}_p\,|m_e f_S\,|u_n)(\bar{u}_e\,|(1+\gamma_5)\,|u_\nu) \qquad \text{(II-20)}$$

which has the form of a scalar interaction of strength $m_e f_S$.

The term in f_P gives a contribution to the total β-decay matrix element of the form

$$-\frac{G_\beta}{\sqrt{2}}(2\pi)^4\,\delta\,(p-n+e^- -\nu_e)(\bar{u}_p\,|\mathrm{i} f_P\gamma_5 k_\lambda\,|u_n)(\bar{u}_e|\gamma_\lambda(1+\gamma_5)|u_\nu) \qquad \text{(II-21)}$$

which, as in the case of the induced scalar interaction, reduces here to an *induced pseudoscalar* interaction

$$-\frac{G_\beta}{\sqrt{2}}(2\pi)^4\,\delta\,(p-n+e^- -\nu_e)(\bar{u}_p\,|\,m_e f_P\gamma_5\,|u_n)(\bar{u}_e|(1+\gamma_5)\,|u_\nu) \qquad \text{(II-22)}$$

of strength $G_P = m_e f_P G_\beta$.

Finally, the f_T term has the form of an *induced tensor interaction* and later (§V-3) will be shown to result from the presence of "second class" currents.

II-2.2. *The effective single nucleon β-decay Hamiltonian*

The discussion so far has been carried on in covariant language and for the decay of an isolated nucleon. On the other hand, in β-decay, the concern is generally with a complex nucleus consisting of many interacting nucleons and for which a covariant description is not available. The next step therefore is to rewrite the previous treatment in a way suitable for use with wave functions of the type occurring in standard nuclear physics.

A suitable starting point is the transition matrix element given in eq. (II-13) with $\langle \mathrm{p}|\mathcal{J}_\lambda(0)|\mathrm{n}\rangle$ replaced by the expressions given in eqs. (II-16) and (II-17) and with $\langle \mathrm{e}^-|\ell_\lambda^{(\mathrm{e})*}(0)|\nu_\mathrm{e}\rangle$ replaced by $-\mathrm{i}(\bar{u}_{\mathrm{e}^-}|\gamma_\lambda(1+\gamma_5)\,|u_{\nu_\mathrm{e}})$ as in eq. (II-14). Thus

$$\mathcal{M} = -\frac{G_\beta}{\sqrt{2}}(2\pi)^4\,\delta\,(p-n+e^- -\nu_e)\,(\bar{u}_p|f_V\gamma_\lambda - f_W\sigma_{\lambda\nu}q_\nu - \mathrm{i} f_S q_\lambda$$

$$- f_A\gamma_\lambda\gamma_5 - \mathrm{i} f_P\gamma_5 q_\lambda - f_T\sigma_{\lambda\nu}q_\nu\gamma_5\,|u_n)(\bar{u}_{e^-}|\gamma_\lambda(1+\gamma_5)\,|u_{\nu_e}) \qquad \text{(II-23)}$$

where, in eq. (II-23), because of the δ-function, $k = p - n$ has been replaced by the lepton momentum transfer $-q$, where $q = e^- - \nu_e$.

Next the 3-dimensional momentum dependent part of the δ-function, i.e. $\delta(\boldsymbol{p} - \boldsymbol{n} + \boldsymbol{e}^- - \boldsymbol{\nu}_e)$ is replaced by the integral

$$\delta(\boldsymbol{p} - \boldsymbol{n} + \boldsymbol{e}^- - \boldsymbol{\nu}_e) = \frac{1}{(2\pi)^3} \int \exp\left[-i(\boldsymbol{p} - \boldsymbol{n} + \boldsymbol{e}^- - \boldsymbol{\nu}_e) \cdot \boldsymbol{r}\right] d^3r \,. \qquad \text{(II-24)}$$

Finally, the different exponentials in the integral are combined with the spinors u_p, u_n, u_{e^-} and u_{ν_e} to form time independent plane wave solutions of the Dirac equation

$$\phi_p(\boldsymbol{p}, \boldsymbol{r}) = u_p \exp(i\boldsymbol{p} \cdot \boldsymbol{r}), \qquad \phi_n(\boldsymbol{n}, \boldsymbol{r}) = u_n \exp(i\boldsymbol{n} \cdot \boldsymbol{r})$$
$$\phi_{e^-}(\boldsymbol{e}^-, \boldsymbol{r}) = u_{e^-} \exp(i\boldsymbol{e}^- \cdot \boldsymbol{r}), \quad \phi_{\nu_e}(\boldsymbol{\nu}_e, \boldsymbol{r}) = u_{\nu_e} \exp(i\boldsymbol{\nu}_e \cdot \boldsymbol{r}) \,. \qquad \text{(II-25)}$$

This gives

$$\mathcal{M} = -\frac{G_\beta}{\sqrt{2}} 2\pi\delta(E_p - E_n + E_{e^-} - E_{\nu_e}) \int \overline{\phi}(\boldsymbol{p}, \boldsymbol{r}) \{ i\,[f_V \gamma_\lambda - f_W \sigma_{\lambda\nu} q_\nu$$
$$- i f_S q_\lambda - f_A \gamma_\lambda \gamma_5 - i f_P \gamma_5 q_\nu - f_T \sigma_{\lambda\nu} q_\nu \gamma_5] L^*_\lambda(\boldsymbol{r}) \} \phi(\boldsymbol{n}, \boldsymbol{r}) \, d^3r \qquad \text{(II-26)}$$

where $L^*_\lambda(\boldsymbol{r}) = -i\overline{\phi}_e(\boldsymbol{e}^-, \boldsymbol{r})\gamma_\lambda(1+\gamma_5)\phi_{\nu_e}(\boldsymbol{\nu}_e, \boldsymbol{r})$ and E_p, E_n etc. are the energies of proton, neutron, etc. The above expression can be written more simply as

$$\mathcal{M} = 2\pi\delta(E_p - E_n + E_{e^-} - E_{\nu_e}) M \qquad \text{(II-27)}$$

where

$$M = \int \phi^\dagger(\boldsymbol{p}, \boldsymbol{r}) H_{\beta^-}(\boldsymbol{r}) \phi(\boldsymbol{n}, \boldsymbol{r}) \, d^3r \qquad \text{(II-28)}$$

and

$$H_{\beta^-}(\boldsymbol{r}) = -\frac{G_\beta}{\sqrt{2}} \gamma_4 \, i\,[f_V \gamma_\lambda - f_W \sigma_{\lambda\nu} q_\nu$$
$$- i f_S q_\lambda - f_A \gamma_\lambda \gamma_5 - i f_P \gamma_5 q_\lambda - f_T \sigma_{\lambda\nu} q_\nu \gamma_5] L^*_\lambda(\boldsymbol{r}) \,. \qquad \text{(II-29)}$$

M can then be identified as the usual β-decay matrix element and $H_{\beta^-}(\boldsymbol{r})$ as the effective Hamiltonian for the process $n + \nu_e \to p + e^-$.

The next step is to reduce $H_{\beta^-}(\boldsymbol{r})$ to a non-covariant form so that it can be used with non-relativistic nuclear wave functions. This can most easily be accomplished by using a Foldy–Wouthuysen transformation (Foldy and Wouthuysen (1950); Rose and Osborn (1954); see also de Vries and Jonker (1968) and the many references therein). The free nucleon Hamiltonian is

$$H_N = \beta M + \boldsymbol{\alpha} \cdot \boldsymbol{P} \tag{II-30}$$

where $\boldsymbol{\alpha}$ and β are the usual Dirac operators and $\boldsymbol{P}$ the nucleon momentum. M is the nucleon mass and since electromagnetic effects are being neglected during this stage of the development we ignore the neutron–proton mass difference. Neglecting terms of order G_β^2, the total Hamiltonian $H = H_N + H_\beta$ can be made "even" (i.e. so as not to involve operators connecting "small" and "large" components of Dirac spinors) to order $1/M$ by means of successive transformations of the form $H' = \exp(\mathrm{i}S) H \exp(-\mathrm{i}S)$ where S is given by

$$\mathrm{i}S = \frac{\beta}{2M}(\boldsymbol{\alpha} \cdot \boldsymbol{P} + H_{\beta^-}^{(o)}) \tag{II-31}$$

and $\boldsymbol{\alpha} \cdot \boldsymbol{P}$ is the "odd" part of H_N and $H_{\beta^-}^{(o)}$ is the "odd" part of H_{β^-}. The transformed Hamiltonian then has the form

$$H' = H_N' + H_{\beta^-}' \tag{II-32}$$

with

$$H_N' = \frac{\beta P^2}{2M} + \beta M \tag{II-33}$$

and

$$H_{\beta^-}' = \frac{1}{2M}\beta\{\boldsymbol{\alpha} \cdot \boldsymbol{P}, H_\beta^{(o)}\}_+ + H_\beta^{(e)} \tag{II-34}$$

where $H_{\beta^-}^{(e)}$ is the "even" part of H_{β^-}.

Writing H_{β^-} (eq. (II-29)) in a form in which oddness and evenness of the Dirac operators is manifest, it is straightforward to show that

$$H_{\beta^-}^{(e)}(\boldsymbol{r}) = -\frac{G_\beta}{\sqrt{2}}[\mathrm{i}f_V L_4^*(\boldsymbol{r}) + \mathrm{i}f_W \beta \boldsymbol{\sigma} \times \boldsymbol{q} \cdot \boldsymbol{L}^*(\boldsymbol{r}) + f_S \beta q_\lambda L_\lambda^*(\boldsymbol{r}) + f_A \boldsymbol{\sigma} \cdot \boldsymbol{L}^*(\boldsymbol{r})$$

$$- \mathrm{i}f_T(\beta \boldsymbol{\sigma} \cdot \boldsymbol{q} L_4^*(\boldsymbol{r}) - \beta \boldsymbol{\sigma} \cdot \boldsymbol{L}^*(\boldsymbol{r}) q_4)] \tag{II-35}$$

$$H_{\beta^-}^{(\mathrm{o})}(\boldsymbol{r}) = -\frac{G_\beta}{\sqrt{2}}[f_\mathrm{V}\boldsymbol{\alpha}\cdot\boldsymbol{L}^*(\boldsymbol{r}) - \mathrm{i}f_\mathrm{W}(\beta\boldsymbol{\alpha}\cdot\boldsymbol{L}^*(\boldsymbol{r})q_4 - \beta\boldsymbol{\alpha}\cdot\boldsymbol{q}L_4^*(\boldsymbol{r}))$$

$$-\mathrm{i}f_\mathrm{A}\gamma_5 L_4^*(\boldsymbol{r}) + f_\mathrm{P}\beta\gamma_5 q_\lambda L_\lambda^*(\boldsymbol{r}) - \mathrm{i}f_\mathrm{T}\beta\boldsymbol{\alpha}\times\boldsymbol{q}\cdot\boldsymbol{L}^*(\boldsymbol{r})] \,. \qquad \text{(II-36)}$$

Using eqs. (II-34), (II-35) and (II-36) H'_{β^-} can then be calculated. Only even terms are retained and these are now correct to order $1/M$. H'_{β^-} is then reduced to two component spinor form by putting $\beta = +1$ and keeping only the "large" components. Thus, finally, the effective non-relativistic β^--decay Hamiltonian has the form (we now drop the prime on H_{β^-}):

$$H_{\beta^-}(\boldsymbol{r}) = -\frac{G_\beta}{\sqrt{2}}\left\{\left[\mathrm{i}(f_\mathrm{V} + f_\mathrm{S}E_0) - \mathrm{i}\left(\frac{f_\mathrm{A}}{2M} + \frac{f_\mathrm{P}}{2M}E_0 + f_\mathrm{T}\right)\boldsymbol{\sigma}\cdot\boldsymbol{q}\right.\right.$$

$$\left. + \mathrm{i}\frac{f_\mathrm{A}}{M}\boldsymbol{\sigma}\cdot\boldsymbol{P} + \mathrm{i}\frac{f_\mathrm{W}}{M}\left(\frac{q^2}{2} - \mathrm{i}\boldsymbol{\sigma}\cdot\boldsymbol{P}\times\boldsymbol{q}\right)\right]L_4^*(\boldsymbol{r})$$

$$+\left[\left(f_\mathrm{A} + f_\mathrm{T}\left(-E_0 + \frac{\boldsymbol{P}\cdot\boldsymbol{q}}{M} - \frac{(\boldsymbol{q})^2}{2M}\right)\right)\boldsymbol{\sigma} - \frac{\mathrm{i}}{2M}(f_\mathrm{V} - f_\mathrm{W}(2M + E_0))\,\boldsymbol{\sigma}\times\boldsymbol{q}\right.$$

$$-\mathrm{i}f_\mathrm{W}\frac{E_0}{M}\boldsymbol{\sigma}\times\boldsymbol{P} + \frac{1}{2M}(f_\mathrm{T} - f_\mathrm{P})\boldsymbol{\sigma}\cdot\boldsymbol{q}\boldsymbol{q} - \frac{f_\mathrm{T}}{M}\boldsymbol{\sigma}\cdot\boldsymbol{q}\boldsymbol{P}$$

$$\left.\left. - \left(\frac{f_\mathrm{V}}{2M} - \frac{f_\mathrm{W}E_0}{2M} - f_\mathrm{S}\right)\boldsymbol{q} + \frac{f_\mathrm{V}}{M}\boldsymbol{P}\right]\cdot\boldsymbol{L}^*(\boldsymbol{r})\right\} \qquad \text{(II-37)}$$

where q_4 has been written $\mathrm{i}E_0$ and E_0 is the energy transfer.

In fact, all energy and momentum dependent terms in eq. (II-37) are small compared with the leading terms. This is because they are of the order q/M or P/M or because, as will be seen later, the corresponding form factors are small. For the moment we shall therefore ignore them although subsequently they will be found to be of importance in discussing the fine details of the β-decay interaction. Neglecting them, gives

$$H_{\beta^-}(\boldsymbol{r}) = -\frac{G_\beta}{\sqrt{2}}[\mathrm{i}f_\mathrm{V}L_4^*(\boldsymbol{r}) + f_\mathrm{A}\boldsymbol{\sigma}\cdot\boldsymbol{L}^*(\boldsymbol{r})] \,. \qquad \text{(II-38)}$$

Finally, if instead of using specific neutron and proton states we use an isospin notation (see App. E) it is clear that $H_{\beta^-}(\boldsymbol{r})$ should be written

$$H_{\beta^-}(\boldsymbol{r}) = -\frac{G_\beta}{\sqrt{2}}[\mathrm{i} f_\mathrm{V} L_4^*(\boldsymbol{r}) + f_\mathrm{A}\, \boldsymbol{\sigma} \cdot \boldsymbol{L}^*(\boldsymbol{r})]\, t_+ \,. \qquad \text{(II-39)}$$

In the above expressions for $H_\beta(\boldsymbol{r})$ the form factors f_V and f_A which are functions of the momentum transfer k should be evaluated for the value of k relevant to the β-decay process under consideration. However, in the spirit of the approximation that terms of the order k/M are negligible it is sufficient to evaluate f_V and f_A at $k = 0$.

Comparing $H_{\beta^-}(\boldsymbol{r})$ (eq. (II-39)), for example, with the usual phenomenological β-decay Hamiltonian for a single nucleon (see eq. (II-1)) it is clear that f_V and f_A can be simply related to the usual polar vector and axial vector coupling constants as follows

$$G_\mathrm{V} = f_\mathrm{V}(0)\, G_\beta \,, \qquad G_\mathrm{A} = f_\mathrm{A}(0)\, G_\beta \qquad \text{(II-40)}$$

where the fact that f_V and f_A are evaluated at $k = 0$ has been made explicit. Thus $H_{\beta^-}(\boldsymbol{r})$ can be written

$$H_{\beta^-}(\boldsymbol{r}) = -\frac{1}{\sqrt{2}}[\mathrm{i}\, G_\mathrm{V} L_4^*(\boldsymbol{r}) + G_\mathrm{A}\, \boldsymbol{\sigma} \cdot \boldsymbol{L}^*(\boldsymbol{r})]\, t_+ \,. \qquad \text{(II-41)}$$

Now the above Hamiltonian was deduced for the symmetrical process

$$\mathrm{n} + \nu_\mathrm{e} \rightarrow \mathrm{p} + \mathrm{e}^-$$

and we have (see eq. (II-26)) $L_\lambda^* = -\mathrm{i}\bar{\phi}_\mathrm{e}(\boldsymbol{e}^-, \boldsymbol{r})\gamma_\lambda(1+\gamma_5)\phi_{\nu_\mathrm{e}}(\boldsymbol{\nu}_\mathrm{e}, \boldsymbol{r})$. We now have to consider briefly the form which it takes for the standard β^-, β^+ and electron-capture processes.

For the β^- decay process $\mathrm{n} \rightarrow \mathrm{p} + \mathrm{e}^- + \tilde{\nu}_\mathrm{e}$ the lepton bilinear term L_λ^* in eqs. (II-38) and (II-39) must be written

$$L_\lambda^* = -\mathrm{i}\bar{\phi}_\mathrm{e}(\boldsymbol{e}^-, \boldsymbol{r})\gamma_\lambda(1+\gamma_5)\phi_{\nu_\mathrm{e}}^{(-)}(-\tilde{\boldsymbol{\nu}}_\mathrm{e}, \boldsymbol{r}) \qquad \text{(II-42)}$$

where $\phi_{\nu_\mathrm{e}}^{(-)}(-\tilde{\boldsymbol{\nu}}_\mathrm{e}, \boldsymbol{r})$ is a negative energy spinor for a neutrino of momentum $-\tilde{\boldsymbol{\nu}}_\mathrm{e}$ and $\tilde{\boldsymbol{\nu}}_\mathrm{e}$ is the momentum of the emitted anti-neutrino. Thus $\phi_{\nu_\mathrm{e}}^{(-)}(-\tilde{\boldsymbol{\nu}}_\mathrm{e}, \boldsymbol{r})$ has the form

$$\phi_{\nu_\mathrm{e}}^{(-)}(-\tilde{\boldsymbol{\nu}}_\mathrm{e}, \boldsymbol{r}) = \upsilon_{\nu_\mathrm{e}}(\tilde{\boldsymbol{\nu}}_\mathrm{e}, \boldsymbol{r})\, \mathrm{e}^{-\mathrm{i}\tilde{\boldsymbol{\nu}}_\mathrm{e} \cdot \boldsymbol{r}} \qquad \text{(II-43)}$$

where υ is defined in App. C. (eq. (C-39)).

For the β^+ decay process, $p \to n + e^+ + \nu_e$, we have to use the Hermitian conjugate of H_{β^-}, thus

$$H_{\beta^+}(\boldsymbol{r}) = -\frac{1}{\sqrt{2}}[\mathrm{i}\, G_\mathrm{V} L_4(\boldsymbol{r}) + G_\mathrm{A}\, \boldsymbol{\sigma} \cdot \boldsymbol{L}(\boldsymbol{r})]\, t_- \tag{II-44}$$

where, in an obvious notation,

$$L_\lambda = -\mathrm{i}\bar{\phi}_{\nu_\mathrm{e}}(\boldsymbol{\nu}_\mathrm{e}, \boldsymbol{r})\gamma_\lambda(1+\gamma_5)\phi_\mathrm{e}^{(-)}(-\boldsymbol{e}^+, \boldsymbol{r})\,. \tag{II-45}$$

Finally, for the electron capture process $p + e^- \to n + \nu_e$ the relevant Hamiltonian takes the form given in eq. (II-44) but with

$$L_\lambda = -\mathrm{i}\bar{\phi}_{\nu_\mathrm{e}}(\boldsymbol{\nu}_\mathrm{e}, \boldsymbol{r})\gamma_\lambda(1+\gamma_5)\,\phi_\mathrm{e}(\boldsymbol{e}, \boldsymbol{r}) \tag{II-46}$$

where $\boldsymbol{e}$ is the momentum of the captured electron.

II-2.3. *The effective many-nucleon β-decay Hamiltonian*

The expressions given above (eqs. (II-39) and (II-44)) for $H_{\beta^\pm}$, the effective β-decay Hamiltonian, refer to single particle processes. To generalise to the case of many (say A) nucleons in a nucleus is straightforward and yields

$$H_{\beta^-}(\boldsymbol{r}_1, ..., \boldsymbol{r}_\mathrm{A}) = -\frac{1}{\sqrt{2}}\sum_{i=1}^{\mathrm{A}}[\mathrm{i}\, G_\mathrm{V} L_4^*(\boldsymbol{r}_i) + G_\mathrm{A}\, \boldsymbol{\sigma}^{(i)} \cdot \boldsymbol{L}^*(\boldsymbol{r}_i)]\, t_+^{(i)} \tag{II-47}$$

$$H_{\beta^+}(\boldsymbol{r}_1, ..., \boldsymbol{r}_\mathrm{A}) = -\frac{1}{\sqrt{2}}\sum_{i=1}^{\mathrm{A}}[\mathrm{i}\, G_\mathrm{V} L_4(\boldsymbol{r}_i) + G_\mathrm{A}\, \boldsymbol{\sigma}^{(i)} \cdot \boldsymbol{L}(\boldsymbol{r}_i)]\, t_-^{(i)}\,. \tag{II-48}$$

The lepton functions L_λ and L_λ^* are now, however, no longer constructed from plane wave solutions of the Dirac equation but are evaluated in terms of spherical wave solutions which take account of the Coulomb field due to the nucleus and the surrounding atomic electrons.

A further complication which occurs in the many-nucleon case has to be considered. In making the generalisations given in eqs. (II-47) and (II-48) the implicit assumption is made that the individual nucleons are completely independent of one another and no allowance has been made for the fact that there are strong interactions *between* the nucleons. These interactions can manifest themselves in two ways. Firstly, the nucleons should no longer be treated as free and moving in plane wave states and account has to be taken of the internucleon potential. From the point of view of conventional nuclear physics this is most readily taken care of by using the beta-decay Hamiltonian with many body nuclear wave functions calculated assuming the presence of nuclear forces. In general this means using some more or less accurate model wave functions and involves all the uncertainties of nuclear structure theory.

The second way in which the internucleon strong interactions can show themselves is through the presence of exchange effects (Bell and Blin-Stoyle (1957)). These effects result in a modification of the actual beta-decay operator because of the proximity of one nucleon to another. Thus, in addition to the single particle terms in $H_{\beta\pm}$ (eqs. (II-47) and (II-48)) there will be two-, three- etc. body terms. Since these terms result from the exchange of mesons between the nucleons they are expected to have ranges typical of nuclear forces. Consideration is limited here to two-body terms only since, as with nuclear forces, three and higher body terms are confidently expected to be unimportant. In addition we take no account of momentum dependent terms. This is, of course, an approximation but because of the uncertainties which emerge in calculating exchange terms this refinement is neither possible nor, at this point in time, meaningful. Thus mesonic exchange terms are written down to the same approximation as the $H_{\beta\pm}$ given in eqs. (II-47) and (II-48).

Two types of term have to be considered: (i) those stemming from the polar vector current and having the same transformation properties as

$$-\frac{1}{\sqrt{2}}\sum_i \mathrm{i}\, G_{\mathrm{V}} L_4^*(r_i) t_+^{(i)}$$

and (ii) those stemming from the axial vector current and transforming like

$$-\frac{1}{\sqrt{2}}\sum G_{\mathrm{A}} \boldsymbol{\sigma}^{(i)} \cdot \boldsymbol{L}^*(r_i) t_+^{(i)} \,.$$

The detailed form of the phenomenological exchange operator has been discussed by Bell and Blin-Stoyle (1957) in the approximation that $L_\lambda^*(r_i)$ can be treated as a constant over the nuclear volume. This approximation is a good one for allowed beta-decay and will be discussed in §II-4. However, in order to achieve maximum generality and also because exchange effects are relevant to the μ-capture process, for which the approximation is not a good one, we here extend the results of Bell and Blin-Stoyle to the case where $L_\lambda^*(r_i)$ is not a constant.

Restricting consideration to the static, two-body exchange operator for the reasons given above, the problem is to write down the most general operator having the transformation properties implicit in (i) and (ii) and constructed from the variables $\boldsymbol{r}_{ij}$ $(=\boldsymbol{r}_i-\boldsymbol{r}_j)$, $L_\lambda(\boldsymbol{r}_i)$, $L_\lambda(\boldsymbol{r}_j)$, $\boldsymbol{\sigma}_i$, $\boldsymbol{\sigma}_j$, $\boldsymbol{t}_i$, $\boldsymbol{t}_j$. It is straightforward to show that the polar vector exchange operator must have the form (for β^- decay)

$$H_{\beta^-\mathrm{V}}^{\mathrm{exch.}} = -\mathrm{i}\frac{G_\mathrm{V}}{\sqrt{2}}\sum_{i<j}\{[f_\mathrm{I}+f_\mathrm{II}\boldsymbol{\sigma}^{(i)}\cdot\boldsymbol{\sigma}^{(j)}+f_\mathrm{III}S_{ij}](t_+^{(i)}+t_+^{(j)})\tfrac{1}{2}(L_4^*(\boldsymbol{r}_i)+L_4^*(\boldsymbol{r}_j))$$

$$+[f'_\mathrm{I}+f'_\mathrm{II}\boldsymbol{\sigma}^{(i)}\cdot\boldsymbol{\sigma}^{(j)}+f'_\mathrm{III}S_{ij}](t_+^{(i)}-t_+^{(j)})\tfrac{1}{2}(L_4^*(\boldsymbol{r}_i)-L_4^*(\boldsymbol{r}_j))\} \tag{II-49}$$

where the f's and f''s are all functions of r_{ij} and where

$$S_{ij}=\left[3\frac{\boldsymbol{\sigma}^{(i)}\cdot\boldsymbol{r}_{ij}\boldsymbol{\sigma}^{(j)}\cdot\boldsymbol{r}_{ij}}{r_{ij}^2}-\boldsymbol{\sigma}^{(i)}\cdot\boldsymbol{\sigma}^{(j)}\right]. \tag{II-50}$$

In the case of the axial vector current the corresponding exchange operator has the form

$$H_{\beta^-\mathrm{A}}^{\mathrm{exch.}} = -\frac{G_\mathrm{A}}{\sqrt{2}}\sum_{i<j}\left\{\left(\left[g_\mathrm{I}\boldsymbol{\sigma}^{(i)}\times\boldsymbol{\sigma}^{(j)}+g_\mathrm{II}\frac{\boldsymbol{r}_{ij}\boldsymbol{r}_{ij}\cdot(\boldsymbol{\sigma}^{(i)}\times\boldsymbol{\sigma}^{(j)})}{r_{ij}^2}\right](\boldsymbol{t}^{(i)}\times\boldsymbol{t}^{(j)})_+\right.\right.$$

$$+\left[h_\mathrm{I}(\boldsymbol{\sigma}^{(i)}-\boldsymbol{\sigma}^{(j)})+h_\mathrm{II}\frac{\boldsymbol{r}_{ij}\boldsymbol{r}_{ij}\cdot(\boldsymbol{\sigma}^{(i)}-\boldsymbol{\sigma}^{(j)})}{r_{ij}^2}\right](t_+^{(i)}-t_+^{(j)})$$

$$\left.+\left[j_\mathrm{I}(\boldsymbol{\sigma}^{(i)}+\boldsymbol{\sigma}^{(j)})+j_\mathrm{II}\frac{\boldsymbol{r}_{ij}\boldsymbol{r}_{ij}\cdot(\boldsymbol{\sigma}^{(i)}+\boldsymbol{\sigma}^{(j)})}{r_{ij}^2}\right](t_+^{(i)}+t_+^{(j)})\right)\cdot\tfrac{1}{2}(\boldsymbol{L}^*(\boldsymbol{r}_i)+\boldsymbol{L}^*(\boldsymbol{r}_j))$$

$$+\left(\left[h'_\mathrm{I}(\boldsymbol{\sigma}^{(i)}-\boldsymbol{\sigma}^{(j)})+h'_\mathrm{II}\frac{\boldsymbol{r}_{ij}\boldsymbol{r}_{ij}\cdot(\boldsymbol{\sigma}^{(i)}-\boldsymbol{\sigma}^{(j)})}{r_{ij}^2}\right](t_+^{(i)}+t_+^{(j)})\right.$$

$$\left.\left.+\left[j'_\mathrm{I}(\boldsymbol{\sigma}^{(i)}+\boldsymbol{\sigma}^{(j)})+j'_\mathrm{II}\frac{\boldsymbol{r}_{ij}\boldsymbol{r}_{ij}\cdot(\boldsymbol{\sigma}^{(i)}+\boldsymbol{\sigma}^{(j)})}{r_{ij}^2}\right](t_+^{(i)}-t_+^{(j)})\right)\cdot\tfrac{1}{2}(\boldsymbol{L}^*(\boldsymbol{r}_i)-\boldsymbol{L}^*(\boldsymbol{r}_j))\right\}$$

(II-51)

where the g's, h's etc. are functions of r_{ij}.

The corresponding operators $H_{\beta^+V}^{\text{exch.}}$ and $H_{\beta^+A}^{\text{exch.}}$ are obtained from eqs. (II-50) and (II-51) by making the replacements $t_+ \to t_-$, $L^* \to L$. It is also to be noted that in the limit that $L_\lambda(\boldsymbol{r})$ is independent of $\boldsymbol{r}$ the terms involving f', h' and j' vanish and the expressions for $H_{\beta\pm}^{\text{exch.}}$ revert back to the forms given by Bell and Blin-Stoyle (1957).

Summarising then, the total many body β-decay Hamiltonian to be used in the following pages in general has two parts to it – a sum of single body terms, $H_{\beta\pm}$, as given in eqs. (II-47) and (II-48) and a sum of two-body terms, $H_{\beta\pm}^{\text{exch.}}$, as given in eqs. (II-50) and (II-51).

II-3. **Lepton wave functions in β-decay**

In order to use the foregoing β-decay Hamiltonian in the framework of a nuclear physics calculation detailed expressions have to be obtained for the lepton functions $L_\lambda(\boldsymbol{r})$ and $L_\lambda^*(\boldsymbol{r})$ for the different possible decay processes. These functions are given in eqs. (II-42), (II-44) and (II-46) in terms of appropriate solutions ϕ_e, ϕ_{ν_e} etc. of the Dirac equation. As pointed out in the last section these solutions are of spherical wave form. Such solutions are discussed in many standard texts (see, in particular, the very full discussion with respect to β-decay by Konopinski (1966)) and have the general form

$$\phi = \phi_\kappa^\mu = \begin{pmatrix} g_\kappa(r)\,\chi_\kappa^\mu(\theta,\phi) \\ -\mathrm{i}f_\kappa(r)\,\chi_{-\kappa}^\mu(\theta,\phi) \end{pmatrix} \tag{II-52}$$

where

$$\chi_\kappa^\mu = \sum_m C(\ell\tfrac{1}{2}j;\mu-m\,m)\,Y_\ell^{\mu-m}(\theta\phi)\,\chi^m\,. \tag{II-53}$$

χ^m $(m=\pm\frac{1}{2})$ is a two component Pauli spinor, $C(\ell\frac{1}{2}j;\mu-m\,m)$ is a Clebsch–Gordan coefficient and κ takes the value ℓ for $j=\ell-\frac{1}{2}$ and $-\ell-1$ for $j=\ell+\frac{1}{2}$. The radial functions $g_\kappa(r)$ and $f_\kappa(r)$ satisfy the following two coupled differential equations

$$\begin{aligned} \frac{\mathrm{d}f_\kappa}{\mathrm{d}r} &= \frac{\kappa-1}{r}f_\kappa - (E-V-1)\,g_\kappa \\ \frac{\mathrm{d}g_\kappa}{\mathrm{d}r} &= -\frac{\kappa+1}{r}g_\kappa + (E-V+1)\,f_\kappa \end{aligned} \tag{II-54}$$

where $E\,(=\sqrt{1+p^2})$ is the total energy of the lepton and $V(r)$ the Coulomb potential experienced by it.

As far as the neutrino is concerned, $V(r) = 0$, and the radial solutions are simply spherical Bessel functions. For the charged lepton, however, the situation is much more complicated. The potential, in general, cannot be given a simple form and it must represent faithfully the effects of the nuclear charge distribution and the screening effects of the surrounding atomic electrons. This means that numerical methods have to be used in order to evaluate f_κ and g_κ. In addition, variations in the calculation of f_κ and g_κ are introduced according as one is dealing with β^- decay, β^+ decay or K-capture. A further difficulty also arises in the case of forbidden transitions which has been pointed out by de Raedt (1968) and taken up by Behrens and Bühring (1970, 1971a). This arises from the usual procedure of expanding the electron or positron radial functions in powers of r, similarly for the neutrino plane wave function, and then subsuming these different powers of r into the nuclear matrix elements (see §II-1). Unfortunately this approach is not completely satisfactory due to convergence difficulties and for this reason Behrens and Bühring (1971) propose a formulation of β-decay theory which involves expanding the radial wave functions "in powers of the mass and energy parameters of the charged lepton and the nuclear charge, while the coefficients of the expansion are still functions of r depending on the shape of the charge distribution". Since our concern in this book is overwhelmingly with allowed transitions we shall not adopt this approach which is, however, important in fitting experimental data in the case of forbidden transitions. Neither shall we embark on the details of the numerical calculations of f_κ and g_κ in the case of allowed transitions although detailed references will be given when we deal with the different specific processes.

II-4. Allowed β-decay

We are now in a position to obtain an expression for the transition probability of a β-transition. By a standard application of time dependent perturbation theory the probability of emission of a $\beta^\mp$ particle with an energy E and momentum p can be written

$$P(E)\,\mathrm{d}E = \frac{2}{\pi} \sum{}' \, |M_\beta|^2 \frac{E}{p} \mathrm{d}E \tag{II-55}$$

where M_β is the relevant matrix element and the sum $\sum{}'$ is taken over all possible lepton angular momentum states and over all final spin orientations of the final nucleus whilst the prime indicates an averaging over the spin orientations of the initial nucleus.

The matrix element M_β has the form

$$M_\beta = \langle \Psi_f | H_\beta | \Psi_i \rangle \tag{II-56}$$

where Ψ_i and Ψ_f are the wave functions for the initial and final nuclear states and where, for the moment, we use either eqs. (II-47) or (II-48) for H_β and do not include exchange effects.

Explicitly, then, for β^+ decay we can write

$$P(E)\,dE = \frac{2}{\pi}\frac{1}{2I_i + 1}$$

$$\times \sum_{\substack{M_i M_f \kappa_e \mu_e \\ \kappa_{\nu_e} \mu_{\nu_e}}} |\langle \Psi_{I_f}^{M_f} | \frac{1}{\sqrt{2}} \sum_i \left[i G_V (L_4(r_i))_{\kappa_{\nu_e}\kappa_e}^{\mu_{\nu_e}\mu_e} + G_A \boldsymbol{\sigma}^{(i)} \cdot (\boldsymbol{L}(r_i))_{\kappa_{\nu_e}\kappa_e}^{\mu_{\nu_e}\mu_e} \right]$$

$$\times\, t_-^{(i)} | \Psi_{I_i}^{M_i} \rangle |^2 \frac{E}{p}\, dE \tag{II-57}$$

where

$$(L_\lambda(r_i))_{\kappa_{\nu_e}\kappa_e}^{\mu_{\nu_e}\mu_e} = i\bar{\phi}_{\kappa_{\nu_e}}^{\mu_{\nu_e}}(r_i)\gamma_\lambda(1+\gamma_5)\,\phi_{\kappa_e}^{(-)\mu_e}(r_i) \tag{II-58}$$

and I_i, M_i and I_f, M_f refer to the spins of the initial and final nuclei. Equivalent expressions can clearly be written down for β^--decay and K-capture.

In the case of an allowed transition the total angular momentum quantum number of the lepton pair is either 1 (Gamow–Teller or axial vector transition) or 0 (Fermi or polar vector transition) and hence in eq. (II-57) only terms with $\kappa = \pm 1$ need be kept. The leading term in the lepton radial functions is then independent of r and it is in the spirit of the allowed approximation to treat the radial functions as constant and to evaluate them at $r = R$ where R is the nuclear radius. As we shall see later (§III-2.4) this approximation is too crude for some purposes. Accepting it for the moment, the expression for the probability of an allowed transition reduces, after some manipulation, to (see e.g. Konopinski (1966)):

$$P(E)\,\mathrm{d}E = \frac{1}{(2\pi)^3}\,[G_V^2\,|\langle t^\pm\rangle|^2 + G_A^2\,|\langle \boldsymbol{\sigma}\, t^\pm\rangle|^2]\,pE(E_0-E)^2\,F(Z,E)\,\mathrm{d}E \quad \text{(II-59)}$$

where the Fermi function $F(Z,E)$ is given by

$$F(Z,E) = \frac{1}{2p^2}\,[g_{-1}^2(R) + f_{+1}^2(R)] \quad \text{(II-60)}$$

and the following notation is used for the nuclear matrix elements *

$$\frac{1}{(2I_i+1)}\sum_{M_f M_i} |\langle \Psi_{I_f}^{M_f}|\sum_i t_\pm^{(i)}|\Psi_{I_i}^{M_i}\rangle|^2$$

$$= \frac{1}{(2I_i+1)}\sum_{M_f M_i} |\langle \Psi_{I_f}^{M_f}|T_\pm|\Psi_{I_i}^{M_i}\rangle|^2 = |\langle t_\pm\rangle|^2$$

$$\frac{1}{(2I_i+1)}\sum_{M_f M_i} |\langle \Psi_{I_f}^{M_f}|\sum_i \boldsymbol{\sigma}^{(i)} t_\pm^{(i)}|\Psi_{I_i}^{M_i}\rangle|^2 = |\langle \boldsymbol{\sigma}\, t_\pm\rangle|^2\,. \quad \text{(II-61)}$$

In eq. (II-59) E_0 is the endpoint energy of the beta-decay and T the total isospin operator.

Finally, integrating over the electron energy we obtain the usual expression

$$ft = \frac{K}{G_V^2\,|\langle t_\pm\rangle|^2 + G_A^2\,|\langle \boldsymbol{\sigma}\, t_\pm\rangle|^2} \quad \text{(II-62)}$$

where $K = 2\pi^3(\ln 2)\hbar^7/m^5c^4 = 1.23063 \times 10^{-94}$ c.g.s. units, t is the half life for the decay and the quantity f is given by

$$f(Z,E_0) = \int_1^{E_0} pE(E_0-E)^2 F(Z,E)\,\mathrm{d}E\,. \quad \text{(II-63)}$$

Eq. (II-62) is the usual expression for the ft-value of an allowed beta-decay.

* An alternative, but closely related, formalism in which β-decay transition amplitudes are expressed in terms of nuclear form-factors has been developed by Stech and Schülke (1964) (see also Schülke (1964)).

Clearly, it is essential to have accurate values for the Fermi function $F(Z, E)$ and over the years there have been a number of tabulations of this function e.g. Dzhelepov and Zyrianova (1956); Bhalla and Rose (1960a, 1961); Bhalla (1964); Bühring (1965); Bahcall (1966); Suslov (1966 and 1967); Blin-Stoyle and Nair (1967); Behrens and Bühring (1968); Behrens and Jänecke (1969). Disagreements exist between the earlier tabulations. For example the corrections due to screening of the nuclear Coulomb field by atomic electrons used by Bhalla and Rose (1961) differ from those used by Dzhelepov and Zyrianova (1956). The former corrections derive from a WKB approximation due to Rose (1936) and the latter from a Thomas–Fermi–Dirac calculation due to Reitz (1950). Subsequently, various authors (Durand III (1964); Brown (1964); Matese and Johnson (1966)) have concluded that Reitz's (1950) results are incorrect. This discrepancy is therefore resolved and all calculations are in essential agreement up to mass $A = 34$. However, for heavier nuclei there is disagreement between Bhalla and Rose (1961, 1964), and the later calculations (Suslov (1966, 1967), Blin-Stoyle and Nair (1967), Behrens and Bühring (1968)) for $f(Z, E_0)$. This disagreement ranges from 0.35% for ^{42}Sc to as much as 1% for ^{54}Co. It seems probable that the discrepancy lies in an inconsistency in the phase conventions adopted by Bhalla and Rose (1961, 1964) and which was pointed out by Bühring (1967) (see also Huffaker and Laird (1967)). The later calculations are all consistent with one another and in this book we, in general, use the f-values calculated by Blin-Stoyle and Nair (1967). The maximum theoretical error is estimated as about 0.3%, but since the evaluation of f is a large scale computing operation part of this error is certainly systematic. In these different calculations it has been the convention to use a uniform charge distribution for the nucleus. Recently it has been suggested (e.g. Halpern and Chern (1968); Dicus and Norton (1970)) that results could be sensitive to the shape assumed for the charge distribution. However, further study by Behrens and Bühring (1972) indicates that this is not the case and that the assumption of a uniform charge distribution is perfectly satisfactory so long as the r.m.s. radius is chosen correctly.

CHAPTER III

THE POLAR VECTOR CURRENT IN β-DECAY

III-1. Conserved vector current theory

The basic theory of β-decay deriving from a fundamental weak interaction Hamiltonian has now been set up and the next step is to investigate the extent to which beta-decay phenomena agree with this theory. First we consider the nature of the polar vector current.

As discussed in Chapt. I it is generally assumed that the polar vector current $\mathcal{V}_\lambda$ is conserved and is a member of an octet of polar vector currents $\mathcal{F}_\lambda^{(i)}$ ($i = 1, \ldots, 8$) i.e.

$$\partial_\lambda \mathcal{V}_\lambda = 0 \tag{III-1}$$

$$\mathcal{V}_\lambda = \mathcal{F}_\lambda^{(1)} + \mathrm{i}\,\mathcal{F}_\lambda^{(2)} . \tag{III-2}$$

The electromagnetic current J_λ is also contained within this octet and is given by

$$J_\lambda = \mathcal{F}_\lambda^{(3)} + \frac{1}{\sqrt{3}} \mathcal{F}_\lambda^{(8)} . \tag{III-3}$$

Further, it will be recalled that the isospin operator T_k ($k = 1, 2, 3$) is given by

$$T_k = -\mathrm{i} \int \mathcal{F}_4^{(k)}(x)\, \mathrm{d}^3x \tag{III-4}$$

so that

$$T_+ = -\mathrm{i} \int (\mathcal{F}_4^{(1)}(x) + \mathrm{i}\,\mathcal{F}_4^{(2)}(x))\, \mathrm{d}^3x = - \int \mathcal{V}_4(x)\, \mathrm{d}^3x \tag{III-5}$$

and

$$T_- = -\mathrm{i} \int (\mathcal{F}_4^{(1)}(x) - \mathrm{i}\,\mathcal{F}_4^{(2)}(x))\, \mathrm{d}^3x = -\mathrm{i} \int \mathcal{V}_4^*(x)\, \mathrm{d}^3x . \tag{III-6}$$

The assumption that $\mathcal{V}_\lambda$ is conserved and is identified with part of the isospin current carries with it a number of important implications for β-decay theory. Previously (eq. (II-16)) the matrix element for $\mathcal{V}_\lambda(0)$ between neutron and proton states has been written down. In an isospin notation and using eq. (III-2) we have

$$\langle N'|\mathcal{V}_\lambda(0)|N\rangle = \langle N'|\mathcal{F}_\lambda^{(1)}(0) + i\,\mathcal{F}_\lambda^{(2)}(0)|N\rangle$$

$$= i(\bar{u}_{N'}|(f_V\gamma_\lambda + f_W\sigma_{\lambda\nu}k_\nu + if_S k_\lambda)t_+|u_N) \qquad \text{(III-7)}$$

where the f's are form-factors, N and N$'$ refer to nucleon states and the four-momentum transfer is given by $k_\lambda = N'_\lambda - N_\lambda$. In addition, from eq. (I-13) we have for the matrix element of the isovector part ($\mathcal{F}_\lambda^{(3)}$) of the electromagnetic current

$$\langle N'|\mathcal{F}_\lambda^{(3)}(0)|N\rangle = i(\bar{u}_{N'}|(F_1^{(V)}\gamma_\lambda - F_2^{(V)}\sigma_{\lambda\nu}k_\nu)\tau_3|u_N) \qquad \text{(III-8)}$$

where $F_1^{(V)}$ and $F_2^{(V)}$ are the nucleon isovector form factors. In the limit of zero momentum transfer, they have the values (see eq. (I-15))

$$F_1^{(V)}(0) = \frac{1}{2}, \qquad F_2^{(V)}(0) = \frac{\kappa_p - \kappa_n}{4M} \qquad \text{(III-9)}$$

where κ_p $(= \mu_p - 1)$ and κ_n $(= \mu_n)$ are the anomalous magnetic moments of the proton and neutron respectively. Since the matrix element in eq. (III-8) is identical with that in eq. (III-7) apart from being rotated from the + to the 3-direction in isospace, it follows that the form factors in the two equations can be identified with one another. Thus, remembering that $\tau_3 = 2t_3$ we obtain

$$f_V = 2F_1^{(V)}, \qquad f_W = -2F_2^{(V)}, \qquad f_S = 0 \qquad \text{(III-10)}$$

and, in the limit $k = 0$, using eq. (III-9)

$$f_V(0) = 1, \qquad f_W(0) = -\frac{\kappa_p - \kappa_n}{2M}, \qquad f_S = 0. \qquad \text{(III-11)}$$

The polar vector form factors in CVC theory therefore take on precise values which can be determined from nucleon electromagnetic data. In particular it should be noted that there is no renormalisation of the polar vector coupling constant G_V due to strong interactions (i.e. $f_V(0) = 1$) and that there is no induced scalar interaction (i.e. $f_S = 0$).

One further result also follows readily from the assumption of CVC theory, namely, that for allowed polar vector matrix elements in a complex nucleus there are no exchange effects *. The allowed matrix element is proportional to $G_V \langle \mathcal{N}' | \int \mathcal{V}_4(x)\, d^3x | \mathcal{N} \rangle$ where $\mathcal{N}'$ and $\mathcal{N}$ refer to nuclear states. By eq. (III-4) *et seq* it follows that

$$G_V \langle \mathcal{N}' | \int \mathcal{V}_4(x)\, d^3x | \mathcal{N} \rangle = G_V \langle \mathcal{N}' | \int (\mathcal{F}_4^{(1)}(x) + i\, \mathcal{F}_4^{(2)}(x))\, d^3x | \mathcal{N} \rangle$$
$$= i\, G_V \langle \mathcal{N}' | T_+ | \mathcal{N} \rangle$$
$$= i\, G_V \langle \mathcal{N}' | \Sigma\, t_+^{(i)} | \mathcal{N} \rangle \qquad \text{(III-12)}$$

which is just the many body generalisation used in eqs. (II-47) and (II-61) and which obviously does not include any two or higher-body exchange terms.

The foregoing implications of CVC theory follow very readily and in the following section the extent to which β-decay experiments support these results will be discussed. A later section will deal with some more subtle implications of CVC theory relevant to higher forbidden corrections and transitions.

III-2. **Superallowed transitions**

Superallowed transitions are defined to be beta-transitions taking place between two members of an isospin multiplet (i.e. between analogue states). Of particular interest for the purpose of studying the polar vector current are decays of the type $0^+ \to 0^+$ between adjacent members of an isospin multiplet for which the axial vector matrix element $\langle \boldsymbol{\sigma}\, t_\pm \rangle$ vanishes. We shall, however, also deal with transitions of a more general type in which there can be a contribution from $\langle \boldsymbol{\sigma}\, t_\pm \rangle$. The crucial importance of such $0^+ \to 0^+$ decays is that, as will be seen, the polar vector matrix element $\langle t_\pm \rangle$ can be evaluated accurately without reference to details of the nuclear wave function. From measured ft-values and using eq. (II-62) it is then possible to determine G_V

* The situation here is completely analogous to that with respect to the total nuclear electric charge which is exactly equal to the sum of the individual proton charges – there is no additional (exchange) contribution.

and hence to test weak interaction and, in particular, CVC theory.

III-2.1. *Value of* $\langle t_{\pm} \rangle$ *for* $0^+ \to 0^+$ *superallowed transitions*

A typical superallowed $0^+ \to 0^+$ decay is that between the 0^+, $T = 1$, $T_3 = +1$ ground state of ^{14}O and the corresponding 0^+, $T = 1$, $T_3 = 0$ first excited state of ^{14}N. The decay is illustrated in fig. III-1. By eq. (II-61) the matrix element for the decay has the form

$$\langle t_- \rangle = \langle ^{14}\mathrm{N}\colon I = 0^+, T = 1, T_3 = 0 | T_- | ^{14}\mathrm{O}\colon I = 0^+, T = 1, T_3 = +1 \rangle . \quad \text{(III-13)}$$

Since both states in the above matrix element are members of the same isospin multiplet differing only in the value of T_3, whilst T_- is an isospin shift operator it follows that we can use the general result from angular momentum theory

$$\langle T, T_3 \pm 1 | T_{\pm} | T, T_3 \rangle = \sqrt{(T \mp T_3)(T \pm T_3 + 1)} . \quad \text{(III-14)}$$

Using this relation then gives $\langle t_- \rangle = \sqrt{2}$ and the result is quite independent of any details of the nuclear wave function and rests only on the assumption that the states involved are pure isospin states. Using this value for $\langle t_- \rangle$ and eq. (II-62) gives

$$ft = \tfrac{1}{2} K G_V^{-2} \quad \text{(III-15)}$$

or, explicitly, inserting appropriate powers of c, $\hbar$ and the electron mass m

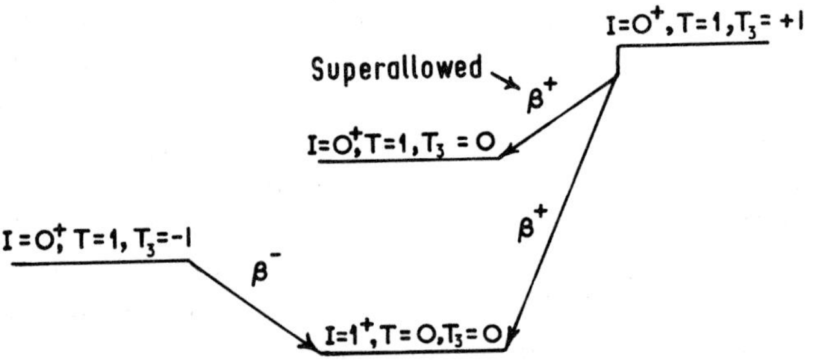

^{14}C ^{14}N ^{14}O

Fig. III-1. Superallowed β-decay of ^{14}O.

$$ft = [G_V^{-2}\pi^3 \ln 2]\hbar^7 m^{-5} c^{-4} \,. \qquad \text{(III-16)}$$

Expressing ft in seconds and G_V in erg cm^3, then

$$ft = (6.1532 \times 10^{-95})\, G_V^{-2} \,. \qquad \text{(III-17)}$$

This same relation holds for all the other $0^+ \to 0^+$ superallowed transitions to be considered.

III-2.2. *Experimental ft-values for* $0^+ \to 0^+$ *superallowed transitions*

During the last few years there has been some very intensive experimental work aimed at obtaining accurate ft-values for $0^+ \to 0^+$ superallowed transitions. The results of this work (see also the recent work of Hardy *et al.* (1972)) are summarised in columns I–VI of table III-1 where the f-functions are taken from the work of Blin-Stoyle and Nair (1967) and where ϵ_K represents the correction due to the competing K-capture process. Clearly all the ft-values are close to one another as is predicted. However, before capitalising on this and setting out to determine the value of G_V a number of small, but nevertheless significant corrections to eq. (III-15) must be considered.

III-2.3. *Electromagnetic radiative corrections*

One of the major uncertainties in the analysis of superallowed β-decay data is the magnitude of electromagnetic radiative corrections *. These arise from, for example, the exchange of virtual photons between the charged particles involved in the β-decay process and lead to corrections to the ft-value of order α (= $e^2/4\pi \approx 1/137$) and higher. Of course, some electromagnetic effects are already taken into account in so far as the evaluation of the Fermi function $F(Z,E)$ (§II-4) allows for the Coulomb interaction between the emitted electron or positron and the daughter nucleus. Because of this some uncertainty has arisen as to whether the correction applied to ft should be *multiplicative* (i.e. of the form $ft(1+\delta_R)$ or *additive* (i.e. of the form $ft + \delta_R$). This point has been touched upon explicitly or implicitly by a number of authors (Chern *et al.* (1967); Logue and Chern (1968); Halpern and Chern (1968); Beg *et al.* (1969); Halpern (1970); Dicus and Norton (1970); Jaus and Rasche (1970)), and will only be finally resolved when $Z\alpha^2$ and higher contributions to electromagnetic radiative corrections can be

* The intimately related phenomenon of inner bremsstrahlung in β-decay has been studied theoretically and experimentally by many authors (see e.g. Schopper (1966)) but in itself is not helpful in giving information about the basic β-decay interaction.

Table III-1

$0^+ \to 0^+$ superallowed β-decay data

I	II	III	IV	V	VI	VII	VIII	IX	X	XI	XII	XIII
Decay	$\mathcal{E}_0$ (keV)	$\mathcal{E}_0$ (mc^2)	ft(raw)	ϵ_K (%)	$ft(1+\epsilon_K)$	Reference	δ'_R (%)	$ft(1+\delta_R)(1+\epsilon_K)$	$\tilde{f}t$(raw)	$\mathcal{F}t = \tilde{f}t(1+\delta'_R)(1+\epsilon_K)\overline{C(E)}$ $r_N=r_P$	$\mathcal{F}t = \tilde{f}t(1+\delta'_R)(1+\epsilon_K)\overline{C(E)}$ $r_N=r_P+0.2$ fm	$\mathcal{F}t = \tilde{f}t(1+\delta'_R)(1+\epsilon_K)\overline{C(E)}$ $r_N=r_P-0.2$ fm
$^{10}C \xrightarrow{\beta^+} {}^{10}B$	888.1 ± 1.8	1.74	3083 ± 39	0.23+	3090 ± 39	Freeman *et al.* (1966a, 1969a, 1969d) Robinson *et al.* (1972)	1.50	3136 ± 39	3083 ± 39	3136 ± 39	3136 ± 39	3136 ± 39
$^{14}O \xrightarrow{\beta^+} {}^{14}N$	1812.6 ± 1.4	3.55	3077 ± 11	0.086 *	3080 ± 11	Bardin *et al.* (1962)	1.30	3120 ± 11	3080 ± 11	3121 ± 11	3122 ± 11	3121 ± 11
	1809.1 ± 1.5	3.54	3039 ± 11 x		3042 ± 11	Freeman *et al.* (1968); Clark *et al.* (1971)		3081 ± 11	3041 ± 11	3081 ± 11	3082 ± 11	3081 ± 11
$^{26}Al^m \xrightarrow{\beta^+} {}^{26}Mg$	3208.6 ± 1.3	6.28	3027 ± 7	0.079 *	3029 ± 7	Freeman *et al.* (1969b)	1.12	3063 ± 7	3034 ± 7	3065 ± 7	3061 ± 7	3069 ± 7
	3211.1 ± 0.6		3037 ± 3		3039 ± 3	De Wit and Van Der Leun (1969)		3073 ± 3	3045 ± 3	3076 ± 3	3072 ± 3	3080 ± 3
$^{34}Cl \xrightarrow{\beta^+} {}^{34}S$	4459.7 ± 4.0	8.73	3092 ± 19	0.071 *	3094 ± 19	Freeman *et al.* (1966b)	1.01	3125 ± 19	3103 ± 19	3127 ± 19	3119 ± 19	3136 ± 19
$^{42}Sc \xrightarrow{\beta^+} {}^{42}Ca$	5409.0 ± 2.3	10.59	3064 ± 9	0.089 *	3067 ± 9	Freeman *et al.* (1966b)	0.95	3096 ± 9	3084 ± 9	3102 ± 9	3088 ± 9	3116 ± 9
$^{46}V \xrightarrow{\beta^+} {}^{46}Ti$	6032.1 ± 2.2	11.80	3072 ± 8	0.091 *	3075 ± 8	Freeman *et al.* (1966b)	0.92	3103 ± 8	3095 ± 8	3111 ± 8	3092 ± 8	3125 ± 8
$^{50}Mn \xrightarrow{\beta^+} {}^{50}Cr$	6609.0 ± 2.6	12.93	3059 ± 9	0.095 *	3062 ± 9	Freeman *et al.* (1966b)	0.88	3089 ± 9	3086 ± 9	3097 ± 9	3074 ± 9	3113 ± 9
$^{54}Co \xrightarrow{\beta^+} {}^{54}Fe$	7227.7 ± 3.8	14.14	3063 ± 17	0.098 *	3066 ± 17	Freeman *et al.* (1966b)	0.85	3092 ± 18	3091 ± 18	3098 ± 18	3070 ± 18	3116 ± 18

+ From Freeman *et al.* (1969a)
* From Behrens and Buhring (1968)
x See Alburger (1972) for a comment on the experimental error.

calculated unambiguously (most recently Jaus (1972) has estimated the $Z^2\alpha^3$ contribution to δ_R). For our purposes we shall take the multiplicative form but it must then be remembered that this includes *some* higher order terms through the product $F(Z, E)\delta_R$ but *not all* and that the results obtained are therefore uncertain to the extent of contributions from the omitted terms. For example, Jaus and Rasche (1970) find additional $Z\alpha^2$ contributions to δ_R of from 0.26% for ^{14}O to 0.93% for ^{54}Co.

Many calculations of electromagnetic radiative corrections in μ-decay, β-decay and other weak interaction processes have been made in the past using perturbation theory and a local point nucleon contact interaction (Berman (1958); Kinoshita and Sirlin (1959); Berman and Sirlin (1962)) and stemming from Feynman diagrams of the type illustrated in fig. III-2. The results are typified by that obtained by Kinoshita and Sirlin (1959), namely *,

$$-\frac{\Delta t}{t} = \delta_R = \frac{\alpha}{2\pi}\left(6 \ln\frac{\Lambda}{M} + 3 \ln\frac{M}{2E_0} - 2.85\right). \qquad \text{(III-18)}$$

$\Delta t/t$ is the fractional change in lifetime due to the radiative corrections, E_0 is the end point energy of the β-decay spectrum and Λ is an ultra-violet cut-off normally taken to be of the order of the nucleon mass M. The effect of such a correction is that in determinations of G_V from experimental ft-values, eq. (III-15) has to be modified to

$$ft(1+\delta_R) = \tfrac{1}{2}KG_V^{-2} \qquad \text{(III-19)}$$

where δ_R is defined by eq. (III-18). Using eq. (III-18) values of δ_R in the approximate range + 1% to + 2% are obtained for the different superallowed decays and the sign of the effect is such as to shorten the observed lifetime.

Subsequent to these calculations a number of developments have taken place involving different treatments of the various interactions involved. Bailin (1964, 1965) and Sirlin (1967a) have taken into account the effects of the weak interaction being mediated by an intermediate vector boson. Källén (1967) takes into account the effect of strong interactions by introducing nucleon form factors. Recently a lot of progress has been made assuming the validity of current algebra at short distances (Bjorken (1966); Abers *et al.* (1967); Cabibbo *et al.* (1967a, 1967b); Johnson *et al.* (1967); Sirlin (1967b, 1967c); Abers *et al.* (1968)).

* Note that the expression given is an asymptotic form and is only accurate for values of $E_0 \gtrsim 2$ MeV (see Källén (1967)).

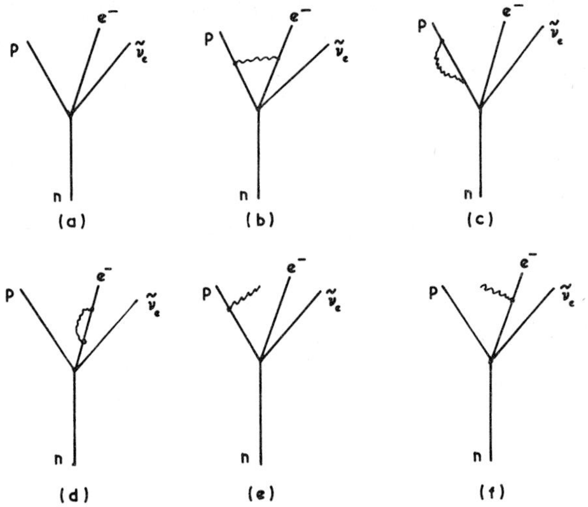

Fig. III-2. Feynman diagrams (b)–(f) contributing to first order electromagnetic radiative corrections in the β-decay of a point nucleon.

As has been pointed out (e.g. Sirlin (1967c, 1969)) the results of all the calculations referred to above can be represented to a good approximation by writing the β-decay spectrum P$\mathrm{d}p$ in the form

$$P\mathrm{d}p = \left[G_V^2\left(1+\frac{\alpha}{2\pi}C\right)|\langle t_\pm\rangle|^2 + G_A^2\left(1+\frac{\alpha}{2\pi}D\right)|\langle \boldsymbol{\sigma} t_\pm\rangle|^2\right] F(Z,E)$$

$$\times \frac{(E_0-E)^2p^2\mathrm{d}p}{2\pi^3}\left[1+\frac{\alpha}{2\pi}g(E,E_0,m)\right] \qquad \text{(III-20)}$$

where p, E, E_0 and m are the momentum, energy, total end point energy and electron mass respectively. C and D are model dependent constants. $F(Z,E)$ is the usual Fermi function and $g(E, E_0 m)$ is a finite universal function given by the following expression

$$g(E,E_0,m) = 3\ln\frac{M}{m} - \frac{3}{4} + 4\left[\frac{\tanh^{-1}\beta}{\beta} - 1\right]\left[\frac{E_0-E}{3E} - \frac{3}{2} + \ln\frac{2(E_0-E)}{m}\right]$$

$$+\frac{4}{\beta}L\left(\frac{2\beta}{1+\beta}\right) + \frac{1}{\beta}\tanh^{-1}\beta\left[2(1+\beta^2) + \frac{(E_0-E)^2}{6E^2} - 4\tanh^{-1}\beta\right] \qquad \text{(III-21)}$$

where $\beta = p/E$ and $L(x)$ is the Spence function defined by

$$L(x) = \int_0^x \frac{\mathrm{d}t}{t} \ln(1-t) \,. \tag{III-22}$$

The effects of electromagnetic radiative corrections can thus be described (Blin-Stoyle and Freeman (1970)) as renormalising the polar vector and axial vector coupling constants according to the relations

$$\begin{aligned} G_\mathrm{V} &\to G'_\mathrm{V} = G_\mathrm{V}\left(1 + \frac{\alpha C}{4\pi}\right) \\ G_\mathrm{A} &\to G'_\mathrm{A} = G_\mathrm{A}\left(1 + \frac{\alpha D}{4\pi}\right) \end{aligned} \tag{III-23}$$

and modifying the overall spectrum by a (known) factor $(1 + (\alpha/2\pi)g(E, E_0, m))$. This means that allowed beta-decay data can sensibly be analysed in terms of effective coupling constants G'_V and G'_A and an effective (known and energy dependent) radiative correction δ'_R given by *

$$\delta'_\mathrm{R}(E_0) = \frac{\alpha}{2\pi}\left[\int (E_0 - E)^2 p^2 g(E, E_0, m)\,\mathrm{d}p \Big/ \int (E_0 - E)^2 p^2\,\mathrm{d}p\right]. \tag{III-24}$$

δ'_R is plotted in fig. III-3 as a function of $\mathcal{E}_0$ (measured in MeV) where $\mathcal{E}_0$ $(= E_0 - mc^2)$ is the maximum kinetic energy of the charged lepton. Here it should be noted that the radiative correction δ_R usually quoted in the literature is related to δ'_R by the relation

$$\delta_\mathrm{R} = \Delta_\mathrm{R} + \delta'_\mathrm{R} \tag{III-25}$$

* $\delta'_\mathrm{R}(E_0)$ is simply related to the function $h(\epsilon)$ tabulated by Källen (1967) as follows:

$$\delta'_\mathrm{R}(E_0) = \frac{\alpha}{2\pi}\left(3 \ln\frac{M}{m} - \frac{3}{4} + 2h(E_0 - 1)\right)$$

where E_0 is measured in units of mc^2 and M is the proton mass. In calculating δ'_R the small effect of the Fermi function $F(Z,E)$ in the integral over the energy spectrum is neglected. Wilkinson and Macefield (1970) have tabulated values of δ'_R calculated without making this approximation.

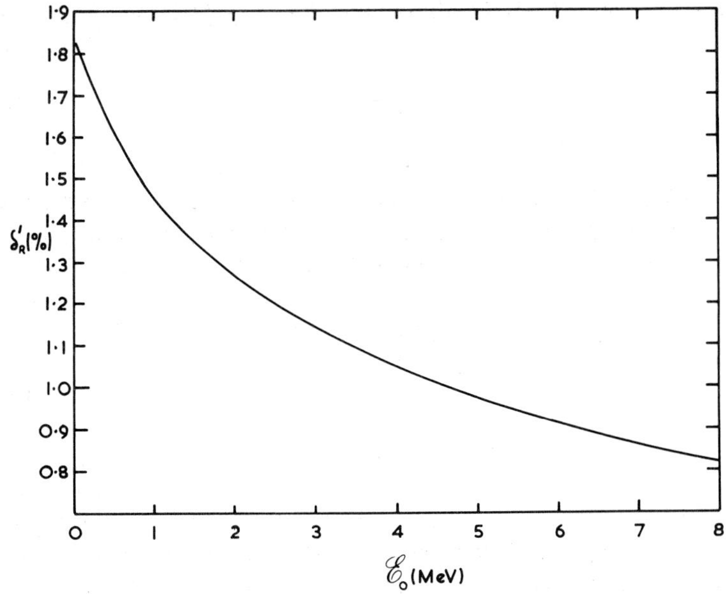

Fig. III-3. Plot of δ'_R (expressed as a percentage) against the end point energy $\mathcal{E}_0$ (measured in MeV).

where

$$\Delta_R = \frac{\alpha}{2\pi} \frac{G_V^2 |\langle t_\pm \rangle|^2 C + G_A^2 |\langle \boldsymbol{\sigma} t_\pm \rangle|^2 D}{G_V^2 |\langle t_\pm \rangle|^2 + G_A^2 |\langle \boldsymbol{\sigma} t_\pm \rangle|^2} \tag{III-26}$$

and is model dependent but, for allowed β-decay, is expected to be energy independent.

In terms of the renormalised coupling constants the β-decay ft-value is now given by (cf. eq. (II-62))

$$ft(1+\delta'_R) = \frac{K}{G_V'^2 |\langle t_\pm \rangle|^2 + G_A'^2 |\langle \boldsymbol{\sigma} t_\pm \rangle|^2} . \tag{III-27}$$

For $0^+ \to 0^+$ superallowed transitions where there is no allowed contribution from the axial vector current, the model dependent constant term in eq. (III-26) reduces to $\Delta_R = (\alpha/2\pi)\, C$.

The following values for Δ_R are obtained for the different models of radiative corrections considered.

(i). Bare nucleon calculations in which strong interactions are neglected (Kinoshita and Sirlin (1959); Berman and Sirlin (1962); Sirlin (1968a)) give

$$\Delta_R = \frac{\alpha}{2\pi}\left[3(1-\lambda)\ln\frac{\Lambda}{M} - \frac{9}{4}\lambda\right] \tag{III-28}$$

where $\lambda = G_A/G_V \approx -1.24$ and Λ is an ultra-violet cut-off. The particular result of Kinoshita and Sirlin (1959) quoted in eq. (III-18) is given by choosing $\lambda = -1$ (i.e. neglecting the renormalisation effect of strong interactions), and, for $\Lambda = M$, corresponds to

$$\Delta_R = 0.26\% . \tag{III-29}$$

(ii). Källén (1967) (see also Jaus (1968); Gustafson (1969)) by introducing nucleon form factors takes some account of strong interactions and his calculations lead to

$$\Delta_R = \frac{\alpha}{2\pi}\left[2\Delta + \tfrac{3}{4}\right]$$

where Δ is defined in his paper. Numerically this corresponds to

$$\Delta_R = (0.64 \pm 0.25)\% \tag{III-30}$$

a rather higher value than that given in eq. (III-29).

(iii). Assuming the validity of current algebra at short distances the following expression for Δ_R is obtained * (see Abers *et al.* (1968); Brene *et al.* (1968)).

$$\Delta_R = \left\{\frac{\alpha}{2\pi}\left[3\ln\frac{\Lambda}{M} + 6\bar{Q}\ln\frac{\Lambda}{M_{A_1}}\right] - 1.2\times 10^{-3}\right\} \tag{III-31}$$

where $\bar{Q}$ is the average charge of the fundamental isodoublet underlying current algebra (see §I-5) and M_{A_1} (= 1070 MeV) is the mass of the A_1 axial vector meson.

* The term -1.2×10^{-3} arises from the induced axial vector correction and relates to the decay of a nucleon. For a $0^+ \to 0^+$ β-decay this term should be omitted (Dicus and Norton (1969)).

Various assumptions about $\bar{Q}$ have been made in the literature. On the usual quark model (e.g. Gell-Mann (1962, 1964b)), $\bar{Q} = \frac{1}{6}$ so that choosing $\Lambda = M$ gives

$$\Delta_R = -0.17\% \tag{III-32}$$

a much smaller value for Δ_R than obtained hitherto. Assuming the results for the algebra of fields (Lee *et al.* (1967)) gives $\bar{Q} = 0$. However, in this case, an additional term of magnitude $+\frac{3}{4}$ has to be added between the square brackets of eq. (III-31) so that

$$\Delta_R = -0.03\%\,. \tag{III-33}$$

A particularly interesting assumption to make about $\bar{Q}$ is that $\bar{Q} = -\frac{1}{2}$ (Cabibbo *et al.* (1967a, 1967b); Johnson *et al.* (1967); Abers *et al.* (1968)) since the result given in eq. (III-31) then becomes finite and independent of the cut-off Λ, namely

$$\Delta_R = \frac{3\alpha}{2\pi}\ln\frac{M_{A_1}}{M} - 1.2 \times 10^{-3} = -0.07\%\,. \tag{III-34}$$

(iv). A final class of models to be considered is that in which the weak interaction is propagated by means of an intermediate vector boson which interacts with the electromagnetic field by the conventional Proca–Wentzel Lagrangian (Sirlin (1967a, b; 1968a, b; 1969); Abers *et al.* (1968); see also Lee (1962); Lee and Wu (1965); Shaffer (1962, 1963); Dorman (1964); Bailin (1964, 1965)). Using this model it is found that in the limit of zero momentum transfer, there is a model independent divergent part renormalising both the μ-decay rate and β-decay rate equally. This means that in discussing the ratio G_V/G_μ as determined from these rates it is sufficient to use the "local" calculation for the μ-decay rate and eq. (III-31) for the β-decay process with Λ replaced by M_W, the mass of the intermediate vector boson.

In table III-2 values for Δ_R are given for $\bar{Q} = \frac{1}{6}$ and $\bar{Q} = 0$, the assumptions already discussed, and for different values of M_W. Of course for $\bar{Q} = -\frac{1}{2}$ the result given in (iii) above still holds for all values of M_W.

It is clear from the foregoing discussion that there is considerable uncertainty about the size of electromagnetic radiative corrections stemming from lack of knowledge about the constant term Δ_R. For this reason in the table of ft-values (table III-1) only the energy dependent corrections embodied in $\delta_R'(\mathcal{E}_0)$ are introduced. These corrections are given in column VIII. In column

Table III-2
Values of Δ_R (expressed as a percentage) for various values of M_W for the two cases $\bar{Q} = \frac{1}{6}, \bar{Q} = 0$.

M_W	Δ_R (%) $\bar{Q} = 1/6$	Δ_R (%) $\bar{Q} = 0$	M_W	Δ_R (%) $\bar{Q} = 1/6$	Δ_R (%) $\bar{Q} = 0$
M	−0.13	−0.12	$40M$	1.58	1.17
$2M$	0.19	0.12	$60M$	1.77	1.31
$4M$	0.51	0.36	$80M$	1.91	1.41
$6M$	0.70	0.50	$100M$	2.01	1.48
$8M$	0.83	0.60	$150M$	2.20	1.63
$10M$	0.94	0.68	$200M$	2.33	1.73
$20M$	1.26	0.92			

IX are listed the ft-values corrected by δ'_R. These corrected ft-values are then only uncertain as far as radiative corrections are concerned to the extent of a further constant but model dependent, correction factor. The value of this constant is of the utmost interest and the light cast upon it by superallowed transitions is discussed in §III-2.6.

III-2.4. *The value of $f(Z, E_0)$ and second forbidden corrections*

It will be remembered that in §II-4 the approximation was made in determining $f(Z, E_0)$ of evaluating the lepton spinors at the nuclear radius R. This is in the spirit of the "allowed" approximation. However, the theoretical and experimental accuracy attempted in dealing with superallowed transitions is such that this approximation is unacceptable. Effectively this means that second forbidden matrix elements must be taken into account. For the decays we are considering, two such matrix elements can contribute, namely, $\langle r^2 t_\pm \rangle$ and $\langle \boldsymbol{\alpha} \cdot \boldsymbol{r} t_\pm \rangle$ in an obvious notation. Rough estimates were made of the size of these matrix elements by Blin-Stoyle *et al.* (1959/60) who obtained corrections to the ft-value of at most a few tenths of a percent. More recently Behrens and Bühring (1968) (see also Bühring (1963a) and Damgaard (1969)) have put these considerations on to a more quantitative footing. They replace ft by $\tilde{f}t\,\overline{C(E)}$ so that in terms of $G_{V'}$ and allowing for radiative corrections eq. (III-19) is now replaced by

$$\tilde{f}t(1+\delta'_R)\,\overline{C(E)} = \tfrac{1}{2} K G_V'^{-2}\,. \qquad \text{(III-35)}$$

The new f function, $\tilde{f}$, is obtained by evaluating the lepton functions at $r = 0$ rather than at $r = R$. Expanding the Dirac functions $g_{-1}(r)$ and $f_{+1}(r)$

(see eq. (II-60)) in powers of r * then leads to an energy dependent correction factor $C(E)$ which has to be averaged over the electron spectrum giving $\overline{C(E)}$. For all practical purposes, $\overline{C(E)}$ is given by

$$\overline{C(E)} = 1 + A_1 \frac{\langle r^2 t_\pm \rangle}{R^2 \langle t_\pm \rangle} \tag{III-36}$$

where the following values of A_1 hold for the various $0^+ \to 0^+$ superallowed decays considered (see Behrens and Bühring (1968))

^{10}C: $A_1 \approx 0$;	^{42}Sc: $A_1 = -7.31 \times 10^{-3}$;
^{14}O: $A_1 = -9.78 \times 10^{-4}$;	^{46}V: $A_1 = -8.67 \times 10^{-3}$;
^{26}Al^m: $A_1 = -2.75 \times 10^{-3}$;	^{50}Mn: $A_1 = -1.02 \times 10^{-2}$;
^{34}Cl: $A_1 = -4.70 \times 10^{-3}$;	^{54}Co: $A_1 = -1.18 \times 10^{-2}$.

It is further straightforward to relate the matrix element ratio in eq. (III-36) to the mean square radii r_n and r_p of the neutron and proton density distributions in the nucleus (Blin-Stoyle (1969a)) giving

$$\frac{\langle r^2 t_\pm \rangle}{R^2 \langle t_\pm \rangle} = 0.3 \left[\left(\frac{r_n^2 + r_p^2}{r_p^2} \right) + \frac{A}{2} \left(\frac{r_n^2 - r_p^2}{r_p^2} \right) \right]. \tag{III-37}$$

An alternative model dependent calculation due to Fayans (1971) leads to rather larger contributions to $\overline{C(E)}$.

In columns XI, XII and XIII of table III-1 the values of $\mathcal{F}t = \tilde{f}t(1+\epsilon_K)(1+\delta_R')\overline{C(E)}$ are given for the cases $r_n = r_p$, $r_n = r_p + 0.2$ fm and $r_n = r_p - 0.2$ fm respectively where the factor $(1+\epsilon_K)$ is the correction for K-capture. It can be seen for the case $r_n = r_p$ that there is little difference between $ft(1+\epsilon_K)(1+\delta_R')$ (as given in column IX) and $\tilde{f}t(1+\epsilon_K)(1+\delta_R')\overline{C(E)}$. For the two cases $r_n \neq r_p$ there is again little effect for the lighter nuclei. However, for the heavier nuclei there is a sizeable effect leading to a tendency for $\mathcal{F}t$ to decrease with increasing A for $r_n = r_p + 0.2$ fm

* There are convergence problems here and a more satisfactory approach in this respect is to make an expansion in powers of $Z\alpha$ (Jaus (1971b)). For practical purposes, however, both approaches seem to be equivalent.

and conversely for $r_n = r_p - 0.2$ fm. We shall see later (§III-2.5) that there are reasons for expecting the latter type of dependence on A thus suggesting that, if anything, the proton distribution for the nuclei considered extends slightly beyond the neutron distribution. This agrees with the general findings of, for example, Khrylin (1968) and Tarburton and Davies (1968) but disagrees with the results obtained by Greenlees *et al.* (1968) (see also Nolen *et al.* (1968); Bethe and Siemens (1968); Nolen and Schiffer (1969); Friedman and Mandelbaum (1969); Auerbach *et al.* (1969a, 1969b); Friedman (1971)). When we come to the point of determining G_V' we shall use data for light nuclei for which the effects considered here are small, and will therefore take the results for the case $r_n = r_p$.

III-2.5. *Isospin impurities*

In the case of superallowed transitions between members of an isospin multiplet the Fermi matrix element was determined to have the value (see eqs. (III-13) and (III-14))

$$\langle t_\pm \rangle = \langle T, T_3 \pm 1 | T_\pm | T, T_3 \rangle = \sqrt{(T \mp T_3)(T \pm T_3 + 1)} \qquad \text{(III-38)}$$

for a transition of the type $\psi(T, T_3) \xrightarrow{\beta^\mp} \psi(T, T_3 \pm 1)$. This result depends on the assumption that the states involved are pure isospin states and are eigenfunctions of a charge independent Hamiltonian. Now in practice the nuclear Hamiltonian is not charge independent. Quite apart from Coulomb forces and the electromagnetic spin-orbit coupling of nucleons, it is clear that the short range nuclear force itself is slightly charge dependent (see §XI-3) and the effect of these terms is to introduce isospin impurities into nuclear states and so modify the value of $\langle t_\pm \rangle$.

To quantify this effect, consider a zero-order charge independent Hamiltonian H_0 whose eigenfunctions and eigenvalues are defined by the following Schrödinger equation:

$$H_0 \psi_\nu(t, t_3) = E_\nu(t) \psi_\nu(t, t_3) \qquad \text{(III-39)}$$

where t and t_3 label the isospin properties of the states and ν labels all other properties. Suppose that H_0 is now perturbed by a charge-dependent term H_{CD} in the Hamiltonian. The initial and final states of a beta-decay which is nominally $\psi(T, T_3) \xrightarrow{\beta^\mp} \psi(T, T_3 \pm 1)$ will now have the form

$$\psi_{\mathrm{f}} = a_0 \psi_0(T, T_3 \pm 1) + \sum_{t,\nu \neq 0} a_\nu^{(t)} \psi_\nu(t, T_3 \pm 1)$$

$$\psi_{\mathrm{i}} = b_0 \psi_0(T, T_3) + \sum_{t,\nu \neq 0} b_\nu^{(t)} \psi_\nu(t, T_3) \tag{III-40}$$

where the a_ν and b_ν are admixture amplitudes and

$$a_0^2 + \sum_{t,\nu \neq 0} a_\nu^{(t)^2} = 1 , \qquad b_0^2 + \sum_{t,\nu \neq 0} b_\nu^{(t)^2} = 1 .$$

Using eq. (III-38) it is then straightforward to calculate the square of the Fermi matrix element for the states given in eq. (III-40) to lowest order in the perturbing terms. The following result is obtained:

$$|\langle t_\pm \rangle|^2 = (T + T_3)(T \pm T_3 + 1)[1 - \delta_c] \tag{III-41}$$

where δ_c is the correction to the zero order matrix element squared and has the value

$$\delta_c = \sum_{\nu \neq 0} \Bigg\{ (a_\nu^{(T)} - b_\nu^{(T)})^2 + \sum_{t \neq T} \left[a_\nu^{(t)^2} - 2 \left(\frac{(t \mp T_3)(t \pm T_3 + 1)}{(T \mp T_3)(T \pm T_3 + 1)} \right)^{1/2} a_\nu^{(t)} b_\nu^{(t)} + b_\nu^{(t)^2} \right] \Bigg\} . \tag{III-42}$$

In the foregoing expressions it is clear that the isospin character of H_{CD}, normally taken to be a combination of isovector ($T = 1$) and isotensor ($T = 2$), will limit t to the range $t = T + 2, T + 1, T, T - 1, T - 2$. A further limitation is imposed by the fact that even with a charge dependent situation, T_3 is still a good quantum number so that $a_\nu^{(t)} = 0$ for $t < T_3 \pm 1$ and $b_\nu^{(t)} = 0$ for $t < T_3$. Thus, for the particular case of superallowed transitions within a $T = 1$ isospin multiplet (MacDonald (1958); Lovitch (1963))

$$\delta_c = \sum_{\nu \neq 0} \{a_\nu^{(0)2} + (a_\nu^{(1)} - b_\nu^{(1)})^2 + (a_\nu^{(2)2} - 2\sqrt{3}\, a_\nu^{(2)} b_\nu^{(2)} + b_\nu^{(2)2})$$

$$+ (a_\nu^{(3)2} - 2\sqrt{6}\, a_\nu^{(3)} b_\nu^{(3)} + b_\nu^{(3)2})\} \,. \qquad \text{(III-43)}$$

It will be noted in eqs. (III-42) and (III-43) that two distinct effects come into play. Firstly there are isospin impurities introduced by H_{CD}, i.e. states of isospin t are admixed for which $t \neq T$. Secondly, there are admixtures of *other* states with the same isospin ($t = T$) as the zero order state. This latter effect has been called dynamic distortion by MacDonald (1958) and inspection of eqs. (III-42) or (III-43) shows that it always reduces the Fermi matrix element (i.e. contributes positively to δ_c).

In order to estimate the size of δ_c the magnitudes of the $a_\nu^{(t)}$ and $b_\nu^{(t)}$ must be calculated. A straightforward application of first-order perturbation theory gives at once

$$a_\nu^{(t)} = \frac{\langle \psi_\nu(t, T_3 \pm 1)\, | H_{\mathrm{CD}} |\, \psi_0(T, T_3 \pm 1)\rangle}{E_0(T) - E_\nu(t)}$$

$$b_\nu^{(t)} = \frac{\langle \psi_\nu(t, T_3)\, | H_{\mathrm{CD}} |\, \psi_0(T, T_3)\rangle}{E_0(T) - E_\nu(t)} \qquad \text{(III-44)}$$

where H_{CD} is the charge dependent perturbation already referred to.

A number of calculations of δ_{c} using one or another nuclear model for the zero order Hamiltonian have been carried out and the general features and results of these calculations will now be described.

The simplest and most clear-cut calculation is due to Bohr *et al.* (1967) who take as a basic model one in which the nuclei under consideration are represented in zero order by a $T_{\mathrm{c}} = 0$ core with two extra-core nucleons in a $T_{\mathrm{n}} = 1$ state. The charge dependent perturbation is taken to be a single particle Coulomb potential generated by a spherical nucleus of radius R and charge Ze, namely,

$$H_{\mathrm{CD}} = V_{\mathrm{C}} = \sum_i \tau_3^{(i)} \frac{Ze^2}{2R} \left[\frac{3}{2} - \frac{1}{2}\left(\frac{r_i}{R}\right)\right]^2 \qquad \text{for } r_i \leq R$$

$$= \sum_i \tau_3^{(i)} \frac{Ze^2}{2r_i} \qquad \text{for } r_i > R \,. \qquad \text{(III-45)}$$

It is here assumed that because of the long range nature of the Coulomb interaction charge dependent effects are dominated by the action of the average Coulomb field resulting from the collective effect of the total nuclear charge. Consideration is restricted to the admixing of a single $T_c = 1$ state in the core which is regarded as a vibrational mode corresponding to the relative density variations of protons and neutrons in the core. The $T_c = 1$ core state can then combine with the $T_n = 1$ state of the extra-core nucleons to give admixtures with $t = 1, 2$ and 3. Including a symmetry term of the form $V_{sym} = (V_1/A)\mathbf{T}_c \cdot \mathbf{T}_n$ in the zero order Hamiltonian, where $V_1 \approx 100$ MeV and A is the nuclear mass number, then leads straightforwardly to

$$\delta_c = \frac{4\alpha^2 V_1}{AE_0} \tag{III-46}$$

where α^2 represents the intensity with which the $T_c = 1$ collective state is admixed into the $T_c = 0$ ground state of the core and E_0 is the excitation of the collective state. Using estimates for α^2 and E_0 obtained from the hydrodynamical model Bohr *et al* (1967) conclude that δ_c is proportional to Z^2 and smaller than 0.1% even for ^{54}Co. As such δ_c is negligible.

Another approach to calculating δ_c is to use the simple Fermi gas model in which the nucleus is approximated by an impenetrable box containing nucleons interacting only through charge dependent forces which are treated as a perturbation. This approach was first used by MacDonald (1955) to estimate total isospin impurities in nuclei and was subsequently developed by Blin-Stoyle *et al.* (1965) (see also Blin-Stoyle (1969b)) in order to calculate δ_c.

In addition to the usual Coulomb potential the calculations also include a short range charge dependent potential of the form discussed in §XI-3.1 (eqs. (XI-33) to (XI-35)), namely

$$V_{CD} = V_0 \sum_{i<j} \{[p + r\boldsymbol{\sigma}^{(i)} \cdot \boldsymbol{\sigma}^{(j)}](\tau_3^{(i)} + \tau_3^{(j)}) + [q + s\boldsymbol{\sigma}^{(i)} \cdot \boldsymbol{\sigma}^{(j)}]\, T^{(i,j)}\}\, e^{-r_{ij}^2/\mu^2} \tag{III-47}$$

where $T^{(i,j)} = \tau_3^{(i)}\tau_3^{(j)} - \frac{1}{3}\boldsymbol{\tau}^{(i)} \cdot \boldsymbol{\tau}^{(j)}$, $V_0 = -50$ MeV and $\mu = 1.73$ fm and the parameters p, r, q, s give a measure of the charge dependence. The energy denominators occurring in the perturbation calculation (see eq. (III-44)) are taken to be the difference in kinetic energy of the zero order and admixed states and no account is taken of any T dependence of these energies. The sum over the various admixed states (Σ_ν in eq. (III-43)) is replaced by an appropriate momentum integral. Proceeding in a standard fashion it is found that; writing $\delta_c = \delta_C + \delta_{CD}$ where δ_C is due only to the Coulomb perturbation and δ_{CD} is due to V_{CD} (including interference with the Coulomb potential), then δ_C varies from 0.26% for the ^{14}O decay to 0.66% for the ^{54}Co decay. No precise statement about δ_{CD} can be made because of the uncertainty about the values of the parameters p, ..., s. Reasonable assumption (see Chapt. XI (eqs. (XI-40) and (XI-41)), however, suggest that δ_{CD} is positive and at most a few tenths of a percent. A calculation due to Fayans (1971) uses the finite Fermi-gas model (Migdal (1967)) and results in rather smaller values for δ_c.

Calculations of δ_c have also been carried out for some of the $0^+ \to 0^+$ superallowed decays using shell model wave functions for the zero order functions (MacDonald (1958); Blin-Stoyle and Le Tourneux (1961, 1962); Weidenmüller (1962); Lovitch (1963); Blin-Stoyle *et al.* (1963); Damgaard (1969); Khadkikar and Warke (1969); see also Blin-Stoyle (1969b)). All calculations limit the admixed states to those which result (i) from a rearrangement of the nucleons within their ground state configuration, (ii) from the excitation of one nucleon through essentially two oscillator spacings and (iii) exciting two nucleons through one oscillator spacing. In general it is found that by far the largest contribution to δ_c comes from single particle excitations. Some calculations restrict consideration to the effects of the Coulomb potential alone, whilst others include short range charge dependent potentials of the type given in eq. (III-47). All calculations lead to positive values for δ_c although the $T = 2$ and $T = 3$ contributions are themselves negative. According to the assumption made about the charge dependent potential V_{CD} and about the precise details of the nuclear wave functions values of δ_c varying from a fraction of a percent to about $\frac{1}{2}$% (in an extreme case) are obtained. As with the Fermi gas model, the value of δ_c tends to increase with increasing nuclear mass.

Finally, mention should be made of a variational shell model calculation due to Chen *et al.* (1966) which aims at calculating δ_c for the ^{14}O β^+ decay. The calculation uses a fairly realistic internucleon potential together with a Coulomb potential and minimises the internal nuclear energy with respect to the range parameters characterising the shell model wave function. It leads to a value for δ_c of the order $\frac{1}{2}$%, rather larger than values obtained by the perturbation approach.

Bearing in mind the different calculations mentioned it is clear that because of the various approximations made there is a considerable measure of uncertainty about the value of δ_c. Even so it is probably reasonable to assume that δ_c is positive, of the order of a few tenths of a percent *, probably at most not much greater than $\frac{1}{2}\%$ for the decays considered and gradually increasing with the nuclear mass.

III-2.6. *The value of G_V, G'_V and the Cabibbo angle θ*

Having discussed the various corrections which need to be made to the *ft*-values of superallowed transitions we can now set about obtaining a value for G_V and the Cabibbo angle θ.

Referring to eq. (III-35) and bearing in mind the corrections due to K-capture, $(1+\epsilon_K)$, and isospin mixing $(1-\delta_c)$, it is clear that we can write:

$$\mathcal{F}t = \widetilde{ft}\,\overline{C(E)}\,(1+\epsilon_K)(1+\delta'_R) = \frac{K}{2G'^2_V(1-\delta_c)}\,. \qquad \text{(III-48)}$$

Assuming that δ_c is small and positive as described in §III-2.5 it follows from eq. (III-48) that considering the various measurements on $0^+ \to 0^+$ super-allowed transitions the "best" experimental result to take for $\mathcal{F}\,t$ in order to determine G'_V is that which has the lowest value, since this must correspond to δ_c having its smallest value. Inspection of table III-1 shows that the value for $^{26}Al^m$ should be taken namely ** $\mathcal{F}t = 3073 \pm 5$ sec. Eq. (III-48) then leads to the result ***

$$G'_V = G_V\left(1+\frac{\alpha C}{4\pi}\right) = (1.4150 \pm 0.0011)\times 10^{-49}\ \text{erg cm}^3\,. \qquad \text{(III-49)}$$

The cabibbo angle relevant to polar vector decays can be determined from β-decay data by using the relations $G_\mu = G$ and $G_\beta = G_V = G\cos\theta^{(V)}$, i.e.

$$G_V = G_\mu \cos\theta^{(V)} \qquad \text{(III-50)}$$

where G_μ is the μ-decay coupling constant and $\theta^{(V)}$ the Cabibbo angle relevant to the polar vector interaction. Taking account of electromagnetic radia-

* Recently Jaus (1971a) has calculated δ_c neglecting the effect of excited nuclear states and treating the nucleus as an elementary particle whose structure manifests itself through an appropriate nuclear form factor. To order $Z^2\alpha^2$ Jaus obtains contributions to δ_c of the order $< 0.1\%$.

** This value was obtained by Blin-Stoyle and Freeman (1970) by taking a weighted mean of the end point energies of the two results quoted and assumes $r_n = r_p$ (see §III-2.4).

*** A slightly different value ($G'_V = (1.4122 \pm 0.0007)\times 10^{-49}$ erg cm^3) has been obtained by Towner and Hardy (1972) using data on both ^{14}O and $^{26}Al^m$.

tive corrections * (e.g. Matsson (1969)), and using the 1969 values of the fundamental constants (Taylor *et al.* (1969)), $G_\mu = (1.4343 \pm 0.0007) \times 10^{-49}$ erg cm^3.

As derived from the experimental values of G'_V, G_V is of course uncertain to the extent of the model dependent renormalisation factor $\alpha C/4\pi$ in eq. (III-49). However, one can take the value of $\theta^{(V)}$ as derived from strange particle leptonic decay data together with β- and μ-decay data and *deduce* the value of $\alpha C/4\pi$ and hence distinguish between different models of electromagnetic radiative corrections. This has been done by Blin-Stoyle and Freeman (1970) using the value $\sin\theta^{(V)} = 0.221 \pm 0.004$ derived from the data of Botterill *et al.* (1968) on K_{e3} decay. They obtain

$$\Delta_R^{(V)} = \frac{\alpha C}{2\pi} = +[2.3 \pm 0.3 - 10.3(\eta_{SU3} + \tfrac{1}{2}\eta_R)]\% \qquad \text{(III-51)}$$

where η_{SU3} and η_R are correction factors for SU3 symmetry breaking effects and the radiative correction in K_{e3} decay respectively. The latter is negligible (Ginsberg (1966, 1967)) whilst the former has been variously estimated (Furlan *et al.* (1965); Srivastava (1965); Jones (1968); Glashow and Weinberg (1968)) to have values such as to give a contribution to $\Delta_R^{(V)}$ of from +0.1% to +1.5%.

Reference to §III-2.3 shows that only those models of radiative corrections involving a heavy intermediate vector boson can account for the value of $\Delta_R^{(V)}$ given in eq. (III-51). Allowing for all uncertainties in the input data Blin-Stoyle and Freeman (1970) conclude that M_W must be *at least* $20M$, consistent with the value $M_W \approx 40M$ suggested on theoretical grounds by Lee (1971) (see also Schwinger (1957); Salam and Ward (1959)). Recently Fischbach *et al.* (1971a) have formulated a theory of K_{e3} decay based on the Kemmer covariant field equations (Kemmer (1939)) rather than the usual Klein–Gordon equation. On this basis Fischbach *et al.* (1971b) obtain a value for $\sin\theta^{(V)}$ from the K_{e3} decay some 20% smaller than that given above so leading to a value for $\Delta_R^{(V)} \approx 0$ to 1% which does not require the existence of an intermediate vector boson. However, it should be noted that there is some disagreement as to whether this work is correct (Nagel and Snellman (1971); Fischbach *et al.* (1972a)).

* As mentioned in §I-5.1 the precise value of these is in a little doubt (Roos and Sirlin (1971)). However, using the Roos–Sirlin value for G_μ ($G_\mu = (1.4354 \pm 0.0003) \times 10^{-49}$ erg cm^3) would not modify the results of this section at all.

III-2.7. *Other superallowed transitions*

The determination of G_V (or G'_V) from superallowed transitions other than $0^+ \rightarrow 0^+$ is complicated by the fact that the axial vector allowed matrix element is non zero and therefore contributes to the decay rate. However, it is possible to determine the ratio of the axial vector and polar vector contributions to a superallowed β-decay of the type $I \rightarrow I$ $(I \neq 0)$ by measuring the angular distributions of electrons (or positrons) emitted from polarised nuclei. The angular distribution can be written in the form (e.g. Jackson *et al.* (1957a, 1957b))

$$W(\theta) = 1 + A \frac{\langle I \rangle}{I} \cdot \frac{\boldsymbol{p}}{E} \tag{III-52}$$

where $\langle \boldsymbol{I} \rangle$ is the expectation value of the spin of the parent nucleus, $\boldsymbol{p}$ is the momentum and E the energy of the emitted electron (or positron). The quantity A is given by

$$A[G_V^2 |\langle t_\pm \rangle|^2 + G_A^2 |\langle \boldsymbol{\sigma} t_\pm \rangle|^2]$$

$$= \mp \frac{1}{I+1} G_A^2 |\langle \boldsymbol{\sigma} t_\pm \rangle|^2 - 2 \left(\frac{I}{I+1}\right)^{\frac{1}{2}} G_V G_A |\langle t_\pm \rangle| |\langle \boldsymbol{\sigma} t_\pm \rangle| \tag{III-53}$$

where as usual the upper sign refers to electron decay and the lower sign to positron decay.

The simplest example of a decay of this kind is the β-decay of the neutron itself for which $I = \frac{1}{2}$, $|\langle t_\pm \rangle| = 1$, $|\langle \boldsymbol{\sigma} t_\pm \rangle| = \sqrt{3}$ so that

$$A = \frac{-2\lambda(\lambda + 1)}{1 + 3\lambda^2} \tag{III-54}$$

where $\lambda = G_A/G_V$ and is, of course, negative. Recently Christensen *et al.* (1969) have measured A accurately for the decay of polarised neutrons and obtained the value $A = -0.115 \pm 0.008$, corresponding to $\lambda = 1.26 \pm 0.02$ *. In conjunction with the measured ft-value for the neutron it now becomes possible to determine G'_V through the relation (see eq. (III-27))

* An equivalent, but much less accurate, measurement of the coefficient B defining the $\tilde{\nu}_e$ angular distribution with respect to neutron polarisation has been carried out by Erozolimsky *et al.* (1970) which gives $B = 0.995 \pm 0.034$ and $\lambda = -[1.16^{+0.41}_{-0.16}]$.

$$ft(1+\delta_R') = \frac{K}{G_V'^2[1+3\lambda'^2]} \tag{III-55}$$

where $\lambda' = G_A'/G_V'$. For the neutron decay, $\mathcal{E}_0 = 0.78$ MeV so that referring to §III-2.3 and fig. III-3, we have $\delta_R' = +1.5\%$. Shann (1971) has shown that radiative corrections to eq. (III-54) are small and that to a good approximation $A = -2\lambda'(\lambda'+1)/(1+3\lambda'^2)$ so that we can take $\lambda' = -1.26 \pm 0.02$. Using $ft = 1081 \pm 16$ (Christensen *et al.* (1967, 1971) and the foregoing values for λ' and δ_R' then gives $G_V' = (1.395 \pm 0.023) \times 10^{-46}$ erg cm^3 which is just about in agreement with the value for G_V' obtained from the $^{26}Al^m$ data (eq. (III-49)).

Other odd A nuclei lend themselves to similar investigation. For example, Freeman *et al.* (1969c) (see also Cramer and Mangelson (1968)) have made a careful measurement of the $\frac{3}{2}^+ \to \frac{3}{2}^+$ β-decay process $^{35}Ar(\beta^+)^{35}Cl$ which in conjunction with an asymmetry measurement due to Calaprice *et al.* (1967) enables G_V to be determined. Here there is an emphatic and unexplained disagreement between the value of G_V so determined and that derived from the $0^+ \to 0^+$ data. Experiments of this kind are all difficult but should be continued with since, in particular, they can enable values of G_V to be determined for electron as well as positron emitters.

III-2.8. *Exchange effects in superallowed transitions*

As pointed out in §III-1, if CVC theory holds then there should be no exchange effects in allowed polar vector β-decays. It is clear from the foregoing discussion that the experimental results are in agreement with this prediction. Sizeable exchange effects, if they did exist, would be expected to vary from nucleus to nucleus and to destroy the good agreement which clearly exists between the ft-values for the different $0^+ \to 0^+$ superallowed β-decays. This point has been investigated in detail by Blin-Stoyle *et al.* (1965) who conclude that there is no evidence for exchange effects and that if they are present then they are certainly smaller in magnitude than crude estimates which have been made as to their size on the assumption that CVC theory does not hold (Blin-Stoyle *et al.* (1959/60)). Further discussion about the magnitude of polar vector exchange effects is given in §III-4.4 which deals with isospin hindered β-transitions.

III-3. Higher forbidden corrections and transitions

We now go on to consider the implications and also the testing of some other consequences of the assumption of CVC theory. As pointed out in §III-1, the CVC hypothesis leads to precise predictions for the values of the form factors f_V, f_W and f_S. We have already seen in the case of superallowed transitions that the experimental data are consistent with no renormalisation of the polar vector coupling constant, i.e. with $f_V(0) = 1$ as given by CVC theory. In this section we consider first the evidence that is available about the value of f_W and second the testing of some relationships between forbidden matrix elements deriving from CVC theory.

III-3.1. *Weak magnetism*

On the basis of CVC theory, the form factor $f_W(0)$ is predicted to have the value (see eq. (III-11))

$$f_W(0) = -\frac{\kappa_p - \kappa_n}{2M}$$

where κ_p and κ_n are the proton and neutron anomalous magnetic moments. Inspection of the general non-relativistic β-decay Hamiltonian given in eq. (II-37) shows that the largest term involving f_W occurs in the third term in the second square bracket namely

$$-\frac{i}{2M}(f_V - f_W(2M+E))\,\boldsymbol{\sigma} \times \boldsymbol{q} \cdot \boldsymbol{L}^* .$$

Neglecting E compared with M and using the CVC values for $f_V(0)$ and $f_W(0)$ then gives for this term

$$H_\beta(\text{weak magnetism}) = -\frac{\kappa_p - \kappa_n + 1}{2M}(\mathrm{i}\boldsymbol{\sigma} \times \boldsymbol{q} \cdot \boldsymbol{L}^*)$$

$$= -\frac{4.7}{2M}(\mathrm{i}\boldsymbol{\sigma} \times \boldsymbol{q} \cdot \boldsymbol{L}^*) . \qquad \text{(III-56)}$$

A similar term arises corresponding to β^+ decay.

The name "weak magnetism" came to be associated with this part of H_β because of its close similarity to the interaction between the spin magnetic dipole moment and a magnetic field. It is small ($\approx q/M$) and has the order of magnitude of a "second forbidden" contribution and since the term is momentum dependent it is most likely to be detected in a β-decay involving a large energy release. Two beta-decays satisfying this criterion and which have been well investigated are those of ^{12}B and ^{12}N to ^{12}C. These are mirror decays from $I = 1^+$, $T = 1$ states to the $I = 0^+$, $T = 0$ ground state of ^{12}C and satisfy the selection rules for "allowed" Gamow–Teller (Axial Vector) transitions and are therefore dominated by the matrix element $G_A \langle t_\pm \boldsymbol{\sigma} \rangle$ (Gell-Mann (1958); Gell-Mann and Berman (1959)). The essential point to note is that the interference between the allowed axial vector term and the weak magnetism term is of opposite sign in the β^+ and β^- decays. Hence in the ratio between the two spectrum shape factors *, whereas the weak magnetism contributions reinforce one another, contributions from other second forbidden terms cancel. Further, because of the assumption of CVC theory the matrix element for the spin contribution to the corresponding $I = 1^+$, $T = 1 \rightarrow I = 0^+$, $T = 0$ M1 γ-transition in ^{12}C is identical (apart from a well-defined constant factor) with the weak magnetism β-decay matrix element. As shown by Weidenmüller (1960) the orbital contribution to the M1 γ-transition is negligible so that the strength of the weak magnetism term can be derived directly from the radiation width of the $T = 1$ state in ^{12}C.

Detailed calculations of the spectrum shape have been carried out by Gell-Mann and Berman (1959), Morita (1959) and Huffaker and Greuling (1962) and all are in agreement that $R(E) = [S(E)]_{^{12}\mathrm{B}} / [S(E)]_{^{12}\mathrm{N}}$ has the essential form $R(E) = 1 + AE$ where the constant A has the value $A = (1.10 \pm 0.17)\%$ MeV. The following experimental values for A have been obtained

$A = (1.07 \pm 0.24)\%$ MeV: Lee *et al.* (1963)

$= (1.30 \pm 0.31)\%$ MeV: Mayer-Kuckuk and Michel (1962)

$= (1.62 \pm 0.28)\%$ MeV: Glass and Peterson (1963).

There is thus good agreement between the experimental and CVC values for A.

Another experiment for investigating weak magnetism and involving a measurement of the β–α angular correlations in the β-decays of ^{8}Li and ^{8}B was suggested by Bernstein and Lewis (1958). This experiment is rather less

* The spectrum shape is taken to be of the form $S(E) P_0(E)$ where $P_0(E)$ is the usual allowed spectrum and $S(E)$ is the shape factor.

conclusive than the foregoing one since its interpretation rests to some extent on nuclear model calculations (Kurath (1960); Weidenmüller (1960)) and also other second forbidden contributions may be important (Soergel *et al.* (1965)). However, with this proviso it is found that the experimental results (Nordberg *et al.* (1960, 1962); Gruhle *et al.* (1963)) are again consistent with CVC theory.

The experimental data on weak magnetism are thus in good agreement with the predictions of CVC theory. It is important, however, to increase the accuracy of this work and to extend it to other situations (e.g. electron–neutrino angular correlations, (Gell-Mann (1958); Morita (1959)) since at present it is difficult to estimate the size of weak magnetism effects should CVC theory not hold. The more exact the agreement between theory and experiment the more confidence will one have in the hypothesis of CVC theory.

III-3.2. *Forbidden transitions*

In general forbidden transitions are extremely complicated to analyse since (a) a number of different terms in the β-decay Hamiltonian contribute and (b) theoretical estimates of the nuclear matrix elements of these different terms are subject to uncertainty because of lack of knowledge about nuclear wave functions. However, some matrix elements deriving from the polar vector current can be simply related to one another if CVC theory holds so that experimental measurement of these matrix elements can lead to further tests of CVC theory. In order to discuss these relations we return to the basic properties of the polar vector current $\mathcal{V}_\lambda$.

Assuming CVC theory we have $\partial_\lambda \mathcal{V}_\lambda = 0$. This relation is exact if the electromagnetic interaction is neglected. Taking it into account a source term is introduced into the continuity equation, which now take the form

$$\left(\partial_\lambda + \mathrm{i}\frac{e}{c}A_\lambda\right)\mathcal{V}_\lambda = 0 \tag{III-57}$$

where A_λ is the vector potential of the electromagnetic field. The above equation results either from a consideration of the field equations or from more general arguments (Eichler (1963); Adler (1965a)).

Writing $\mathcal{V}_4 = \mathrm{i}\rho$ eq. (III-57) can be written

$$\operatorname{div} \boldsymbol{\mathcal{V}} + \frac{\partial \rho}{\partial t} = -\mathrm{i}\frac{e}{c}A_\lambda \mathcal{V}_\lambda \,. \tag{III-58}$$

At this point it is convenient to label explicitly the isospin character of $\mathcal{V}_\lambda$

(which transforms like the + component of an isovector; see §III-1) and to write $\mathcal{V}_+, \rho_+$ for its four components. Bearing in mind that the 3-component in isospace of this same current is the isovector component of the electromagnetic current ot follows that $eA_\lambda \mathcal{V}_\lambda$ is simply related to the electromagnetic interaction (eq. (I-5)). Thus eq. (III-58) can be written

$$\operatorname{div} \mathcal{V}_+ + \frac{\partial \rho_+}{\partial t} = \mathrm{i}\,\mathcal{H}^{(\mathrm{V})}_{(\mathrm{e.m})_+} \tag{III-59}$$

where $\mathcal{H}^{(\mathrm{V})}_{(\mathrm{e.m})_+}$ is the Hamiltonian density for the isovector part of the electromagnetic interaction rotated into the + direction in isospin space. Similarly for $\mathcal{V}^*_\lambda \equiv (\mathcal{V}_-, \mathrm{i}\rho_-)$

$$\operatorname{div} \mathcal{V}_- + \frac{\partial \rho_-}{\partial t} = -\mathrm{i}\,\mathcal{H}^{(\mathrm{V})}_{(\mathrm{e.m})_-}\,. \tag{III-60}$$

By considering the identity

$$\int \mathcal{V}_\pm \,\mathrm{d}^3 r = -\int r \operatorname{div} \mathcal{V}_\pm \,\mathrm{d}^3 r \tag{III-61}$$

and using the Heisenberg relation

$$\frac{\partial \rho_\pm}{\partial t} = \mathrm{i}\,[H, \rho_\pm] \tag{III-62}$$

where H is the total Hamiltonian, it follows from eqs. (III-59) and (III-60) that

$$\int \mathcal{V}_\pm \,\mathrm{d}^3 r = -\mathrm{i} \int r \,\{[\rho_\pm, H] \pm \mathrm{i}\,\mathcal{H}^{(\mathrm{V})}_{(\mathrm{e.m})_\pm}\}\,\mathrm{d}^3 r\,. \tag{III-63}$$

On the basis of *CVC* theory, $\rho_\pm(r)$ for a nucleus can be written

$$\rho_\pm(r) = \sum_i t_\pm^{(i)} \delta(r - r_i) \tag{III-64}$$

in complete analogy with the isovector component of the nuclear charge distribution ($\rho^{(\mathrm{e.m})} = \Sigma\,(\tfrac{1}{2} + t_3^{(i)})\,\delta(r - r_i)$); this result is a simple application of the Siegert Theorem (Siegert (1937); Sachs (1953)). It is then easy to verify that

$$\mathcal{H}^{(V)}_{(e.m)\pm} = \mp [\rho_\pm(r), H_{e.m.}] \tag{III-65}$$

where $H_{e.m.}$ is the electromagnetic part of the nuclear Hamiltonian (see §XI-2) i.e.

$$H_{e.m.} = \sum_{i<j} (\tfrac{1}{2} + t_3^{(i)})(\tfrac{1}{2} + t_3^{(j)})\frac{e^2}{r_{ij}} + \sum_i t_3^{(i)}(M_p - M_n) \tag{III-66}$$

and a term has been included to allow for the neutron–proton mass difference.

Inserting eq. (III-65) into eq. (III-63) gives

$$\int \mathcal{V}_\pm \, d^3r = -i \int r[\rho_\pm, H - H_{e.m.}] \, d^3r \, . \tag{III-67}$$

Taking matrix elements of eq. (III-67) between initial and final states of a β-decay process which are eigenfunctions of H gives

$$\langle f | \int \mathcal{V}_\pm \, d^3r | i \rangle = -i \langle f | \sum_i r_i t_\pm^{(i)} | i \rangle \, [(E_i - E_f) \mp \tfrac{1}{2}(M_n - M_p)]$$

$$+ \, i \langle f | [\sum_i r_i t_\pm^{(i)}, V_C] | i \rangle \tag{III-68}$$

where use has been made of eqs. (III-64) and (III-66) and the Coulomb part of $H_{e.m.}$ has been denoted by V_C.

The last term can be evaluated reasonably accurately by inserting a complete set of states between the two terms in the commutator bracket and retaining only diagonal matrix elements of V_C (an approximation due to Ahrens and Feenberg (1952) and usually estimated to be good to about 10% – but see later). This leads to the final result

$$\langle f | \int \mathcal{V}_\pm \, d^3r | i \rangle = -i \langle f | \sum_i r_i t_\pm^{(i)} | i \rangle \, [(E_i - E_f) \mp \tfrac{1}{2}(M_n - M_p) \pm E_C] \tag{III-69}$$

where E_C is the magnitude of the Coulomb energy difference between the initial and final nuclear states. It is usual to replace E_C by the semi-classical value $\frac{6}{5} Z\alpha/R$ where R is the nuclear radius.

The above result was first obtained by Fujita (1962a, 1962b) and Eichler (1963) and similar relations can be obtained for higher forbidden matrix elements (Fujita (1962b); Eichler (1963); see also Stech and Schülke (1964);

Blin-Stoyle and Nair (1966); Schopper (1966)). Other aspects of this relation have been discussed by Spector (1963a) and Fujii and Fujita (1965). It is usually written

$$\int \boldsymbol{\alpha} = -\mathrm{i} \int r\,[(E_{\mathrm{i}} - E_{\mathrm{f}}) \mp \tfrac{1}{2}(M_{\mathrm{n}} - M_{\mathrm{p}}) \pm E_{\mathrm{C}}] \qquad \text{(III-70)}$$

in terms of a more conventional terminology for the two nuclear matrix elements *.

Investigating the extent to which eq. (III-70) and other similar relations hold obviously provides a test of CVC theory. Most experimental and theoretical work in this direction has been carried out on the first forbidden $1^- \to 0^+$ β-decay of Ra*E* (^{210}Bi) since, due to an accidental cancellation, the spectrum shape differs considerably from the "allowed" shape. The situation here is that the *ft*-value, spectrum shape and the electron longitudinal polarisation are determined by the magnitudes of three nuclear matrix elements: $\int \boldsymbol{\alpha}$, $\int \boldsymbol{r}$ and $\int \boldsymbol{\sigma} \times \boldsymbol{r}$. The analysis of the experimental results and the associated theoretical work is very complicated (see e.g. Ullman (1962); Daniel (1962); Bühring (1963b); Deutsch and Lipnik (1965) for experimental analyses and Newby and Konopinski (1959); Banerjee and Zeh (1960); Fujita (1962a); Spector and Blin-Stoyle (1962); Spector (1963b); Kim and Rasmussen (1963); Sodemann and Winther (1965) for theoretical work). Wu and Moszkowski (1966, p. 87 *et seq*) also give a fairly detailed discussion of this decay and the general conclusion seems to be that agreement with eq. (III-70) is obtained.

However, as with other tests of CVC theory it is not clear how eq. (III-70) would be modified if CVC theory did not hold. Crude estimates for this situation have been made by Pursey (1951) and Ahrens and Feenberg (1952) but the results are not in agreement and the situation is very uncertain (see Blin-Stoyle (1964)) because of the approximations made. As with weak magnetism it is most important in order to establish the correctness of CVC theory that the relation given in eq. (III-70) should be checked as accurately as possible. It should also be remembered that in eq. (III-70) the approximation has been made of neglecting off-diagonal matrix elements of the Coulomb interaction and this approximation has been shown by Damgaard and Winther (1965) to be bad in some cases (see also Fujita (1967b); Fayans and Khodel (1969)).

It is also important to use and test eq. (III-70) in connection with other

* See Wu and Moszkowski (1966) p. 91 for a discussion of the sign conventions used with this notation.

forbidden decays and theoretical and experimental work in this direction has been carried out, for example, by Damgaard and Winther (1964)-decay of ^{207}Tl, ^{209}Pb; Lipnik and Sunier (1964)-decay of ^{124}Sb, ^{152}Eu, ^{154}Eu, ^{72}Ga and ^{140}La; Camp *et al.* (1965)-decay of ^{72}Ga and ^{124}Sb; Simms (1965)-decay of ^{86}Rb and ^{84}Rb; Bogdan and Vâtă (1968)-decay of ^{170}Tm; Bogdan *et al.* (1968)-decay of ^{186}Re, ^{184}Re, ^{170}Tm; Hocquenghem and Berthier (1968)-decay of ^{94}Nb. Most of this work assumes CVC theory and uses eq. (III-70) in order to help analyse the decays under consideration. It is confused to a considerable extent by theoretical and experimental uncertainties and much more work in this area is needed.

III-4. **Isospin hindered Fermi transitions**

In this section we discuss the situation in which the Fermi (polar vector) allowed matrix element, $\langle t_\pm \rangle$, is measured for a β-transition between members of two *different* isospin multiplets. As will emerge, if there are no charge dependent terms present in the nuclear Hamiltonian and if CVC theory holds, then such matrix elements should be identically zero. Any deviation from zero is related to the magnitude of isospin mixing and can therefore give information about isospin purity and the charge dependence of nuclear forces. It can also serve as a measure of the validity of CVC theory.

From the point of view of determining $\langle t_\pm \rangle$ two classes of transition have to be distinguished. Firstly, there are transitions of the type $0^+ \to 0^+$, Δ Multiplet = "Yes" where, in general, a change in isospin multiplet (Δ Multiplet) implies a change in isospin (i.e. $\Delta T \neq 0$) but not necessarily so. Secondly, there are transitions of the type $I^+ \to I^+$, $(I \neq 0)$, Δ Multiplet = "Yes". The important difference between these two categories lies in the fact that in the former there can be no "allowed" contribution from the axial vector current whereas there can be such a contribution in the latter. This means that in the case of $0^+ \to 0^+$ transitions, since $\langle t_\pm \rangle$ is expected to be small the ft-value is much larger than the "allowed" range and $\langle t_\pm \rangle$ can be deduced from it. In the case of $I^+ \to I^+$ $(I \neq 0)$ transitions, there is an allowed contribution from the axial vector current so that a typical "allowed" ft-value is expected. Furthermore, in this case, since two matrix elements now contribute, $\langle t_\pm \rangle$ cannot be determined from the ft-value alone and recourse has to be made to the measurement of electron decay asymmetries or of β–γ circularly polarised (CP) angular correlations.

III-4.1. *The theory of isospin hindered Fermi transitions*

The basic situation is represented in fig. III-4 which is modelled on the diagrams of Atkinson *et al.* (1968).

In the case of β^--decay the transition is between the parent state Ⓟ with quantum numbers I^π, T, T_3 (denoted by $|P:I^\pi, T, T_3\rangle$) and the state labelled $(T_<)$ with quantum numbers I^π, T', T_3+1 (denoted by $|T:I^\pi, T', T_3+1\rangle$). The state Ⓟ by definition is the state in the β-decay with the lowest value of T_3. It is the initial state for β^- decay and the final state for β^+ decay. The state Ⓐ denoted by $|A:I^\pi, T, T_3+1\rangle$ is the analogue of the parent state Ⓟ and is related to it by

$$|A:I^\pi, T, T_3+1\rangle = NT_+|P:I^\pi, T, T_3\rangle \tag{III-71}$$

where the normalisation constant N is given by

$$N = \frac{1}{\sqrt{(T-T_3)(T+T_3+1)}}\,. \tag{III-72}$$

The state Ⓖ is included to indicate a possible γ-decay following the β-decay.

If *CVC* theory holds the Fermi matrix for the β-decay is given precisely by

$$\langle t_+\rangle = \langle T_<:I^\pi, T', T_3+1|T_+|P:I^\pi, T, T_3\rangle = 0 \tag{III-73}$$

since the initial state Ⓟ and final state $(T_<)$ belong to different isospin

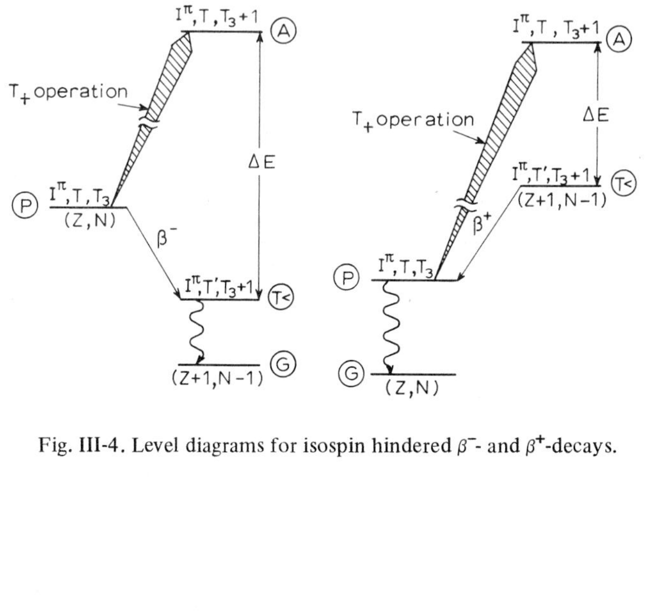

Fig. III-4. Level diagrams for isospin hindered β^-- and β^+-decays.

multiplets. However, because of charge dependent effects, Ⓟ and $(T_<)$ are not completely pure isospin states and, in particular, the state Ⓐ is admixed into $(T_<)$ with an amplitude α (say). Considering this admixture only (since other admixtures lead only to second order effects) we have for the initial and final states of the β-decay,

$$
\begin{aligned}
|\mathrm{i}\rangle &= |P:I^\pi, T, T_3\rangle \\
|\mathrm{f}\rangle &= |T:I^\pi, T', T_3+1\rangle + \alpha|A:I^\pi, T, T_3+1\rangle
\end{aligned}
\tag{III-74}
$$

which then leads to

$$
\begin{aligned}
\langle t_+\rangle &= \langle \mathrm{f}|T_+|\mathrm{i}\rangle \\
&= \alpha\langle A:I^\pi, T, T_3+1|T_+|P:I^\pi, T, T_3\rangle \\
&= \alpha\sqrt{(T-T_3)(T+T_3+1)}
\end{aligned}
\tag{III-75}
$$

by eqs. (III-71) and (III-72). An identical result is obtained for β^+ decay.

Thus measurement of $\langle t_\pm\rangle$ allows α to be determined where α is the amplitude with which the analogue state is admixed into either the initial or final state of the β-decay. It should be noted that since there are other admixtures (which do not, however, contribute in first order to $\langle t_\pm\rangle$), $|\alpha|^2$ does not represent the *total* isospin impurity. For all decays of this type which have been studied, it turns out that $|T_3|$ takes its maximum value T, in which case for both β^- and β^+ decay,

$$\langle t_\pm\rangle = \alpha\sqrt{2T}. \tag{III-76}$$

Consider now the quantity α. If H_{CD} represents all charge dependent terms in the nuclear Hamiltonian then a straightforward application of first order perturbation theory gives

$$
\begin{aligned}
\alpha &= -\frac{\langle T:I^\pi, T', T_3+1|H_{\mathrm{CD}}|A:I^\pi, T, T_3+1\rangle}{\Delta E} \\
&= -\frac{\langle H_{\mathrm{CD}}\rangle}{\Delta E}
\end{aligned}
\tag{III-77}
$$

where ΔE is the magnitude of the energy separation between the state $(T_<)$ and Ⓐ (see fig. III-4).

ΔE can be simply related to the total energy release $E_0(\beta^\pm)$ in the $\beta^\pm$-decays and the energy difference ΔE_{CD} between the parent Ⓟ and analogue Ⓐ states due to charge dependent forces. The appropriate relations are obvious (see fig. III-4) and are as follows:

$$\beta^-\text{-decay}: \Delta E = \Delta E_{\text{CD}} + E_0(\beta^-) - (M_\text{n} - M_\text{p})c^2 \qquad \text{(III-78)}$$

$$\beta^+\text{-decay}: \Delta E = \Delta E_{\text{CD}} - E_0(\beta^+) - (M_\text{n} - M_\text{p})c^2 \,. \qquad \text{(III-79)}$$

Theoretical and experimental values for ΔE_{CD} are discussed by Anderson *et al.* (1965) but very accurate values are not necessary. Bloom (1964), for example, has used the relation

$$\Delta E_{\text{CD}} = E_\text{C}(Z+1) - E_\text{C}(Z) = \frac{1.36}{A^{1/3}} \{(Z+\tfrac{1}{2}) - 0.51 Z^{-1/3}\} \text{ MeV} \qquad \text{(III-80)}$$

where $E_\text{C}(Z)$ is the Coulomb energy of a nucleus of charge Ze and mass number A. The first term is the classical Coulomb displacement energy whilst the second term comes from the exchange and self-energy terms in the Coulomb energy (see Swamy and Green (1958); Sengupta (1961)). The above expression neglects short range charge dependent contributions to ΔE_{CD} but these are expected to be small compared with the long range contribution from the Coulomb potential. Certainly eq. (III-80) agrees well with experimental data where this is known (Anderson *et al.* (1965)). Bhattacherjee *et al.* (1967) have used a similar relation based on a semi-empirical fit to the experimentally observed energies of the isobaric analogue states in a wide mass region due to Anderson *et al.* (1965), namely

$$\Delta E_{\text{CD}} = E_\text{C}(Z+1) - E_\text{C}(Z) = \left[(1.444 \pm 0.005)\frac{Z+\frac{1}{2}}{A^{1/3}} - (1.13 \pm 0.04)\right] \text{MeV} \,. \qquad \text{(III-81)}$$

In making use of the foregoing relationships there is firstly the problem of determining $\langle t_\pm \rangle$ and $\langle H_{\text{CD}} \rangle$ from experimental data and secondly the problem of giving a theoretical interpretation of the value of $\langle H_{\text{CD}} \rangle$ in terms of nuclear model calculations assuming specific forms for H_{CD} and assuming *CVC* theory.

III-4.2. *The experimental determination of* $\langle t_\pm \rangle$, α *and* $\langle H_{\mathrm{CD}} \rangle$

III-4.2.1. *Basic theory*

For an allowed β-decay the ft-value is related to $\langle t_\pm \rangle$ and $\langle \boldsymbol{\sigma} t_\pm \rangle$ by the following expression (see eq. (II-62))

$$ft = \frac{K}{G_{\mathrm{V}}^2 |\langle t_\pm \rangle|^2 + G_{\mathrm{A}}^2 |\langle \boldsymbol{\sigma} t_\pm \rangle|^2} \tag{III-82}$$

where $K = 2\pi^3 \ln 2$.

For the case of a $0^+ \to 0^+$ superallowed transition within a $T = 1$ isospin multiplet $|\langle t_\pm \rangle|^2 = 2$, $|\langle \boldsymbol{\sigma} t_\pm \rangle|^2 = 0$, so that

$$(ft)_{\mathrm{superallowed}} = \frac{K}{2G_{\mathrm{V}}^2}\,. \tag{III-83}$$

Comparing eqs. (III-82) and (III-83) gives

$$|\langle t_\pm \rangle|^2 = \frac{2(ft)_{\mathrm{superallowed}}\, y^2}{ft(1+y^2)} \tag{III-84}$$

where $y = (G_{\mathrm{V}} \langle t_\pm \rangle)/(G_{\mathrm{A}} \langle \boldsymbol{\sigma} t_\pm \rangle)$. Finally, using eq. (III-76) for $\langle t_\pm \rangle$ the following expression is obtained for α^2

$$\alpha^2 = \frac{(ft)_{\mathrm{superallowed}}\, y^2}{ft(1+y^2)\,T}\,. \tag{III-85}$$

In the particular case of a $0^+ \to 0^+$ isospin hindered transition, $\langle \boldsymbol{\sigma} t_\pm \rangle = 0$ and $y \to \infty$ so that

$$\alpha^2 = \frac{1}{T} \frac{(ft)_{\mathrm{superallowed}}}{ft} \tag{III-86}$$

and the magnitude of α can be determined directly from the measured ft-value. In the case of an $I^+ \to I^+$ $(I \neq 0)$ transition information is also needed about the magnitude of y. y is a quantity which appears in theoretical expressions for phenomena involving interference between the polar vector and axial vector interactions in a β-decay. Two experimental measurements sensitive to such interference are (i) the angular distribution of leptons from oriented nuclei and (ii) the $\beta-\gamma$ (CP) * angular correlation from unoriented nuclei.

For case (i) assuming a V–A time reversal invariant theory, the angular distribution of the electrons (or positrons) has the form

$$W_\beta(\theta) = 1 + PA_{I\beta} \cos\theta \qquad \text{(III-87)}$$

where $P = (\langle I_3 \rangle / I)$ is the nuclear polarisation, and the asymmetry parameter $A_{I\beta}$ is given by (e.g. Jackson *et al.* (1957a))

$$A_{I\beta} = \frac{1}{1+y^2}\left[\mp \frac{1}{I+1} - 2|y|\left(\frac{I}{I+1}\right)^{1/2}\right]\frac{p}{E}. \qquad \text{(III-88)}$$

Here the upper sign refers to β^- decay and the lower to β^+ decay; p and E are the magnitudes of the lepton's momenta and energy respectively.

For a $\beta-\gamma$ (CP) angular correlation in which the β-decay between two states of spin I is followed by a mixed γ-transition to a state of spin I', the angular correlation has the form (e.g. Alder *et al.* (1957); Mann *et al.* (1965))

$$W_{\beta\gamma}(\theta) = 1 + \tau(v/c) A_{\beta\gamma} \cos\theta \qquad \text{(III-89)}$$

where

$$A_{\beta\gamma} = \frac{\sqrt{3}/6}{1+\delta^2}\left(\frac{1}{1+y^2}\right)\left\{\mp \frac{2}{[I(I+1)]^{1/2}} + 4y\right\}$$
$$\times\, [F_1(\lambda,\lambda,I',I) + \delta^2 F_1(\lambda+1,\lambda+1,I',I) + 2\delta F_1(\lambda,\lambda+1,I',I)]\,.$$

The upper sign refers to β^--decay and the lower to β^+-decay; δ is the usual ratio of the matrix elements for a mixed $2^{\lambda+1}$-pole, 2^λ-pole γ-transition; $\tau = +1$ or -1 corresponds to positive or negative circular polarisation respectively; v is the speed of the lepton; the functions F are those usually given in angular correlation theory and are tabulated, for example, by Jackson *et al.* (1957b) and more recently by Frauenfelder and Steffen (1965a).

Thus, from asymmetry experiments of the type (i) or (ii) (usually the latter in practice) it is possible to determine y and hence, using eq. (III-85) and the measured ft-value, α can be determined.

* (CP) signifies Circularly Polarised.

III-4.2.2. *Experimental results for* $\langle t_\pm \rangle$, α *and* $\langle H_{CD} \rangle$

During the last few years a large number of measurements have been made of $\langle t_\pm \rangle$ (and hence α and $\langle H_{CD} \rangle$) in various isospin hindered beta-decays. These measurements involve the procedures outlined in the last section. The experimental results have been periodically gathered together (e.g. Bloom (1964 and 1966); Daniell and Schmitt (1965); Bhattacherjee *et al.* (1967); Behrens (1967)). Unfortunately in many cases there are large disagreements between different experimental groups. These need resolving. Meanwhile we content ourselves by presenting a diagram (based on one due to Bloom (1964)) showing the general variation of the matrix element $\langle H_{CD} \rangle$ with mass number A and by giving a selection of detailed results for cases in which corresponding theoretical investigations have been carried.

The overall situation is summarised in fig. III-5 in which $\langle H_{CD} \rangle$ is plotted on a logarithmic scale against A. Note that experiment does not in general determine the sign of $\langle H_{CD} \rangle$. The values of $|\langle H_{CD} \rangle|$ lie mainly in the region between 1 keV and 40 keV although a few exceptions at both extremes will be noted. In table III-3 which is based on the compilation of Bhattacherjee *et al.* (1967)), the details of selected isospin hindered transitions are given. Column I gives the decay; II and III the initial and final isospins (T_i and T_f);

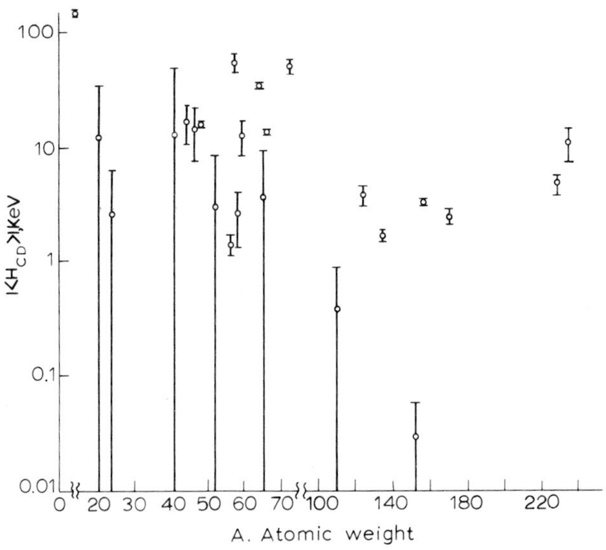

Fig. III-5. Logarithmic plot of $|\langle H_{CD} \rangle|$ against A.

Table III-3
Isospin hindered β-decay (based on Bhattacherjee *et al.* (1967)) *.

I	II	III	IV	V	VI	VII
β-decay	T_i	T_f	$\|M_V\| \cdot 10^3$	$\alpha \cdot 10^3$	ΔE (MeV)	$\|\langle H_{CD}\rangle\|$ (keV)
$^{20}F \xrightarrow{\beta^-} {}^{20}Ne$	1	0	2.0 ± 3.5	1.4 ± 2.5	8.56	12 ± 22
$^{24}Na \xrightarrow{\beta^-} {}^{24}Mg$	1	0	0.7 ± 1.0	0.5 ± 0.7	5.24	2.6 ± 3.7
$^{24}Al \xrightarrow{\beta^+} {}^{24}Mg$	1	0	0.3 ± 3.7	0.2 ± 2.5	5.41	1.1 ± 13.5
$^{41}A \xrightarrow{\beta^-} {}^{41}K$	$\frac{5}{2}$	$\frac{3}{2}$	4.0 ± 11.4	1.8 ± 5.1	7.04	12.7 ± 36.0
$^{44}Sc \xrightarrow{\beta^+} {}^{44}Ca$	1	2	8.4 ± 3.0	4.2 ± 1.5	3.98	16.7 ± 6.0
$^{52}Mn \xrightarrow{\beta^+} {}^{52}Cr$	1	2	1.0 ± 1.8	0.5 ± 0.9	5.98	3.0 ± 5.4
$^{56}Co \xrightarrow{\beta^+} {}^{56}Fe$	1	2	0.5 ± 0.1	0.25 ± 0.05	5.57	1.4 ± 0.3
$^{57}Ni \xrightarrow{\beta^+} {}^{57}Co$	$\frac{1}{2}$	$\frac{3}{2}$	14.4 ± 2.6	8.3 ± 1.5	6.51	54 ± 10
$^{65}Ni \xrightarrow{\beta^-} {}^{65}Cu$	$\frac{9}{2}$	$\frac{7}{2}$	1.3 ± 1.9	4.0 ± 6.3	9.01	3.6 ± 5.7
$^{156}Eu \xrightarrow{\beta^-} {}^{156}Gd$	15	14	1.02 ± 0.05	0.19 ± 0.01	17.30	3.3 ± 0.2
$^{170}Lu \xrightarrow{\beta^+} {}^{170}Yb$	14	15	1.03 ± 0.15	0.19 ± 0.03	13.05	2.5 ± 0.4
$^{234}Np \xrightarrow{\beta^+} {}^{234}U$	24	25	$4.2\,^{+1.6}_{-1.0}$	0.6 ± 0.2	17.94	10.8 ± 3.6

* Note the error in their entry for ^{20}F.

IV the Fermi matrix element $|\langle t_\pm \rangle| \times 10^3$; V the isospin admixture parameter $\alpha \times 10^3$; VI the energy separation ΔE between the analogue state Ⓐ and the state $(T_<)$ (see eqs. (III-80) and (III-81)); VII the magnitude of $|\langle H_{CD}\rangle|$.

Before commenting on the values of $|\langle H_{CD}\rangle|$ obtained from experiment we consider the various relevant theoretical calculations that have been carried out.

III-4.3. *The theoretical calculation of* $\langle H_{CD}\rangle$

The basic theoretical problem is to calculate the matrix element of $\langle H_{CD}\rangle$ which has the essential form (see eq. (III-77))

$$\langle H_{\mathrm{CD}}\rangle = \langle T_{<}:I^{\pi}, T', T_3+1\,|H_{\mathrm{CD}}|\,A:I^{\pi}, T, T_3+1\rangle\,. \qquad \text{(III-90)}$$

The earliest calculations of $\langle H_{\mathrm{CD}}\rangle$ were those of Kelly and Moszkowski (1960) and Bouchiat (1960), using the simple nuclear shell model. Bouchiat used very simple jj-configurations for the decays of ^{24}Na, ^{44}Sc and ^{52}Mn and for H_{CD} used the usual Coulomb potential. The calculations were straightforward and yielded a result for $\langle H_{\mathrm{CD}}\rangle$ in agreement with the experimental data for ^{52}Mn but in disagreement by an order of magnitude for ^{44}Sc and ^{24}Na. These early discrepancies led Blin-Stoyle and Novakovic (1964), Blin-Stoyle and Yap (1966) and Yap (1967) to reinvestigate these cases and some additional ones taking into account isospin mixing due to both the Coulomb potential, V_{C}, and the phenomenological charge dependent potential V_{CD} introduced in §XI-3.1. (see also §III-2.5), namely

$$V_{\mathrm{CD}} = V_0 \sum_{i<j} \{[p + r\boldsymbol{\sigma}^{(i)} \cdot \boldsymbol{\sigma}^{(j)}](\tau_3^{(i)} + \tau_3^{(j)}) + [q + s\boldsymbol{\sigma}^{(i)} \cdot \boldsymbol{\sigma}^{(j)}]\, T^{(i,j)}\}\, \mathrm{e}^{-r_{ij}^2/\mu^2} \qquad \text{(III-91)}$$

with $V_0 \approx -50$ MeV and $\mu = 1.73$ fm. The terms in p and r are isovector in character and those in q and s, isotensor ($T^{(i,j)} = \tau_3^{(i)}\tau_3^{(j)} - \frac{1}{3}\boldsymbol{\tau}^{(i)} \cdot \boldsymbol{\tau}^{(j)}$). The nuclear states were described by a closed shell core together with extra-core nucleons coupled to total angular momentum I^{π} in a jj-coupling scheme classified by seniority and reduced isobaric spin (Flowers (1951)). The results (see table III-4) were found (Yap (1967)) to be in agreement with experiment (except in the case of the $A = 24$ decays, *q.v.*) subject to p and r being of the order a fraction of a percent, and q and s being limited as follows: $2\% \lesssim q \lesssim 4\%$; $\frac{1}{2}\% \lesssim s \lesssim 1\frac{1}{2}\%$. These relations are consistent with what is known about the charge dependence of nuclear forces from nucleon–nucleon scattering data (§XI-3) and indeed we have here a means of seeking further information about the detailed nature of this charge dependence. Thus Yap, Tee and Yalcin (Yalcin and Yap (1970); Yap and Tee (1971a, b) and Tee and Yap (1971a, 1971b)) have made direct calculations of isospin hindered Fermi matrix elements for decays in ^{20}F, ^{24}Na, ^{41}Ar, ^{44}Sc and ^{52}Mn using explicit forms for the charge dependent potential derived from one boson exchange models (see §XI-3.1) including ρ–ω–ϕ and η–π mixing. General agreement between theory and experiment is again obtained.

In the case of the $A = 24$ decays it has been suggested that the unsatisfactory result stems from the fact that $A = 24$ nuclei are distorted. For this reason, further calculations were carried out by Blin-Stoyle and Yap (1966) using Nilsson (1955) wave functions to describe the nuclear states. Because of

Table III-4
Theoretical and experimental values for $\langle H_{CD} \rangle$.

Decay	$\langle H_{CD} \rangle$theory (keV)	$\lvert\langle H_{CD} \rangle\rvert$expt. (keV)
$^{20}F \rightarrow {}^{20}Ne$	$-(44p - 160r - 35)$	12 ± 22
$^{24}Na \rightarrow {}^{24}Mg$	$-(59p + 180r + 46)$	2.6 ± 3.7
$^{24}Al \rightarrow {}^{24}Mg$	$+(59p - 180r - 46)$	1.1 ± 13.5
$^{41}A \rightarrow {}^{41}K$	$-(22(p-q) - 54(r-s))$	12.7 ± 36.0
$^{44}Sc \rightarrow {}^{44}Ca$	$-(11(p-q) - 33(r-s))$	16.7 ± 6.0
$^{52}Mn \rightarrow {}^{52}Cr$	$-(-19(p+q) + 44(r+s) + 30)$	3.0 ± 5.4
$^{56}Co \rightarrow {}^{56}Fe$	$-(2p - q + 3r + s + 3)$	1.4 ± 0.3

a K-selection rule the theoretical value for $\langle H_{CD} \rangle$ is reduced by two orders of magnitude bringing theory and experiment into agreement. This effect does, however, indicate the sensitivity of calculations of this type to the nuclear structure adopted (Tee and Yap (1971a)).

Deformed shell model calculations have also been carried out in another region of the periodic table by Damgaard (1966). He considers the decays of ^{156}Eu, ^{170}Lu and ^{234}Np using wave functions corresponding to a distorted core with two extra-core nucleons in Nilsson orbitals. For H_{CD} he uses the Coulomb potential generated by the distorted core. His theoretical results are in reasonable agreement with experiment.

III-4.4. *General consideration of isospin hindered β-decays*

The foregoing analysis of isospin hindered β-decays shows that in principle there is the possibility of exploring details of the charge dependence of nuclear forces and also, insofar as there is agreement between theory and experiment, of implicitly testing *CVC* theory. Unfortunately the analysis does depend upon calculations using model wave functions and the case of ^{24}Na illustrates the extent to which the theoretical value for $\langle H_{CD} \rangle$ can be sensitive to the model used. In this connection it has been pointed out by various authors (e.g. Mann *et al.* (1967); Atkinson *et al.* (1968)) that a small admixtures of the anti-analogue state * into the state $T_<$ by the ordinary nuclear

* Basically an anti-analogue state is one belonging to the same shell model configuration as the analogue state, having the same spin and parity, but different isospin and whose wave function is simply related to the analogue state. See Lane and Soper (1962a, b); Sherr *et al.* (1965); De Toledo Piza *et al.* (1966).

force can lead to large values for $\langle H_{CD} \rangle$. Atkinson *et al.* (1968) also point out that anomalous values for $\langle H_{CD} \rangle$ can result as in ^{24}Na, from the operation of a model selection rule. They quote the case of the decay of ^{65}Ni for which $\langle H_{CD} \rangle_{expt} = 3.6 \pm 5.7$ keV and attribute the small value of $\langle H_{CD} \rangle$ to the fact that the $(T_<)$ and (A) states in ^{65}Cu have a phonon content differing by one.

Such uncertainties as these mean that much more detailed theoretical study of these isospin hindered transitions is necessary *. It is also important to make experimental and theoretical studies of beta-decays involving particularly simple configurations. One such has been suggested by Damgaard (1968) namely the β^- decay of the $I = \frac{3}{2}^-$, $T = \frac{9}{2}$ ground state of ^{49}Ca to the $I = \frac{3}{2}^-$, $T = \frac{7}{2}$ excited state of ^{49}Sc. Both nuclei have a single nucleon outside the double-closed shell of ^{48}Ca and so are particularly easy to deal with from the theoretical point of view. Unfortunately so far no experimental measurements have been made for this decay.

There is also a further ambiguity with respect to isospin hindered transitions in that so far no account has been taken of contributions from second forbidden matrix elements. In general, for an allowed transition, these can be neglected. But in the present case we are interested in the value of a Fermi matrix element which is non-zero only by virtue of charge dependent effects. It is therefore relatively small so that contributions to it from second forbidden terms could be important. This particular point has been looked into by Van Neste *et al.* (1966, 1967) and by Coussement and Van Neste (1967) who conclude that in cases where $\langle t_\pm \rangle$ is particularly small (e.g. for ^{152}Gd where $\langle H_{CD} \rangle = (30 \pm 30)$ eV) second forbidden effects become quite significant. They stress the importance of determining these matrix elements by experimental means (e.g. by detailed analysis of the spectrum shape and/or measurements of longitudinal polarization) so as to remove some of the uncertainty. Holstein *et al.* (1971) have also pointed out that it may be important to take recoil effects into account.

As far as CVC theory is concerned, the fact that theory and experiment are in essential agreement means that the assumption of CVC theory has not been invalidated. However, this is not a very strong statement and one can, in fact, make more progress. It was pointed out by Bouchiat (1959) in connection with the mirror isospin hindered beta-decays of ^{24}Na and ^{24}Al to ^{24}Mg that, assuming CVC theory and that the transitions proceeds only by charge dependent effects, then the Fermi matrix elements for the two decays

* Fujita and Ikeda (1965, 1966a, 1966b) have developed (although, not used) a method which would seem to give more reliable estimates of hindered matrix elements.

should be equal, i.e.

$$\langle t_+\rangle_{^{24}\mathrm{Na}} - \langle t_-\rangle_{^{24}\mathrm{Al}} = 0\,. \qquad \text{(III-92)}$$

Any deviation from zero is a measure of the breakdown in CVC theory and represents a contribution due to exchange effects (see §II-2.3) which have opposite signs for the two decays. Defining the Fermi matrix element due to exchange effects (M_{exch}) as

$$M_{\mathrm{exch}} = \tfrac{1}{2}\left[\langle t_+\rangle_{^{24}\mathrm{Na}} - \langle t_-\rangle_{^{24}\mathrm{Al}}\right] \qquad \text{(III-93)}$$

then experiment gives

$$M_{\mathrm{exch}} = (1 \pm 3) \times 10^{-3} \quad \text{(Haase } et\ al. \text{ (1963))}$$

$$M_{\mathrm{exch}} = (0.4 \pm 3) \times 10^{-3} \quad \text{(Bloom } et\ al. \text{ (1964))}\,.$$

Both results are consistent with $M_{\mathrm{exch}} = 0$ and therefore with CVC theory.

However, it is important to know the size of M_{exch} should CVC theory not hold. Yap (1968) has estimated this quantity for various isospin hindered decays using the earlier crude calculation of the β-decay exchange operator which results if CVC theory does not hold (Blin-Stoyle *et al.* (1959/60)). He obtains $M_{\mathrm{exch}} \approx (0.5 \text{ to } 1.0) \times 10^{-3}$ which value is not excluded by the experimental results. It is obviously important that more accurate experiments should be carried out.

As a general point it should be noted that if CVC theory does not hold, then exchange contributions would lead to contributions to $\langle t_\pm\rangle$ for all the decays discussed. Yap (1968) points out that these effects would seem to be of the same order of magnitude as charge dependent effects and would seriously disturb the analysis of the values of $\langle t_\pm\rangle$ in terms of charge dependent effects alone.

There is obviously every incentive in the case of these isospin hindered transitions to increase experimental accuracy and to develop the theoretical treatment to a more precise state.

CHAPTER IV

THE AXIAL VECTOR CURRENT IN β-DECAY

IV-1. Partially conserved axial vector current theory

As is discussed in Chapt. I it is generally assumed that the axial vector current $\mathcal{A}_\lambda$ is partially conserved (PCAC theory) and that it is a member of an SU3 octet of axial vector currents $\mathcal{F}_\lambda^{5(i)}$ ($i = 1, ..., 8$). Specifically

$$\mathcal{A}_\lambda = \mathcal{F}_\lambda^{5(1)} + \mathrm{i}\mathcal{F}_\lambda^{5(2)} . \tag{IV-1}$$

For the purposes of our discussions we use the field-theoretical statement of PCAC due to Gell-Mann and Levy (1960) (see also Bernstein *et al.* (1960a, b); Adler and Dashen (1968)) namely

$$\partial_\lambda \mathcal{A}_\lambda = c\phi^\dagger_{\pi^+} = c\phi_{\pi^-} \tag{IV-2}$$

where c is a constant (q.v.) and the field operators $\phi^\dagger_{\pi^+}$, ϕ_{π^-} create a π^+ and annihilate a π^- respectively.

The main reason for introducing PCAC theory is that it gives a simple explanation of the so-called Goldberger–Treiman relation (Goldberger and Treiman (1958)). This remarkable expression for the charged pion decay amplitude is discussed in the next section. Other implications of PCAC theory for nuclear β-decay are dealt with in subsequent sections.

IV-2. The Goldberger–Treiman relation

The Goldberger–Treiman relation connects the β-decay axial vector coupling constant G_{A}, the pion leptonic decay amplitude F and the renormalised pion–nucleon pseudoscalar coupling constant $G_{\pi\mathrm{NN}}$ and we now consider the determination of this relation in some detail.

We first have to determine the magnitude of the constant c in eq. (IV-2). Consider the matrix element of $\mathcal{A}_\lambda(0)$ between neutron and proton states, namely (see eq. (II-17))

$$\langle \mathrm{p}|\,\mathcal{A}_\lambda(0)\,|\mathrm{n}\rangle = \mathrm{i}(\bar{u}_\mathrm{p}\,|-f_\mathrm{A}\gamma_\lambda\gamma_5 + \mathrm{i}f_\mathrm{P}\gamma_5 k_\lambda + f_\mathrm{T}\sigma_{\lambda\nu}k_\nu\gamma_5\,|\,u_\mathrm{n})\,. \tag{IV-3}$$

Correspondingly, the matrix element of $\partial_\lambda \mathcal{A}_\lambda$ is given by

$$\begin{aligned}\langle \mathrm{p}|\,\partial_\lambda \mathcal{A}_\lambda(0)\,|\mathrm{n}\rangle &= -\mathrm{i}k_\lambda\,\langle \mathrm{p}|\,\mathcal{A}_\lambda(0)\,|\mathrm{n}\rangle \\ &= -\mathrm{i}(2Mf_\mathrm{A} - f_\mathrm{P}k^2)(\bar{u}_\mathrm{p}|\gamma_5|\,u_\mathrm{n})\end{aligned} \tag{IV-4}$$

where use has been made of eq. (IV-3) and the Dirac equation to simplify the first term. Further, the matrix element of $c\,\phi^\dagger_{\pi^+}$ between neutron and proton states can be written

$$\begin{aligned}\langle \mathrm{p}|\,c\,\phi^\dagger_{\pi^+}|\,\mathrm{n}\rangle &= \frac{c}{k^2+m_\pi^2}\,\langle \mathrm{p}|(-\Box + m_\pi^2)\,\phi^\dagger_{\pi^+}|\,\mathrm{n}\rangle \\ &= \frac{c}{k^2+m_\pi^2}\,\langle \mathrm{p}|\,j_{\pi^+}|\,\mathrm{n}\rangle\end{aligned} \tag{IV-5}$$

where j_{π^+} is the current generating the pion field and satisfies the equation

$$(\Box - m_\pi^2)\,\phi^\dagger_{\pi^+}(x) = -j_{\pi^+}(x)\,. \tag{IV-6}$$

But

$$\langle \mathrm{p}|\,j_{\pi^+}|\,\mathrm{n}\rangle = \mathrm{i}\sqrt{2}\,G_{\pi\mathrm{NN}}K_{\pi\mathrm{NN}}(k^2)\,(\bar{u}_\mathrm{p}|\gamma_5|\,u_\mathrm{n}) \tag{IV-7}$$

where $G_{\pi\mathrm{NN}}$ is the rationalised, renormalised pion–nucleon coupling constant $[G^2_{\pi\mathrm{NN}}/4\pi \approx 14.6]$ and $K_{\pi\mathrm{NN}}(k^2)$ is the invariant pion–nucleon vertex function normalised so that $K_{\pi\mathrm{NN}}(-m_\pi^2) = 1$.

Thus, equating the neutron–proton matrix elements of both sides of the PCAC equation (eq. (IV-2)) and using eqs. (IV-4), (IV-5) and (IV-7) it follows that

$$-2Mf_\mathrm{A}(k^2) + f_\mathrm{P}(k^2)\,k^2 = \frac{c\sqrt{2}G_{\pi\mathrm{NN}}\,K_{\pi\mathrm{NN}}(k^2)}{k^2+m_\pi^2}\,. \tag{IV-8}$$

Taking $k = 0$ then gives

$$c = -\frac{\sqrt{2}Mm_\pi^2 f_A(0)}{G_{\pi NN}K_{\pi NN}(0)} = -\frac{\sqrt{2}Mm_\pi^2 G_A}{G_\beta G_{\pi NN}K_{\pi NN}(0)} \quad \text{(IV-9)}$$

so that substituting for c in eq. (IV-8) we have

$$2Mf_A(k^2) - f_P(k^2)k^2 = 2Mf_A(0)\frac{m_\pi^2}{k^2+m_\pi^2}\left(\frac{K_{\pi NN}(k^2)}{K_{\pi NN}(0)}\right). \quad \text{(IV-10)}$$

We now turn briefly to the π-decay processes $\pi^- \to \ell^- + \tilde{\nu}_\ell$ where $\ell = e$ or μ. The decay rate for this process is given by (e.g. Lee and Wu (1965); Marshak *et al.* (1968))

$$\Gamma_\pi = \left(\frac{G}{\sqrt{2}}\right)^2 \cos^2\theta \frac{f_\pi^2}{4\pi}\frac{m_\ell^2}{m_\pi^3}(m_\pi^2 - m_\ell^2)^2 = \frac{F_\pi^2}{4\pi}\frac{m_\ell^2}{m_\pi^3}(m_\pi^2 - m_\ell^2)^2 \quad \text{(IV-11)}$$

where G is the basic weak interaction coupling constant, f_π is the charged pion decay constant given by

$$\langle 0|\mathcal{A}_\lambda(0)|\pi^-\rangle = -iq_\lambda(2\omega_\pi)^{-1/2}f_\pi \quad \text{(IV-12)}$$

and

$$F_\pi = \frac{G\cos\theta\, f_\pi}{\sqrt{2}}. \quad \text{(IV-13)}$$

In eq. (IV-12) ω_π $(=\sqrt{m_\pi^2+q^2})$ is the pion energy and q_λ is the four momentum of the decaying pion.

Consider now the matrix element of the PCAC continuity equation (eq. (IV-2)) taken between the π^- state and the vacuum, i.e.

$$\langle 0|\partial_\lambda \mathcal{A}_\lambda|\pi^-\rangle = \langle 0|c\,\phi_{\pi^-}|\pi^-\rangle. \quad \text{(IV-14)}$$

Using $\langle 0|\partial_\lambda\mathcal{A}_\lambda|\pi^-\rangle = +iq_\lambda\langle 0|\mathcal{A}_\lambda|\pi^-\rangle$, eq. (IV-12) and the normalisation properties of ϕ_{π^-} (i.e. $\langle 0|\phi_{\pi^-}|\pi^-\rangle = (2\omega_\pi)^{-1/2}$) we obtain *

* Note that for a conserved axial vector current (i.e. $c = 0$) since for a real decaying pion on the mass shell $q^2 = -m_\pi^2$ it follows from eq. (IV-15) that $f_\pi = 0$ and that $\pi^\pm_{\ell 2}$ decay is forbidden contrary to experiment. This argument against the axial vector current being completely conserved was first given by Taylor (1958).

$$q^2 f_\pi = c\,. \tag{IV-15}$$

But for a pion on the mass shell $q^2 = -m_\pi^2$ so that substituting for c from eq. (IV-9) gives

$$f_\pi = \frac{\sqrt{2}MG_A}{G_\beta G_{\pi NN} K_{\pi NN}(0)} \quad \text{or} \quad F_\pi = \frac{MG_A}{G_{\pi NN} K_{\pi NN}(0)}\,. \tag{IV-16}$$

If it is assumed that $K_{\pi NN}(q^2)$ is a slowly varying function of q^2, so that $K_{\pi NN}(0) \approx K_{\pi NN}(-m_\pi^2) = 1$ then we obtain the celebrated Goldberger–Treiman (1958) relation *

$$F_\pi = MG_A/G_{\pi NN}\,. \tag{IV-17}$$

This relation agrees with experiment to within 10% and, as mentioned earlier, is one of the basic reasons for introducing PCAC theory.

By treating the nucleus as an elementary particle, Kim and Primakoff (1965a) have obtained a Goldberger–Treiman relation equivalent to that given in eq. (IV-17) for any nuclear beta-decay of the type $N_i(I_i = \frac{1}{2}^{\pm}, T_i = \frac{1}{2}) \to N_f(I_f = \frac{1}{2}^{\pm}, T_f = \frac{1}{2})$. They write the matrix element of the axial vector current as follows (c.f. eq. (II-17)):

$$\langle N_f | \mathcal{A}_\lambda(0) | N_i \rangle = i(\bar{u}_{N_f} | t_\pm [-f_A^{N_i \to N_f}(k^2)\gamma_\lambda\gamma_5 + i f_P^{N_i \to N_f}(k^2)\gamma_5 k_\lambda] | u_{N_i}) \tag{IV-18}$$

where u_{N_i} and u_{N_f} are spinors describing the motion as a whole of the initial and final nuclei, $f_A^{N_i \to N_f}$ and $f_P^{N_i \to N_f}$ are form factors for the transition $N_i \to N_f$ and k is the four-momentum transfer.

By exactly the same arguments which led to eq. (IV-17), it then follows that

$$F_\pi = \frac{M_{N_i} + M_{N_f}}{2}\, G_A(N_i \to N_f)/G_{\pi N_i N_f} \tag{IV-19}$$

where $G_A(N_i \to N_f)$ is the effective axial vector coupling constant for the

* This relation can also be written in terms of renormalised (by the electromagnetic interaction) coupling constants F'_π, G'_A along the lines described in §III-2.3 (Sirlin (1972)).

decay considered and $G_{\pi N_i N_f}$ is the pion-initial nucleus – final nucleus coupling constant. The ft-value for such a transition is, of course, given by the same expression as for neutron decay but with G_A replaced by $G_A(N_i \to N_f)$, namely *

$$(ft)_{N_i \to N_f} = \frac{K}{G_V^2 + 3\,G_A^2(N_i \to N_f)} \,. \tag{IV-20}$$

In combination with the usual Goldberger–Treiman relation (eq. (IV-17)), eq. (IV-19) yields

$$\frac{G_A(N_i \to N_f)}{G_A} = \frac{G_{\pi N_i N_f}}{(M_{N_i} + M_{N_f})} \bigg/ \frac{G_{\pi NN}}{2M} = f_{\pi N_i N_f} / f_{\pi NN} \tag{IV-21}$$

where $f_{\pi N_i N_f}$ and $f_{\pi NN}$ are equivalent pseudovector coupling constants. $G_{\pi NN}$ is known and if $G_{\pi N_i N_f}$ could be determined (e.g. by a pological analysis of the process $n + N_f \to p + N_i$) a value for the corresponding ft value could be obtained independent of details of the nuclear wave function, exchange effects etc. Unfortunately no values of $G_{\pi N_i N_f}$ other than $G_{\pi NN}$ are available as yet, although Grismore (1968) has suggested an investigation of the process $n + {}^3He \to p + {}^3H$ as a means of determining $G_{\pi {}^3H {}^3He}$ and hence a value for the triton ft-value. It would obviously be of great interest to exploit the relation (eq. (IV-21) since it tests PCAC theory and efforts to determine $G_{\pi N_i N_f}$ are to be encouraged. Kim and Primakoff (1965a) used an extremely crude approach in which the $G_{\pi N_i N_f}$ are related to the nuclear magnetic moments. However, the limited agreement which they obtain between theory and experiment is probably not significant.

We now go on to consider various implications of PCAC for nuclear beta-decay.

IV-3. The axial vector coupling constants G_A and G_A'

As discussed in §III-2.3 model dependent radiative corrections make it difficult to determine the fundamental β-decay coupling constants G_V and G_A from experimental data and make it more reasonable to determine the renormalised constants G_V' and G_A' defined in eq. (III-23). The value of G_V'

* Note that if *CVC* theory holds $f_V^{N_i \to N_f}(0) = f_V(0) = 1$, so that $G_V(N_i \to N_f) = G_V$.

has already been obtained (§III-2.6, eq. (III-49)) from the data on super-allowed transitions and in this section we combine this data with that on the neutron decay * to determine a value for $\lambda' = G'_A/G'_V$.

Two ft-values are quoted in the literature for neutron β-decay, namely,

$(ft)_n = 1190 \pm 31$ sec. (Sosnovskii *et al.* (1959); Bhalla (1966))

$(ft)_n = 1081 \pm 16$ sec. (Christensen *et al.* (1967, 1972))

where, in both cases, electromagnetic radiative corrections are *not* included.

Referring to eq. (III-27) we have for neutron decay ($|\langle t_\pm \rangle|^2 = 1$; $|\langle \boldsymbol{\sigma} t_\pm \rangle|^2 = 3$)

$$(ft)_n(1+\delta'_R(n)) = \frac{K}{G'^2_V + 3G'^2_A} \,. \tag{IV-22}$$

On comparing with eq. (III-48) for the case $\delta_c = 0$ this gives

$$\frac{(\mathcal{F}t)_{0^+\to 0^+}}{(ft)_n(1+\delta'_R(n))} = \frac{1}{2}(1+3\lambda'^2) \,. \tag{IV-23}$$

Taking $\delta'_R(n) = +1.5\%$ from fig. III-3 (for the neutron decay $\mathcal{E}_0 = 0.78$ MeV) and the $^{26}Al^m$ value for $(\mathcal{F}t)_{0^+\to 0^+}$ namely $(\mathcal{F}t)_{0^+\to 0^+} = 3073 \pm 5$ sec then gives, using the two values of $(ft)_n$,

$|\lambda'| = 1.17 \pm 0.02$ (Sosnovskii *et al.* (1959))

$|\lambda'| = 1.239 \pm 0.011$ (Christensen *et al.* (1967, 1972)).

As described in §III-2.7, Christensen *et al.* (1969) have made a measurement of the electron asymmetry coefficient A in the decay of polarised neutrons thereby deducing an independent result for the ratio of the coupling constants. Since, according to Shann (1971), radiative corrections are small in such an asymmetry measurement the result of this measurement can be written

$$|\lambda'| = 1.26 \pm 0.02 \tag{IV-24}$$

* Note that in §III-2.7 we used an inverse procedure using an experimental value for λ' in conjunction with the neutron β-decay data in order to determine the value of G'_V.

and so certainly favours the second value for $|\lambda'|$ based on the work of Christensen *et al.* (1967, 1972).

From the foregoing results it is clear that some renormalisation of G_A takes place and that $f_A(0)$ is not equal to unity. Thus, accepting the result of Christensen *et al.* (1967, 1972), implies that $|f_A(0)| \approx 1.239 \pm 0.011$ where $\approx$ signifies the uncertainty in the model dependent radiative correction Δ_R.

Theoretically $|f_A(0)|$ has been evaluated assuming PCAC theory and that the axial and polar vector charges defined by (see §I-5)

$$A_+ = -\mathrm{i}\int \mathcal{A}_4(x)\,\mathrm{d}^3x = -\mathrm{i}\int (\mathcal{F}_4^{5(1)}(x) + \mathrm{i}\,\mathcal{F}_4^{5(2)})\,\mathrm{d}^3x$$

$$A_- = -\mathrm{i}\int \mathcal{A}_4^*(x)\,\mathrm{d}^3x = -\mathrm{i}\int (\mathcal{F}_4^{5(1)}(x) - \mathrm{i}\,\mathcal{F}_4^{5(2)}(x))\,\mathrm{d}^3x \qquad \text{(IV-25)}$$

$$T_3 = -\mathrm{i}\int \mathcal{F}_4^{(3)}(x)\,\mathrm{d}^3x$$

satisfy the current commutation relation (see §I-5, eq. (I-28))

$$[A_+, A_-] = 2T_3\,. \qquad \text{(IV-26)}$$

The method of evaluating $|f_A(0)|$ was suggested by Fubini and Furlan (1965) and applied by Adler (1965b, 1965c) and Weisberger (1965, 1966). Essentially what is involved is to take the matrix element of eq. (IV-26) between physical one proton states, thus:

$$\langle \mathrm{p}_2|[A_+, A_-]|\mathrm{p}_1\rangle = 2\langle \mathrm{p}_2|T_3|\mathrm{p}_1\rangle = (2\pi)^3\delta(\boldsymbol{p}_2 - \boldsymbol{p}_1) \qquad \text{(IV-27)}$$

where, as usual, p_1 and p_2 are the four-momenta of the protons and the last step follows since $|\mathrm{p}_1\rangle$ and $|\mathrm{p}_2\rangle$ are eigenstates of T_3 with eigenvalue $+\frac{1}{2}$. Inserting a complete set of states between the operators A_+ and A_- gives

$$\left[\sum_{\text{spin}}\int \frac{\mathrm{d}n}{(2\pi)^3}\langle \mathrm{p}_2|A_+|\mathrm{n}\rangle\langle \mathrm{n}|A_-|\mathrm{p}_1\rangle + \sum_{j\neq n}\langle \mathrm{p}_2|A_+|\mathrm{j}\rangle\langle \mathrm{j}|A_-|\mathrm{p}_1\rangle\right]$$

$$-(A_+ \leftrightarrow A_-) = \delta(\boldsymbol{p}_2 - \boldsymbol{p}_1) \qquad \text{(IV-28)}$$

where the one nucleon (only the neutron, n, contributes) contribution has been separated out. The first term can obviously be related to $|f_A(0)|^2$ since it just involves matrix elements of the axial vector current with respect to nucleon states. Using the PCAC hypothesis the second term can be related to an integral over pion–nucleon scattering cross sections and the final result is:

$$|f_A(0)|^2 = \left|\frac{G_A}{G_V}\right|^2 = \left[1 - \frac{4M^2}{G_{\pi NN}^2 K_{\pi NN}^2(0)} \frac{1}{\pi} \int_{M+m_\pi}^{\infty} \frac{W\,dW}{W^2-M^2} [\sigma_0^+(W) - \sigma_0^-(W)]\right]^{-2} \tag{IV-29}$$

where $\sigma_0^\pm(W)$ is the total cross section for the scattering of *zero mass* $\pi^\pm$ on protons and W is the centre of mass energy.

There are, of course, uncertainties in PCAC theory (e.g. the Goldberger–Treiman relation does not agree exactly with experiment) and, furthermore, in evaluating eq. (IV-29) uncertainties arise in going from the cross section for zero mass pions to experimental cross sections. The following values have been obtained for $|f_A(0)|$:

$$|f_A(0)| = 1.18 \quad \text{(Weisberger (1965, 1966))}$$
$$= 1.24 \quad \text{(Adler (1965b, 1965c))}$$
$$= 1.17 \quad \text{(Höhler and Strauss (1967))}$$

and their spread reflects the theoretical uncertainty. Even so the agreement with experiment $|f_A(0)| \approx 1.239 \pm 0.011$ can be considered satisfactory and important supporting evidence for PCAC theory and for the integrated current commutation relation given in eq. (IV-26).

An exactly similar relation to eq. (IV-29) has been obtained by Kim and Primakoff (1966) using the elementary particle treatment of nuclei (see §IV-2) relating the effective axial vector coupling constant $G_A(N_i \to N_f)$ to integrals over pion energy of pion-final nucleus total cross sections. An experimental check of these sum rules would provide a further test of the correctness of PCAC theory and the assumed equal time commutation relations.

IV-4. The pseudoscalar coupling constant G_P

Referring to §II-2.1 it will be recalled that an induced pseudoscalar interaction of strength $G_P = m_e f_P G_\beta$ derives from the axial vector current. PCAC theory makes a clear prediction as to the size of G_P. This follows at once from eq. (IV-10) which can be rewritten as follows:

$$f_P(k^2) = \frac{2M}{k^2}\left[f_A(k^2) - \left(\frac{K_{\pi NN}(k^2)}{K_{\pi NN}(0)}\right)\frac{f_A(0)\,m_\pi^2}{k^2+m_\pi^2}\right]. \tag{IV-30}$$

Neglecting the k^2 dependence of $K_{\pi NN}$, f_P and f_A which is certainly a valid approximation for β-decay transitions, gives *

$$G_P = m_e f_P G_\beta = \frac{2MG_A m_e}{k^2+m_\pi^2} \tag{IV-31}$$

or, using the Goldberger–Treiman relation (eq. (IV-17))

$$G_P = \frac{2F_\pi G_{\pi NN} m_e}{k^2+m_\pi^2}. \tag{IV-32}$$

For β-decay these results correspond to $G_P \approx \frac{1}{20} G_A$. On the other hand for μ-capture (see Chapt. VII) a much larger value is obtained since m_e in the expressions for G_P has to be replaced by m_μ.

It is interesting to see from the experimental side what upper limit has been imposed on the magnitude of G_P in β-decay processes.

Let it be said at once that it is extremely difficult to measure even a large value of G_P in β-transitions. This is because the pseudoscalar operator can only contribute in $\Delta J = 0$ first forbidden or higher order transitions. In such transitions, ambiguities in interpretation arise both from the need to go to a non-relativistic form for a relativistic operator and also because other contributions can arise from the main axial vector interaction and from retardation terms (i.e. from $\langle\gamma_5 t_\pm\rangle$ and $\langle\boldsymbol{\sigma}\cdot \boldsymbol{r} t_\pm\rangle$). An analysis ** of the $0^+ \to 0^+$ β-decay of ^{144}Pr, for example, by Bhalla and Rose (1960b) is only able to give $|G_P/G_A| < 90$! Daniel *et al.* (1964) by analysing the energy spectrum of this same decay concluded that $|G_P/G_A| < 5$ but later found an error in their

* Kim and Primakoff (1965), treating the nucleus as an elementary particle have obtained a similar relation namely

$$G_P(N_i \to N_f) = \frac{(M_{N_i} + M_{N_f})\, G_A(N_i \to N_f)\, m_e}{k^2 + m_\pi^2}$$

** A general discussion and formulae relevant to the effect of the induced pseudoscalar interaction on spectra etc. have been given by Bühring and Schülke (1965).

analysis (Daniel (1968)). More recently Krmpotić and Tadić (1969) have analysed the decay of ^{144}Pr (and also of ^{166}Ho) in more detail but are unable to come to any precise upper limit for $|G_P/G_A|$. They see no disagreement, however, with the theoretical result $|G_P/G_A| \sim \frac{1}{20}$. This work illustrates the difficulty of determining G_P from β-decay transitions (unless it is anomalously large) and it seems unlikely that it will ever be possible to disentangle the effects of a pseudoscalar interaction as small as that predicted theoretically.

IV-5. Exchange effects in allowed axial vector transitions

Exchange effects in β-decay were discussed in general terms in §II-2.3 and although CVC theory means that they do not occur for allowed polar vector transitions, there is no reason to suppose that they do not contribute to allowed axial vector transitions. The most convincing experimental evidence for the existence of these effects comes from a consideration of the beta-decay * of ^{3}H (Blin-Stoyle *et al.* (1959); Blin-Stoyle (1964b); Blin-Stoyle and Papageorgiou (1965a)).

Using eq. (II-62) and including electromagnetic radiative corrections a comparison of the ft-values for the β-decay of ^{3}H and the neutron gives

$$\frac{[(ft)(1+\delta_R)]_{n\to p}}{[(ft)(1+\delta_R)]_{^3H\to {}^3He}} = \frac{G_V^2 + |\langle \boldsymbol{\sigma} t_+\rangle|^2 G_A^2}{G_V^2 + 3G_A^2} \tag{IV-33}$$

where $\langle \boldsymbol{\sigma} t_+\rangle$ is the axial vector matrix element for the $^3H \to {}^3He$ decay. If δ_R is written in the form $\delta_R = \Delta_R + \delta_R'$ (see eq. III-25) and it is assumed that Δ_R is the same for both decays ** then we have

$$\frac{(ft)_{n\to p}}{(ft)_{^3H\to {}^3He}} = \frac{1 + |\langle \boldsymbol{\sigma} t_+\rangle|^2 \lambda^2}{1+3\lambda^2} (1 - \delta_R'^{\,n\to p} + \delta_R'^{\,^3H\to {}^3He}) . \tag{IV-34}$$

Experimentally (excluding radiative corrections),

* We shall see in §XI-4 that the magnetic moments of ^{3}H and ^{3}He indicate the presence of *electromagnetic* exchange effects.

** This is a good assumption since, as will be seen $|\langle \boldsymbol{\sigma} t_+\rangle|^2$ does not differ greatly from 3 (its value for the neutron decay) so that Δ_R, as defined in eq. (III-26), is practically the same for both decays.

$$(ft)_{n\to p} = 1081 \pm 16 \text{ sec} \quad \text{(Christensen } et\ al. \text{ (1967, 1972))}$$

$$(ft)_{^3H\to{}^3He} = (1121 \pm 6) \text{ sec} \quad \text{(Lewis * (1970))}$$

$$\lambda \approx \lambda' \text{ (see §IV-3)} = 1.239 \pm 0.011$$

and from fig. III-3 $\delta_R'^{n\to p} = 1.5\%$, $\delta_R'^{^3H\to{}^3He} = 1.8\%$. Substituting into eq. (IV-34) gives

$$|\langle \boldsymbol{\sigma} t_+ \rangle|^2 = 2.86 \pm 0.06\,. \tag{IV-35}$$

Now it has been shown by Blatt (1952) that for the decay of ^{3}H,

$$\begin{aligned}|\langle \boldsymbol{\sigma} t_+ \rangle|^2 = 3\{&p(\text{I}^2\text{S}) - \tfrac{1}{3}p(\text{I}^2\text{P}) - \tfrac{1}{3}p(\text{II}^2\text{S}) \\ &+ \tfrac{1}{9}[p(\text{II}^2\text{P}) - 5p(\text{II}^4\text{P}) - 6\alpha(\text{II}^2\text{P})\,\alpha(\text{II}^4\text{P})] \\ &+ \tfrac{1}{3}p(\text{II}^4\text{D}) - \tfrac{5}{3}p(\text{III}^2\text{S}) + \tfrac{5}{9}(\text{II}^2\text{P})\}^2\end{aligned} \tag{IV-36}$$

where p is the probability and α the amplitude with which a particular state occurs in the ^{3}H or ^{3}He wave functions. The symbols I, II and III refer to states which are spatially symmetric, of mixed symmetry, or antisymmetric respectively. It is clear from a study of the three nucleon problem (see e.g. Delves and Phillips (1969) for a recent review of the problem and also Delves and Hennell (1971)) that the I^{2}S state dominates, followed by the II^{4}D state (D state) and with a small contribution from the II^{2}S$'$ state (S$'$ state). The remaining states contribute negligibly so that to a very good approximation we can write

$$|\langle \boldsymbol{\sigma} t_+ \rangle|^2 = 3\{p(\text{S}) - \tfrac{1}{3}p(\text{S}') + \tfrac{1}{3}p(\text{D})\}^2\,. \tag{IV-37}$$

Values of $|\langle \boldsymbol{\sigma} t_+ \rangle|^2$ for various assumptions about $p(\text{D})$ and $p(\text{S}')$ are given in table IV-1.

* This value is based on a suggestion by Lewis (1970) that the "best" ^{3}H end point energy to use is $E_0 = 18.610 \pm 0.035$ keV. He concludes this in the light of the following measurements of E_0

$E_0 =$	18.54 ± 0.095 keV	(Lewis (1970))
	18.57 ± 0.075 keV	(Daris and St. Pierre (1969))
	18.61 ± 0.10 keV	(Porter (1959))
	18.63 ± 0.05 keV	(Bergkvist (1969))
	18.72 ± 0.05 keV	(Salgo and Staub (1969)) .

Table IV-1
Theoretical values for $|\langle \boldsymbol{\sigma} t_+ \rangle|^2$ for ^{3}H β-decay.

p(D) (%)	p(S′) (%)	$\|\langle \boldsymbol{\sigma} t_+ \rangle\|^2$	$\delta_{ex.}$ (%)
4	0	2.84	0.4
6	0	2.77	1.6
6	2	2.61	4.8
9	2	2.50	7.2

The minimum percentage of D-state is probably around 6% (e.g. Blatt and Delves (1964)) and recent calculations by Delves and Hennell (1971) suggest $p(\mathrm{D}) \approx 9\%$ and $p(\mathrm{S}') \approx 2\%$ (see also Hennell and Delves (1972); Samaranayake and Wilk (1972)). On the other hand, Gibson (1965) has obtained $p(\mathrm{D}) \approx 6\%$ and $p(\mathrm{S}') \approx 2\%$ and Nunberg *et al.* (1972), who give further references, suggest $p(\mathrm{D}) \lesssim 4\%$. There is clearly some confusion here but even so there is some indication that the theoretical and experimental values for $|\langle \boldsymbol{\sigma} t_+ \rangle|^2$ are not in agreement for reasonable D- and S′-state admixtures.

This discrepancy can be resolved by appealing to exchange effects and writing

$$\langle \boldsymbol{\sigma} t_+ \rangle = \langle \boldsymbol{\sigma} t_+ \rangle_0 [1 + \delta_{ex}] \qquad \text{(IV-38)}$$

where $\langle \boldsymbol{\sigma} t_+ \rangle_0$ is given by eq. (IV-37) and where δ_{ex} represents the corrections due to exchange effects. The necessary values of δ_{ex} (excluding experimental error) to reconcile theory and experiment under different assumptions about $p(\mathrm{D})$ and $p(\mathrm{S}')$ are given in table IV-1 and it can be seen that although δ_{ex} may be very small it could be as high as 7% or more.

We now go on to consider briefly the attempts which have been made to calculate δ_{ex}.

IV-5.1. *The calculation of exchange effects*

Referring to eq. (IV-38) and to the discussion in §II-2.3 it is clear that δ_{ex} is given by

$$\delta_{ex} = \frac{\langle \mathrm{f} | H_{\beta\text{-}\mathrm{A}}^{\mathrm{exch}} | \mathrm{i} \rangle}{G_{\mathrm{A}} \langle \mathrm{f} | \sum_i \boldsymbol{\sigma}^{(i)} t_+^{(i)} | \mathrm{i} \rangle} \qquad \text{(IV-38)}$$

where $H_{\beta^-A}^{\mathrm{exch}}$ is the nucleon part of the axial vector exchange operator defined in eq. (II-51). In the allowed approximation that $\boldsymbol{L}(\boldsymbol{r})$ is constant over the nuclear volume $H_{\beta^-A}^{\mathrm{exch}}$ is given explicitly by

$$H_{\beta^-A}^{\mathrm{exch}} = G_A \sum_{i<j} \left\{ \left[g_I \boldsymbol{\sigma}^{(i)} \times \boldsymbol{\sigma}^{(j)} + g_{II} \frac{\boldsymbol{r}_{ij}\boldsymbol{r}_{ij} \cdot (\boldsymbol{\sigma}^{(i)} \times \boldsymbol{\sigma}^{(j)})}{r_{ij}^2} \right] (\boldsymbol{t}^{(i)} \times \boldsymbol{t}^{(j)}) \right.$$

$$+ \left[h_I (\boldsymbol{\sigma}^{(i)} - \boldsymbol{\sigma}^{(j)}) + h_{II} \frac{\boldsymbol{r}_{ij}\boldsymbol{r}_{ij} \cdot (\boldsymbol{\sigma}^{(i)} - \boldsymbol{\sigma}^{(j)})}{r_{ij}^2} \right] (t_+^{(i)} - t_+^{(j)})$$

$$\left. + \left[j_I (\boldsymbol{\sigma}^{(i)} + \boldsymbol{\sigma}^{(j)}) + j_{II} \frac{\boldsymbol{r}_{ij}\boldsymbol{r}_{ij} \cdot (\boldsymbol{\sigma}^{(i)} + \boldsymbol{\sigma}^{(j)})}{r_{ij}^2} \right] (t_+^{(i)} + t_+^{(j)}) \right\} \tag{IV-39}$$

where the g, h and j are functions of r_{ij}.

For the purposes of calculating δ_{ex} it is usually assumed that the ^{3}H and ^{3}He wave functions are 100% space symmetric $^2S_{1/2}$ states. The exchange matrix element then takes the form (Blin-Stoyle *et al.* (1959); Blin-Stoyle and Papageorgiou (1965a))

$$|\langle f | H_{\beta^-A}^{\mathrm{exch}} | i \rangle| = 4\sqrt{3}\, G_A \langle \phi(1,2,3) | G(r_{12}) | \phi(1,2,3) \rangle \tag{IV-40}$$

where $G(r_{12}) = [g_I(r_{12}) + \frac{1}{3} g_{II}(r_{12}) + h_I(r_{12}) + \frac{1}{3} h_{II}(r_{12})]$ and $\phi(1,2,3)$ is the fully space symmetric three-body wave function for ^{3}H and ^{3}He. To obtain the magnitude of this matrix element then requires that the radial functions g and h be calculated.

The first attempt to calculate g and h was by Blin-Stoyle *et al.* (1959) who used fourth order no-recoil perturbation theory to estimate the size of pion exchange effects. Such a calculation can at most indicate the order of magnitude of the effects and it was therefore encouraging that a value for δ_{ex} of $\approx +(3-7)\%$ was obtained for reasonable values of the various wave-function and meson parameters involved. However, as pointed out by Cheng (1966), the quantitative success of the calculation can only be regarded as partial because of the critical dependence of δ_{ex} on the high momentum cut-off of the pion–nucleon interaction and on the values at small internucleon distances of the ^{3}H and ^{3}He wave function.

Cheng (1966) set out to calculate δ_{ex} by schematising the pion–nucleon or multi–pion system present in the intermediate state of a pion-exchange

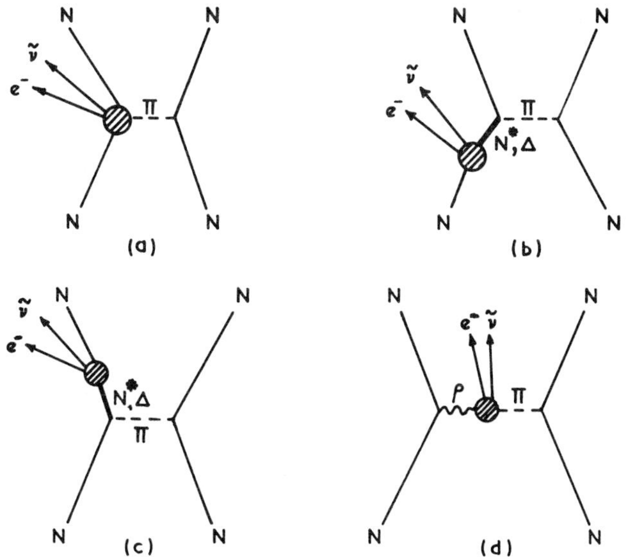

Fig. IV-1. Meson exchange diagrams in β-decay.

diagram (fig. IV-1(a)) by an appropriate "resonance", the resonance being treated as a stable particle. Specifically, diagrams of the type shown in fig. IV-1 (b–d) were taken into account where only the $N^*(I=\frac{1}{2}^+, T=\frac{1}{2})$, $N^*(I=\frac{3}{2}^-, T=\frac{1}{2})$ and $\Delta(I=\frac{3}{2}^+, T=\frac{3}{2})$ resonances contribute. Phenomenological Hamiltonians were used for the various strong vertices with coupling constants determined from experiment (e.g. from the decay width for $\Delta \to p + \pi^+$, $G_{\pi N\Delta}$ can be determined). The details of the baryon weak vertices were obtained by assuming PCAC and obtaining Goldberger–Treiman relations (see eq. (IV-17)) of the form

$$F_\pi \triangleq -\frac{M G_A^{N\to N*}}{G_{\pi NN*}}. \tag{IV-41}$$

In conjunction with the usual Goldberger–Treiman relation (eq. (IV-17)) this gives

$$\frac{G_A^{N\to N*}}{G_A} \triangleq -\frac{G_{\pi NN*}}{G_{\pi NN}} \tag{IV-42}$$

so enabling $G_A^{N\to N^*}$, the axial vector coupling constant for the $N\to N^*$ weak vertex to be determined. This procedure was used for the $\Delta(I=\frac{3}{2}^+, T=\frac{3}{2})$ and the $N^*(I=\frac{3}{2}^-, T=\frac{1}{2})$. Because of experimental uncertainty an Adler–Weissberger sum-rule approach (cf. §IV-3) had to be used to obtain $G_A^{N\to N^*}$ for $N^*(I=\frac{1}{2}^+, T=\frac{1}{2})$. In the case of the ($\pi$–$\rho$) exchange diagram (fig. IV-1(d)) PCAC theory together with other reasonable assumptions about ρ-nucleon coupling again enabled the weak and strong vertex strength to be evaluated. Given the strength of the various vertices the calculation of $H_\beta^{(2)}$ is then straightforward. A value for $\delta_{ex} \approx +10\%$ was obtained which seems to err on the large side.

More recently Chemtob and Rho (1969, 1971) have approached the problem in a more sophisticated manner in which the pion production vertex (fig. IV-1(a)) is evaluated using the low-energy theorem (Adler (1968); Adler and Dothan (1966)) which follows from PCAC theory. The "blob" in fig. IV-1(a) requires the evaluation of a matrix element of the form

$$M_\lambda = \langle N\pi | \mathcal{A}_\lambda | N\rangle \tag{IV-43}$$

which through the complexities of PCAC theory can be related to off-shell pion–nucleon scattering amplitudes. In terms of numerical values for these amplitudes $H_{\beta\text{-}A}^{\text{exch}}$ can then be calculated. Using a Gaussian form for the three-body function the authors obtain $\delta_{ex} \approx (0.053\,^{+\,0.18}_{-\,0.043})\%$. One reason for this reduced value for δ_{ex} compared with the work of Cheng (1966) is that Cheng omitted certain contributing diagrams of the type given in fig. IV-1(b) and IV-1(c), but with N^* replaced by N. This (Born) contribution is found to be sizeable and negative hence leading to the reduction.

A related semi-phenomenological calculation of δ_{ex} has been carried out by Blin-Stoyle and Tint (1967). If PCAC theory holds then taking matrix elements of the divergence equation (eq. (IV-2)) for $\mathcal{A}_\lambda$ between two nuclear states i and f gives

$$\langle f | \partial_\lambda \mathcal{A}_\lambda(0) | i\rangle = c\langle f | \phi_{\pi^+}^\dagger(0) | i\rangle = \frac{c}{k^2+m_\pi^2}\langle f | j_\pi(0) | i\rangle \tag{IV-44}$$

as in eq. (IV-5). But,

$$\langle f | j_\pi(0) | i\rangle = (2\pi)^{3/2} T(f\to i+\pi^+) \tag{IV-45}$$

where $T(f\to i+\pi^+)$ is the transition amplitude for the process $f\to i+\pi^+$ so that

$$\langle \mathrm{f} | \partial_\lambda \mathscr{A}_\lambda(0) | \mathrm{i} \rangle = \frac{c(2\pi)^{3/2}}{k^2 + m_\pi^2} T(\mathrm{f} \to \mathrm{i} + \pi^+) . \tag{IV-46}$$

If the π-production process is described by an "effective" Hamiltonian H_π of the form

$$H_\pi = \int S_\mu(r)\, \partial_\mu \phi_\pi(r)\, \mathrm{d}^3 r \tag{IV-47}$$

then use of eqs. (IV-44) to (IV-47) leads, after some approximation, to the result

$$H_\beta = \frac{c}{m_\pi^2} G_\beta \int \boldsymbol{S}(r)\, \mathrm{d}^3 r \tag{IV-48}$$

where H_β is the effective allowed axial vector β-decay operator.

It is thus possible to relate one- and two-body (i.e. exchange) terms in the beta-decay Hamiltonian to equivalent terms (in $\boldsymbol{S}(r)$) occurring in the pion-production Hamiltonian. The form of the two-body terms in $\boldsymbol{S}(r)$ was determined by a phenomenological analysis of the low energy π production process $\mathrm{p} + \mathrm{p} \to \mathrm{d} + \pi^+$ assuming a Yukawa radial dependence for the functions $g(r)$, $h(r)$, $j(r)$ occurring in eq. (IV-39) and the results seemed to imply that δ_{ex} is negative. However, recently (Blin-Stoyle (1969)) it has been shown in the light of new experimental data and because of theoretical uncertainties about the size of the *one-body* contribution to $\boldsymbol{S}(r)$ that δ_{ex} could be positive – but not very. In addition, Cheng and Fischbach (1969) have shown that if more singular terms than Yukawa arise in the radial dependence of the exchange operator as suggested by Cheng's (1966) work, then it is much easier to reconcile the ^{3}H decay and pion production data.

All of the calculations referred to so far evaluate the exchange effect using only the S-state part of the 3-body wave function. Recently Blomqvist (1970) and Riska and Brown (1970) (see also Fischbach *et al.* (1972b)) have shown that, in fact, the largest contribution to δ_{ex} comes from matrix elements of tensor like parts of the exchange operator (terms involving g_{II}, h_{II}, j_{II} etc. in eq. (IV-39)) between the 3-body S and D states. In particular tensor contributions from (non-Born) diagrams of the type (c) and to a lesser extent (d) in fig. IV-1 are important and, according to Riska and Brown (1970), can lead to values of δ_{ex} of the order $\gtrsim 10\%$. These calculations involve some rather "rough" approximations, however, and further do not take into account Born terms or wave function renormalisation (needed to take account of the fact that there are wave function components with meson presents – see Chemtob and Rho (1971) and Green and Schucan (1972)). The latter authors derive an expression equivalent to that given in eq. (IV-37) which includes a probability amplitude for an N^* being present in the 3-body wave function. With reasonable assumption about the different probability amplitudes, they obtain $|\langle \boldsymbol{\sigma} t_+\rangle|^2 \approx 2.71$ which to agree with experiment ($|\langle \boldsymbol{\sigma} t_+\rangle|^2 = 2.86 \pm 0.06$) still requires some contribution from other diagrams than those involving an N^*.

The situation with respect to the theoretical evaluation of exchange effects is therefore still uncertain and much more work is needed. Even so, within the (wide) limits of theoretical and experimental error there is no violent discrepancy between theory and experiment.

IV-5.2. *Exchange effects in heavier nuclei*

Given that exchange effects contribute to the ^3H decay it is important to estimate their size in heavier nuclei. Considering the β-decay of a nucleus consisting of an odd nucleon or nucleon hole together with a spin and isospin saturated core (e.g. ^{17}O, ^{15}N, ^{41}Sc etc.), it has been shown (Bell and Blin-Stoyle (1957)) that in the zero-range limit for the exchange operator, an additional effective one-body operator is introduced of the form

$$[H_{\beta A}^{\text{exch}}]_{\text{one-body}} = G_A [\tfrac{1}{2} \int G(r')\, d^3r'] \rho(r) \boldsymbol{\sigma} t_{\pm} \tag{IV-49}$$

where $\rho(r)$ is the density of core nucleons and $G(r)$ is the function occurring in the ^3H decay (eq. IV-40). Using a phenomenological form $\Gamma e^{-\alpha r}/\alpha r$ for $G(r)$ and taking for $\rho(r)$ the density of nuclear matter, namely $\rho_0 = 3/4\pi r_0^3$ where $r_0 \approx 1.2$ fm, gives

$$[H_{\beta A}^{\text{exch}}]_{\text{one-body}} \approx \delta\, G_A \boldsymbol{\sigma} t_{\pm} \tag{IV-50}$$

where $\delta = \frac{3}{2} \Gamma/(\alpha r_0)^3$. If the same phenomenological form for $G(r)$ is used for the ^3H decay together with $\phi(1,2,3) \propto \exp[-\frac{1}{2}(r_{12} + r_{23} + r_{13})]$ and $k = 0.78\ \text{fm}^{-1}$ as given by electron scattering data, it is found that

$$\delta_{\text{ex}}(^3H) = a(\alpha)\, \Gamma \tag{IV-51}$$

where $a(0.7\ \text{fm}^{-1}) \approx 1.2$, $a(1.4\ \text{fm}^{-1}) \approx 0.3$. Thus eliminating Γ between eq. (IV-50) and eq. (IV-51) gives

$$\delta \approx A\,\delta_{\rm ex}(^3\text{H}) \tag{IV-52}$$

where $A = 3/2a(\alpha)(\alpha r_0)^3$ and takes the values $A \approx 2$ for $\alpha = 0.7\ \text{fm}^{-1}$ (one pion exchange) and $A \approx 1$ for $\alpha = 1.4\ \text{fm}^{-1}$ (two pion exchange). Thus on the basis of this calculation exchange effects of at least the order of magnitude of those occurring in the ^{3}H decay deriving from the *S*-state part of the wave function should arise in heavier nuclei.

However, the zero range approximation and the assumption of constant density are suspect and more detailed calculations are necessary. Such calculations were carried out by Blin-Stoyle and Papageorgiou (1965b) and showed that in fact the situation is qualitatively unchanged. It is therefore important to search for exchange effects in other decays than that of ^{3}H. Unfortunately, in heavier nuclei, wave function uncertainties greatly complicate the situation and it is difficult to disentangle exchange contributions – but as our knowledge of nuclear structure increases there is every incentive to do so.

CHAPTER V

UNUSUAL HADRON CURRENTS AND INTERACTIONS IN BETA-DECAY

V-1. Normal and regular currents

In the foregoing discussions quite specific and restrictive assumptions have been made implicitly about the nature of the polar and axial vector currents. Thus the assumption of *CVC* theory and the identification of $\mathcal{V}_\lambda$ with the isospin current implies that it is "normal" *, i.e. odd, under time reversal and "regular" ** i.e. odd, under the charge symmetry operation. Similarly the axial vector current, having the transformation properties implied by PCAC theory is also "normal" and "regular" in the above sense. Briefly, the assumptions about the transformation properties of $\mathcal{V}_\lambda$ and $\mathcal{A}_\lambda$ under time reversal and charge symmetry are that $\mathcal{V}_\lambda$ transforms like $\mathrm{i}\,\overline{\psi}_\mathrm{p}\gamma_\lambda\psi_\mathrm{n}$ and $\mathcal{A}_\lambda$ like $\mathrm{i}\,\overline{\psi}_\mathrm{p}\gamma_\lambda\gamma_5\,\psi_\mathrm{n}$ where ψ is a Dirac spinor. We now study other possibilities for the transformation properties $\mathcal{V}_\lambda$ and $\mathcal{A}_\lambda$ in detail and the extent to which β-decay processes can test for "unusual" currents.

V-2. Behaviour of currents under the time reversal operation

The behaviour of Dirac spinors and operators under the time reversal operation is set out in App. C where, in particular, it is shown that under time reversal a Dirac spinor $\psi(\boldsymbol{x}, t)$ transforms according to the relation

$$\begin{aligned} \psi(\boldsymbol{x}, t) &\to \psi^t(\boldsymbol{x}, -t) = \eta B \psi^\times(\boldsymbol{x}, t) \\ \overline{\psi}(\boldsymbol{x}, t) &\to \overline{\psi}^t(\boldsymbol{x}, -t) = \eta \overline{\psi}^\times(\boldsymbol{x}, t) B^{-1} \end{aligned} \qquad \text{(V-1)}$$

where the symbol X signifies complex conjugation, where the matrix B

* We here use a terminology due to Kim and Primakoff (1969).

** This terminology is due to Cabibbo (1964).

satisfies the relations

$$B^{-1}\gamma_\lambda B = \tilde{\gamma}_\lambda; \qquad B^\dagger = B^{-1}; \qquad \tilde{B} = -B\,. \tag{V-2}$$

and $\eta = \pm 1$.
Under this transformation the quantity $\mathrm{i}\bar{u}_\mathrm{p}\gamma_\lambda u_\mathrm{n}$ transforms as follows:

$$\begin{aligned}\mathrm{i}\bar{u}_\mathrm{p}\gamma_\lambda u_\mathrm{n} \to \mathrm{i}\bar{u}^t_\mathrm{p}\gamma_\lambda u^t_\mathrm{n} &= \mathrm{i}\bar{u}^\times_\mathrm{p} B^{-1}\gamma_\lambda B u^\times_\mathrm{n} \\ &= \mathrm{i}\tilde{u}_\mathrm{p}\tilde{\gamma}_4\tilde{\gamma}_\lambda u^\times_\mathrm{n} \\ &= \mathrm{i}u^\dagger_\mathrm{n}\gamma_\lambda\gamma_4 u_\mathrm{p} \quad \text{by taking the transpose} \\ &= -(\mathrm{i}\bar{u}_\mathrm{p}\gamma_\lambda u_\mathrm{n})^\dagger\,. \end{aligned} \tag{V-3}$$

Similarly

$$\mathrm{i}\bar{u}_\mathrm{p}\gamma_\lambda\gamma_5 u_\mathrm{n} \to \mathrm{i}\bar{u}^t_\mathrm{p}\gamma_\lambda\gamma_5 u^t_\mathrm{n} = -(\mathrm{i}\bar{u}_\mathrm{p}\gamma_\lambda\gamma_5 u_\mathrm{n})^\dagger\,. \tag{V-4}$$

It is in this sense that the matrix elements of the currents $\mathcal{V}_\lambda$ and $\mathcal{A}_\lambda$ that we have so far dealt with are said to be odd (i.e. normal) under time reversal. The currents themselves must satisfy equivalent relations and we can write

$$\begin{aligned} U_\mathrm{T}\mathcal{V}^{(n)}_\lambda U_\mathrm{T}^{-1} &= -\mathcal{V}^{(n)\dagger}_\lambda \\ U_\mathrm{T}\mathcal{A}^{(n)}_\lambda U_\mathrm{T}^{-1} &= -\mathcal{A}^{(n)\dagger}_\lambda \end{aligned} \tag{V-5}$$

where the suffix (n) signifies "normal" and U_T denotes a time reversal operator acting on the currents $\mathcal{V}^{(n)}_\lambda$ and $\mathcal{A}^{(n)}_\lambda$ (see Appendix D).

Similar manipulations show that the lepton current $\ell_\lambda = -\mathrm{i}\bar{u}_{\nu_\mathrm{e}}\gamma_\lambda(1+\gamma_5)u_\mathrm{e}$ is normal as are the terms $\mathrm{i}\bar{u}_\mathrm{p}\sigma_{\lambda\nu}k_\nu u_\mathrm{n}$, $\mathrm{i}\bar{u}_\mathrm{p}\mathrm{i}k_\lambda u_\mathrm{n}$, $\mathrm{i}\bar{u}_\mathrm{p}\mathrm{i}\gamma_5 k_\lambda u_\mathrm{n}$ and $\mathrm{i}\bar{u}_\mathrm{p}\sigma_{\lambda\nu}k_\nu\gamma_5 u_\mathrm{n}$ occurring in the expressions (II-16) and (II-17) for the matrix elements of $\mathcal{V}_\lambda$ and $\mathcal{A}_\lambda$.

Now the total amplitude for the β-decay process is of the form $\mathcal{M} + \mathcal{M}^\dagger$ where $\mathcal{M}$ is given by eq. (II-23) and is a product of nucleon and lepton bilinear terms of the type just discussed, each nucleon term being prefixed by a form factor $f\,(f_\mathrm{V}, f_\mathrm{W}, f_\mathrm{S}, f_\mathrm{A}, f_\mathrm{P}, f_\mathrm{T})$. Clearly, these latter terms are all normal *provided that the corresponding f's are real*. In this case under time reversal $\mathcal{M} \to \mathcal{M}^\dagger$ and $\mathcal{M}^\dagger \to \mathcal{M}$. Thus with normal currents the various form factors are real and the β-decay interaction is invariant under time reversal.

We can now introduce the concept of abnormal currents (denoted by (a))

which transform as follows under time reversal:

$$U_T \mathcal{V}_\lambda^{(a)} U_T^{-1} = + \mathcal{V}_\lambda^{(a)\dagger}$$
$$U_T \mathcal{A}_\lambda^{(a)} U_T^{-1} = + \mathcal{A}_\lambda^{(a)\dagger} \qquad \text{(V-6)}$$

and correspondingly lead to imaginary form factors. The simultaneous presence of normal and abnormal currents can then lead to a β-decay interaction and to physical effects which are odd under time reversal. The nature and significance of these effects as they arise in β-decay will be discussed in §V-4.

V-3. Behaviour of currents under the charge symmetry operation

With CVC theory it is clear that $\mathcal{V}_\lambda$ and $\mathcal{V}_\lambda^*$ transform as the $T_3 = +1$ and $T_3 = -1$ members of an isospin triplet (see eq. (III-4)). Similarly with PCAC theory, since $\partial_\lambda \mathcal{A}_\lambda$ is proportional to the pion field (which is itself an isospin triplet), the $\mathcal{A}_\lambda$ and $\mathcal{A}_\lambda^*$ also transform as the $T_3 = +1$ and $T_3 = -1$ members of an isospin triplet. This means that $\mathcal{V}_\lambda$ and $\mathcal{V}_\lambda^*$ are related by a 180° rotation about the 2-axis in isospin space (Lee and Yang (1962); see App. E), namely,

$$\mathcal{V}_\lambda^* = - U_S \mathcal{V}_\lambda U_S^{-1}$$

and similarly (V-7)

$$\mathcal{A}_\lambda^* = - U_S \mathcal{A}_\lambda U_S^{-1} .$$

Currents transforming in this way are defined to be regular and will be denoted by a suffix (r). Thus a *regular* current $\mathcal{J}_\lambda^{(r)}$ is defined by the equation

$$\mathcal{J}_\lambda^{(r)*} = - U_S \mathcal{J}_\lambda^{(r)} U_S^{-1} . \qquad \text{(V-8)}$$

Similarly, an *irregular* current $\mathcal{J}_\lambda^{(i)}$ is defined by the converse equation

$$\mathcal{J}_\lambda^{(i)*} = + U_S \mathcal{J}_\lambda^{(i)} U_S^{-1} . \qquad \text{(V-9)}$$

Consider now the transformation properties of the matrix elements of $\mathcal{V}_\lambda$ and $\mathcal{A}_\lambda$ with respect to nucleon states under the charge symmetry operation. We have from eqs. (II-16) and (II-17)

$$\langle p|\mathcal{V}_\lambda|n\rangle = i(\bar{u}_p|f_V\gamma_\lambda + f_W\sigma_{\lambda\nu}k_\nu + if_S k_\lambda|u_n)$$
$$\langle p|\mathcal{A}_\lambda|n\rangle = i(\bar{u}_p|-f_A\gamma_\lambda\gamma_5 + if_P\gamma_5 k_\lambda + f_T\sigma_{\lambda\nu}k_\nu\gamma_5|u_n)\,. \tag{V-10}$$

Similarly

$$\langle n|\mathcal{V}^*_\lambda|p\rangle = i(\bar{u}_n|f^\times_V\gamma_\lambda + f^\times_W\sigma_{\lambda\nu}k_\nu - if^\times_S k_\lambda|u_p)$$
$$\langle n|\mathcal{A}^*_\lambda|p\rangle = i(\bar{u}_n|-f^\times_A\gamma_\lambda\gamma_5 + if^\times_P\gamma_5 k_\lambda - f^\times_T\sigma_{\lambda\nu}k_\nu\gamma_5|u_p) \tag{V-11}$$

where in both sets of equations k_λ is the four-momentum transfer and X signifies "complex conjugate".

Now, using eq. (V-8) and eq. (E8) it is easy to show for a regular current that

$$\langle n|\mathcal{J}^{(r)*}_\lambda|p\rangle = +\langle p|\mathcal{J}^{(r)}_\lambda|n\rangle \tag{V-12}$$

and using eq. (V-9) for an irregular current that

$$\langle n|\mathcal{J}^{(i)*}_\lambda|p\rangle = -\langle p|\mathcal{J}^{(i)}_\lambda|n\rangle\,. \tag{V-13}$$

Comparing the explicit expressions for these matrix elements given in eqs. (V-10) and (V-11) the following statements can then be made about the form factors (see Lee and Wu (1965) for further discussion)

Regular currents

f_V, f_W, f_A, f_P	real
f_S, f_T	imaginary .

Irregular currents

f_V, f_W, f_A, f_P	imaginary
f_S, f_T	real .

Obviously these restrictions on the reality or otherwise of the various form factors impinge on the conditions imposed by whether the currents are normal or abnormal with respect to time reversal. To make the situation clear we introduce a double labelling of a current $\mathcal{J}_\lambda$ according to its normality and regularity. An arbitrary current $\mathcal{J}_\lambda$ can clearly be decomposed into four parts $\mathcal{J}_\lambda^{(n)(r)}$, $\mathcal{J}_\lambda^{(n)(i)}$, $\mathcal{J}_\lambda^{(a)(r)}$ and $\mathcal{J}_\lambda^{(a)(i)}$ and by virtue of the discussion in this and the previous paragraph each type of current can lead to only certain non-vanishing form factors which are either real or imaginary. The correspondences between currents and form factors are given in table V-1. Here it will be seen that currents of the type $\mathcal{J}_\lambda^{(n)(r)}$ and $\mathcal{J}_\lambda^{(a)(i)}$ have been designated "first class" and currents of the type $\mathcal{J}_\lambda^{(n)(i)}$ and $\mathcal{J}_\lambda^{(a)(r)}$ have been designated "second class". This is an alternative classification due to Weinberg (1958) based on the transformation properties of the currents under the operation (G-parity) $G = U_C U_S$ where U_C is the charge conjugation operator (see App. E).

Having related currents to form factors we can now consider the extent to which β-decay processes are sensitive to the presence of "unusual" currents.

V-4. The search for T-violation in β-decay

The experimental search for time reversal non-invariance effects in β-decay involves the measurement of quantities which are odd under time reversal. In a beta-decay experiment there are essentially four vector quantities that can be measured namely $\langle \boldsymbol{I} \rangle / I$, the polarization of the decaying nucleus; $\boldsymbol{\sigma}$, the electron spin polarization; $\boldsymbol{p}_e$, the electron momentum; and $\boldsymbol{p}_\nu$, the neutrino momentum.

Each of these vectors changes sign under time reversal and so the scalar triple product of any three of them gives a term odd under time reversal but invariant under rotations. The detection of such a term in the angular distribution of a beta-decay process could then constitute evidence for time reversal non-invariance of the beta-decay interaction *. The precise form of these terms has been studied by Jackson *et al.* (1957a) for the general case of a four fermion beta-decay interaction incorporating scalar (S), polar vector (V), tensor (T), axial vector (A) and pseudo-scalar (P) terms. The only time reversal non-invariant term which occurs if only the V and A terms are present is that which arises when the electron and neutrino momenta are measured for decay from a polarised nucleus. The resulting angular correlation has the form

* Such terms can also arise from final state Coulomb interactions. This point is discussed later.

Table V-1

Correspondence between currents and form factors

Polar vector current (form factors f_V, f_W, f_S)				Axial vector current (form factors f_A, f_P, f_T)			
First class currents		Second class currents		First class currents		Second class currents	
$\mathcal{V}_\lambda^{(n)(r)}$	$\mathcal{V}_\lambda^{(a)(i)}$	$\mathcal{V}_\lambda^{(n)(i)}$	$\mathcal{V}_\lambda^{(a)(r)}$	$\mathcal{A}_\lambda^{(n)(r)}$	$\mathcal{A}_\lambda^{(a)(i)}$	$\mathcal{A}_\lambda^{(n)(i)}$	$\mathcal{A}_\lambda^{(a)(r)}$
f_V, f_W real	f_V, f_W imag.	$f_V = f_W = 0$	$f_V = f_W = 0$	f_A, f_P real	f_A, f_P imag.	$f_A = f_P = 0$	$f_A = f_P = 0$
$f_S = 0$	$f_S = 0$	f_S real	f_S imag.	$f_T = 0$	$f_T = 0$	f_T real	f_T imag.

$$W \propto 1 + D \frac{\langle I \rangle}{I} \cdot \frac{p_e \times p_\nu}{E_e E_\nu} + \ldots \tag{V-14}$$

where E refers to the energy of the particle indicated. For an allowed beta decay, $I \to I'$, the coefficient D is given explicitly by

$$D = \delta_{I'I} \frac{\mathrm{i}(f_V f_A^\times - f_A f_V^\times)|\langle t_\pm \rangle||\langle \boldsymbol{\sigma} t_\pm \rangle|}{|f_V|^2|\langle t_\pm \rangle|^2 + |f_A|^2|\langle \boldsymbol{\sigma} t_\pm \rangle|^2} \sqrt{\frac{I}{I+1}}\,. \tag{V-15}$$

Obviosuly D depends on interference between the V and A parts of the interaction and, because of the selection rule for allowed polar vector transitions there can be no spin change. If f_V and f_A are both real as is required for a T-invariant theory, then $D = 0$. A finite value for D thus gives a measure of the extent to which f_V and f_A differ in phase.

In table V-2 recent experimental values for the value of D are given. Also given is the value of the angle ϕ defined by

$$\frac{f_A}{f_V} = r\mathrm{e}^{\mathrm{i}\phi} \tag{V-16}$$

where ϕ takes the value $\phi = 180^\circ$ for a T-invariant theory.

It should also be mentioned here that a phase difference between f_A and f_V can show up through modifying terms in the energy spectrum and electron polarisation in some β-decays. Of particular interest is the decay of RaE (see §III-3.2), which, because of accidental destructive interference, allows such effects to be significant. Even so the experimental and theoretical analysis is very complicated. Daniel (1968) (see also Sodemann and Winther (1965)) concludes that the deviation of ϕ from 180° is at most about 6°. It is clear that the present results do not indicate any T non-invariance.

Before commenting on the significance of this there are two points to be made. As pointed out by Jackson *et al.* (1957a) and elaborated by them (Jackson *et al.* (1957b)), terms odd under time reversal can in general be introduced into a β-decay angular correlation because of final state electromagnetic (Coulomb) interactions. Nevertheless, if only the form factors f_A and f_V are taken into account, then in lowest order, $(\alpha Z m/p_e)$, there is no contribution to the coefficient D. However, Callan and Treiman (1967) (see also Chen (1969a, 1969b); Brodine (1970, 1971)) have shown that a contribution can occur from momentum dependent terms in the β-decay amplitude (i.e. from terms proportional to f_W, f_S, f_P and f_T). Assuming only first class

Table V-2
β-decay tests of time reversal non-invariance

Process	Reference	D	ϕ
$^{19}Ne \rightarrow {}^{19}F + e^+ + \nu_e$	Calaprice *et al.* (1967, 1969) Girvin (1972)	+0.002 ± 0.014 +0.001 ± 0.003	180.2 ± 1.6° 180.1 ± 0.3°
$n \rightarrow p + e^- + \tilde{\nu}_e$	Erozolimsky *et al.* (1968)	+0.01 ± 0.01	178.7 ± 1.3°

currents ($f_S = f_T = 0$) Callan and Treiman (1967) show that the contribution from f_P is negligible but that from f_W is of the order $D \approx (2.6 \times 10^{-4}) p_e/(p_e)_{max}$ for the ^{19}Ne decay. This is some fifty times smaller than the best experimental result and so does not confuse the situation at this time. Hopefully, however, experimental accuracy will improve and then this spurious T non-invariant effect will have to be taken into account.

The second point to note was first made by Kim and Primakoff (1969) and generalised by Rosen (1971) and is simply that the expression for D and the two experiments given in table V-2 are sensitive only to a phase difference between f_V and f_A. But f_V and f_A both result from first class currents whereas theories (see Chapt. I) which, attribute T-violation to the weak interaction and which would lead to T-violating effects in β-decay (e.g. Cabibbo (1964, 1965); Maiani (1968)) depend on the properties of second class axial vector currents. In other words, in these theories f_T is expected to have an imaginary part. The magnitude of the effect to be expected in the case of neutron-decay can be estimated roughly as follows (see eq. (V-15))

$$|D| \simeq \frac{2|f_V||\operatorname{Im} f_T|(M_n - M_p)}{|f_V|^2 + 3|f_A|^2} \tag{V-17}$$

where the magnitude of the momentum transfer k_ν occurring in the induced tensor term $f_T \sigma_{\lambda\nu} k_\nu$ (see eq. (II-17)) has been put equal to $M_n - M_p$. The order of magnitude of f_T is expected to be $\approx 1/M$ by analogy with the weak magnetism term so that

$$|D| \approx \frac{M_n - M_p}{M} \approx 10^{-3} - 10^{-4} \tag{V-18}$$

which is obviously hard to detect at this stage.

However, as is pointed out by Kim and Primakoff (1969) much larger effects might be expected for certain allowed β-decays not between mirror states (as is the case for $^{19}Ne \rightarrow {}^{19}F$ and $n \rightarrow p$). They argue that the various reality conditions imposed on f_A, f_V etc. and set out in table V-1 depend on the fact that the $n \rightarrow p$ process is between mirror states (see §V-3). A decay between non-mirror states (e.g. in particle physics $\Sigma \rightarrow n$) would not involve the same conditions and, for example, an abnormal regular axial vector current ($\mathcal{A}^{(a)(r)}$) could lead to an imaginary term in the total axial vector form factor for the decay. In the nucleus, this term would only appear in the exchange interaction, however, since the usual sum of single particle terms $f_A \Sigma \boldsymbol{\sigma}^{(i)} t_{\pm}^{(i)}$ has as its coefficient the form factor relevant to the (mirror) $n \rightarrow p$ decay.

Thus, a transition is needed in which the usual axial vector matrix element $\langle \boldsymbol{\sigma} t_{\pm} \rangle$ is very small so that axial vector exchange effects can substantially interfere with the polar vector terms (in this case the weak magnetism term, since the main polar vector matrix element $\langle t_{\pm} \rangle$ will vanish for a transition of the type being considered here).

The magnitude of the effect under these circumstances is then of the order

$$D \triangleq \frac{2|f_W||\langle \sigma_{\lambda\nu} k_\nu \rangle||\mathrm{Im}\langle H_{\beta A}^{\mathrm{exch}} \rangle|}{|f_A|^2 |\langle \boldsymbol{\sigma} t_{\pm} \rangle|^2} \qquad \text{(V-19)}$$

where $\langle H_{\beta A}^{\mathrm{exch}} \rangle$ is the nuclear matrix element of the axial vector exchange interaction stemming from $\mathcal{A}_\lambda^{(a)(r)}$ and $\langle \sigma_{\lambda\nu} k_\nu \rangle$ is the weak magnetism matrix element. Kim and Primakoff (1969) suggest the decay $^{32}P(T=1) \rightarrow {}^{32}S(T=0) + e^- + \tilde{\nu}_e$ as suitable for investigation since $|\langle \boldsymbol{\sigma} t_{\pm} \rangle|^2$ is small ($\approx 6 \times 10^{-3}$). Taking $|f_W| \approx 1/M$, $\mathrm{Im}\langle H_{\beta A}^{\mathrm{exch}} \rangle \approx 1$ (i.e. assuming maximum T-invariance violation) and $|k| \approx M_{^{32}P} - M_{^{32}S}$ gives $D \approx 0.3$ which should be observable with present accuracies. Here it should be noted that Brodine (1970) estimates a contribution to D from final state interactions of the order $10^{-2}\, p_e/(p_e)_{max}$.

We thus have a situation in which a substantial T non-invariance could be present in the basic weak interaction through the presence of $\mathcal{A}_\lambda^{(a)(r)}$ but which would not show up in β-decay tests so far carried out. It is therefore important to investigate the decay of ^{32}P and other decays which are more sensitive to the presence of abnormal currents (see also Holstein (1971) who has suggested other angular correlation tests for T-violating second class currents).

Morita and Morita (1957, 1958) have also shown that a study of β–γ

angular correlations from oriented nuclei can indicate the presence of *T*-violating terms in the β-decay interaction. So far, however, no significant experimental studies have been carried out.

V-5. *T*-invariant second class currents and the induced tensor interaction

As discussed in §V-3 and explicitly stated in table V-1, second class currents in β-decay only show themselves in the nuclear matrix element through the form factors f_S and f_T. Now, if the polar vector current is conserved then, whether or not it has second class components, $f_S = 0$ (see App. F). Thus it is likely that if second class currents do play a role in the weak interaction then they will show up in β-decay mainly through the induced tensor form factor f_T. This term $(f_T \sigma_{\lambda\nu} k_\nu \gamma_5)$ is momentum dependent and will be most important in a β-decay process where there is large momentum transfer. Even so, it would, in general, be difficult to be sure about the presence of such a term since it would have to be disentangled from all the other (2nd forbidden etc.) contributions to the β-decay.

There is, however, one particularly profitable approach to this problem (Weinberg (1958)) and that is to make a comparison of the energy spectra or ft-values of axial vector mirror transitions where effects of such a term might be partially singled out. This is because the matrix elements of second class currents have opposite signs for mirror-transitions (compare the f_T terms in eqs. (V-10) and (V-11)). Referring to eq. (II-37) it can be seen that as far as f_A and f_T are concerned the leading contribution to the nuclear Hamiltonian for β^- decay has the form

$$H_{\beta^-} = -\frac{G_\beta}{\sqrt{2}} \boldsymbol{\sigma} \cdot \{(f_A - f_T E_0) \boldsymbol{L}^* - \mathrm{i} f_T \boldsymbol{q} L_4^*\} \qquad \text{(V-20)}$$

where E_0 is the energy transfer, $\boldsymbol{q}$ the momentum transfer, and L_λ^* is an electron–neutrino function (see eq. (II-26)). A similar expression holds for H_{β^+} with the sign of the f_T term changed and detailed calculation then gives for the relative difference in the ft-values for a pair of mirror decays (Huffaker and Greuling (1963))

$$\delta \simeq \frac{4}{3} \frac{f_T}{|\lambda|} (E_0^+ + E_0^-) \qquad \text{(V-21)}$$

where $\lambda = G_A/G_V$ and $E_0^{+,-}$ refer to the end point energies (in units of mc^2)

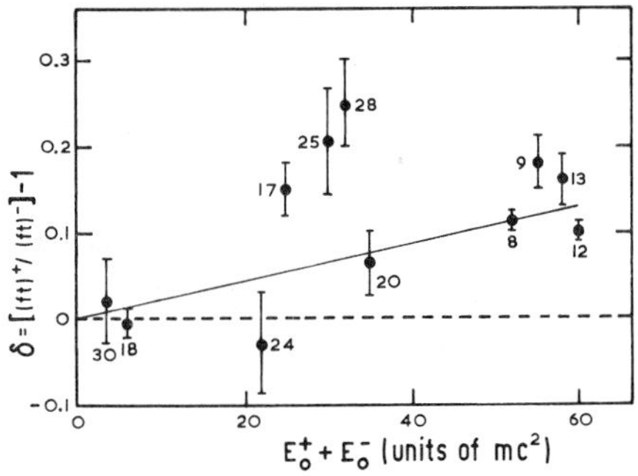

Fig. V-1. The asymmetry $\delta = [(ft)^+/(ft)^-] - 1$ in mirror axial vector β-decays. The sloping straight line corresponds to $f_T \approx 2 \times 10^{-3}$.

for the mirror β^+ and β^- decays respectively. In the above expression δ is defined as follows

$$\delta = \frac{(ft)^+}{(ft)^-} - 1 . \tag{V-22}$$

The experimental situation has recently received a lot of attention from Wilkinson and his collaborators (Wilkinson (1970); Alburger and Wilkinson (1970); Wilkinson and Alburger (1970); Wilkinson *et al.* (1971); Hardy *et al.* (1971); Wilkinson and Alburger (1971); Wilkinson (1971b)) and is summarised in fig. V-1 (Wilkinson (1971b).

The straight line drawn in fig. V-1 corresponds to $f_T \approx 2 \times 10^{-3}$ which is of the order of magnitude $1/M$ and might be expected on dimensional grounds. There is clearly a rough fit.

However, there are several points to mention before accepting this interpretation of the data. Firstly, as pointed out by Weinberg (1958) and re-emphasised by Lipkin (1971a) (see also Wolfenstein and Henley (1971)), for nucleons on the mass shell Hermiticity ensures that there is no asymmetry. In the foregoing treatment off-mass shell effects have been taken into account crudely by using the impulse approximation and the experimental end point energies. However, off-mass-shell effects are brought about by interaction with other nucleons in the nucleus and a proper treatment should allow for all such effects. In other words it should take full account of exchange effects. Assuming a divergenceless second-class current Delorme and Rho (1971 a, b) have in fact shown that exchange effects and the assumptions made in deducing eq. (V-21) are essentially equivalent. However, this assumption is arbitrary and, for example, Lipkin (1971b) using a specific meson exchange model including the "second class" decay $\omega \to \pi + e + \nu$ (cf. the "first class" decay $\rho \to \pi + e + \nu$ included in usual axial vector exchange effects (fig. IV-1(d)) obtains a value for δ roughly proportional to A. Thus there are theoretical reasons for suspecting the simple relation given in eq. (V-21) and much further study is needed of the fundamental theoretical issues involved (Delorme and Rho (1971b); Kim and Fulton (1971); Bég and Bernstein (1972)).

Secondly, the energy dependence evident in fig. V-1 rests heavily on the points for A = 18 and A = 30 and it is interesting to note that these result from a comparison of successive β^+ decays and not of β^+ with β^-. This oddity derives from Coulomb effects which lift the $T = 0$ ground state of the central (isospin-wise) nucleus into a position *between* the $T = 1$, $T_3 = \pm 1$ states. These points might therefore tentatively be excluded from the analysis in which case a line corresponding to δ = const. would equally well fit the data (Vatai (1971)). In this connection an experiment by Wilkinson and Alburger (1971) on the mirror decays $^8\text{Li} \to {}^8\text{Be}$ and $^8\text{B} \to {}^8\text{Be}$ supports such a hypothesis. The decay in question is to a very broad (many MeV) state in ^{8}Be and so the energy dependence can be studied within the same nucleus. Suffice it to say here that no such energy dependence is found.

The question then arises as to whether nuclear structure effects can account for the experimental data. A first full study of this point was made by Blin-Stoyle and Rosina (1965) for the A = 12 mirror system. They obtained the following contributions to δ from the effects indicated:

(i) Second forbidden contributions *: $\delta \approx -0.01$ to -0.03;

(ii) Corrections involving electromagnetic interactions between the charged leptons and the nucleus (including the usual electromagnetic radiative corrections): $\delta \approx -0.02$.

(iii) Isospin mixing effects in the initial and final states: $\delta \approx -0.03$ to $+0.01$;

(iv) The effect of different binding energies in ^{12}B and ^{12}N. Here account is taken of the fact that because of binding energy differences the nuclear wave functions will be different in ^{12}B and ^{12}N: $\delta \approx +0.035$ to 0.055.

The main effect is obviously (iv) and it is in the right direction to account for the value of δ_{exp} for the A = 12 system ($\delta_{\text{exp}} \approx +0.1$). It has also been investigated for A = 12 by Eichler *et al.* (1964) [$\delta \approx +0.1$], Mafethe and

* A recent more detailed calculation by Wilkinson (1971c) shows that the second forbidden contribution to δ due to weak magnetism is quite negligible.

Hodgson (1966) [$\delta \approx 0.05$], Laverne and Do Dang (1971) [$\delta \approx +0.085$], Blomqvist (1971) [$\delta \approx 0.14$ to 0.19], Wilkinson (1971a) [$\delta \approx 0.098$ to 0.146]. Some of these later calculations are much more sophisticated and elaborate. They have also been carried out for other nuclei (Blomqvist (1971); Wilkinson (1971a); Blin-Stoyle *et al.* (1971); Wilkinson (1972)) so that the effect (iv) (and in some cases (iii)) can be considered for the totality of decays studied. Wilkinson (1971a, b) concludes on the basis of these calculations that in considering the total situation for nuclei in the 1p shell the binding energy effect "fails systematically to reproduce the experimental δ-values". In support of this statement the averaged values over the 1p shell for the theoretical and experimental values of δ are as follows

$$\overline{\delta_{\text{theo}}} \approx 0.042 \text{ to } 0.054$$
$$\overline{\delta_{\text{exp}}} \approx 0.144\,. \qquad \text{(V-23)}$$

He also shows that a similar situation exists for the (2s, 1d) shell nuclei.

There is therefore an unresolved problem here. The experimental values of δ may derive from a second class current effect which is largely energy independent or there is still the chance that other nuclear structure effects not fully considered so far (e.g. configurational changes across an isospin multiplet) or higher order β-decay and relativistic corrections (Blin-Stoyle and Rosina (1965); Strubbe and Callebaut (1970)) may help. In both areas further study is clearly needed.

Other possible nuclear ways of investigating this problem have been suggested. For example, as indicated earlier, the f_T term can also affect the shape of the β-spectrum and so, conversely, the detailed study of suitable spectra might give information about f_T. Theoretical work along these lines has been carried out for $0^- \to 0^+$ transitions (Daniel and Kaschl (1966); Krmpotić and Tadić (1966)) and for unique beta transitions (Eman *et al.* (1967)). The theoretical analysis is extra ordinarily complicated and the experimental data not always clear cut, and the interpretation of the results is consequently confused and uncertain (Krmpotić and Tadić (1969)).

Another approach is to make accurate comparative measurement of electron–neutrino and electron–nuclear spin angular correlations. Here Kim (1971), using the "elementary particle" model of the nucleus, has suggested that such angular correlations could be sensitive to the presence of second class currents (giving a contribution ≈ 1–10%). Similarly Holstein and Treiman (1971) and Holstein (1971a,b; 1972) (see also Holstein *et al.* (1971)) have shown that a study of the above correlations and of β–γ angular correlations could reveal the presence of recoil terms stemming from second class currents when the β-decay is between members of a common isospin multiplet.

Finally it should be mentioned that there is some slight evidence from μ-capture data for the presence of effects due to second class currents. This is discussed in Chapt. VII.

V-6. Possible scalar (S) or tensor (T) interactions in β-decay

Throughout the discussion of β-decay phenomena it has been assumed that the basic interaction is V–A in nature and obviously the success with which such a theory is able to account for the main features of β-decay strongly supports the assumption. Nevertheless it is important to see what upper limits can be imposed on possible scalar or tensor interactions of, for example, the form

$$H_\beta^{(\text{scalar})} \sim \frac{G_\text{S}}{\sqrt{2}}(\bar{u}_\text{p}u_\text{n})(\bar{u}_\text{e}(1+\gamma_5)u_{\nu_\text{e}})$$

$$H_\beta^{(\text{tensor})} \sim \frac{G_\text{T}}{\sqrt{2}}(\bar{u}_\text{p}\gamma_\lambda\gamma_\nu u_\text{n})(\bar{u}_\text{e}\gamma_\lambda\gamma_\nu(1+\gamma_5)u_{\nu_\text{e}}) \qquad \text{(V-24)}$$

playing a role in the total β-decay interaction.

If either of the above terms are present then they can lead to modifications of the β-decay energy spectrum through what are known as Fierz interference terms (Fierz (1937)). In addition the angular correlation between electron and neutrino in a β-decay is sensitive to the presence of a scalar or tensor interaction. Experiments in this latter field are part of the well established part of beta decay studies and will not be discussed here. Schopper (1966) gives a full treatment of the situation and assuming scalar and tensor interactions of the form given in eq. (V-24) he obtains

$$\frac{G_\text{T}}{G_\text{A}} \lesssim 0.005\,, \qquad \frac{G_\text{S}}{G_\text{V}} \lesssim 0.07\,.$$

As far as Fierz interference is concerned this shows itself up as a modifying term in the allowed β-decay spectrum which, assuming that all coupling constants are real, now takes the form (cf. eq. (II-59) and see e.g. Schopper (1966) or Wu and Moszkowski (1966))

$$P(E)\,\mathrm{d}E = \frac{1}{(2\pi)^3}\Big[(G_V^2+G_S^2)|\langle t_\pm\rangle|^2 + (G_A^2+G_T^2)|\langle \boldsymbol{\sigma} t_\pm\rangle|^2 \pm \frac{2m}{E}(G_S G_V|\langle t_\pm\rangle|^2 + G_A G_T|\langle \boldsymbol{\sigma} t_\pm\rangle|^2)\Big]\, pE(E_0-E)^2F(Z,E)\,\mathrm{d}E \tag{V-25}$$

where the + and – signs refer to β^- and β^+ decay respectively.

The magnitude of the interference term can be determined by careful comparison of experimental allowed spectra and the theoretical spectrum given in eq. (V-25). Such comparisons have been made by various authors and the following results obtained

$$\frac{G_S}{G_V} = 0.00 \pm 0.12 \qquad \text{(Gerhart (1958))}$$

$$\frac{G_T}{G_A} \lesssim 0.04 \qquad \text{(Daniel (1958))}$$

$$\frac{G_T}{G_A} = 0.008 \pm 0.006 \text{ (Leutz (1961))}\,.$$

Another way of determining the Fierz interference term is by measurement of the K capture to β^+-decay ratio for a decaying nucleus. This ratio is given by (de Groot and Tolhoek (1950); see also Schopper (1966))

$$\frac{\lambda_K}{\lambda_{\beta^+}} = \left(\frac{\lambda_K}{\lambda_{\beta^+}}\right)_0 \frac{1+2b/E_K}{1-2b\langle E^{-1}\rangle} \tag{V-26}$$

where

$$b = m\,\frac{G_S G_V|\langle t_\pm\rangle|^2 + G_A G_T|\langle \boldsymbol{\sigma} t_\pm\rangle|^2}{(G_V^2+G_S^2)|\langle t_\pm\rangle|^2 + (G_A^2+G_T^2)|\langle \boldsymbol{\sigma} t_\pm\rangle|^2}\,.$$

E_K is the K-electron binding energy in units of mc^2 and $\langle E^{-1}\rangle$ is the average value of E^{-1} over the β^+ spectrum. $(\lambda_K/\lambda_{\beta^+})_0$ is the theoretical value for $\lambda_K/\lambda_{\beta^+}$ neglecting Fierz interference. The following results have been obtained using this method (see Berenyi (1968) for a general review of electron capture phenomena)

$$\frac{G_S}{G_V} = -0.02 \pm 0.09 \qquad \text{(Scobie and Lewis (1957))}$$

$$\frac{G_T}{G_A} = -0.012 \pm 0.022 \quad \text{(Konijn \textit{et al.} (1958))}$$

$$\frac{G_T}{G_A} = -0.004 \pm 0.013 \quad \text{(Ramaswamy (1959))}$$

$$\frac{G_T}{G_A} = -0.025 \pm 0.006 \quad \text{(Williams (1964))}$$

$$\frac{G_T}{G_A} = -0.024 \pm 0.009 \quad \text{(Leutz and Wenninger (1967))} .$$

There is evidently some disagreement here between the different results and also with values of G_T/G_A determined from e–ν angular correlations. One of the main sources of uncertainty lies in the application of electron exchange corrections (Bahcall (1963a, b, c)) to the theoretical expression for λ_K. This point is discussed by various authors (Murthy and Ramaswamy (1963); Leutz and Wenninger (1967); Vatai *et al.* (1968); Williams (1968); Ledingham *et al.* (1971)) and at present the uncertainty is such as to make the disagreements probably insignificant.

Paul (1970) has made a least squares study of beta-decay coupling constants using all available data and obtains $G_S/G_V = -0.001 \pm 0.006$ and $G_T/G_A = -0.0004 \pm 0.0003$. Thus, in summary, it can be said that there is at present no evidence for the presence of a fundamental S or T interaction but that there are some uncertainties still to be cleared up.

CHAPTER VI

THE LEPTON CURRENT AND THE NEUTRINO

VI-1. Introduction

Throughout the discussion of beta-decay phenomena it has been implicitly assumed that lepton number (see §I-5) is conserved and that the electron–neutrino has zero mass and negative helicity (i.e. is left handed) and described by 2-component theory. These various assumptions are related to one another and, in particular, it should be noted that only if both the two component theory and the conservation law of lepton number are valid is the neutrino mass obliged to be zero. These assumptions are fundamental, far reaching and aesthetically pleasing and it is clearly of importance that they should be checked with extreme accuracy. Any slight deviation would have profound implications for our understanding of leptons. In this chapter we shall look into the present limits on the mass and helicity of the electron–neutrino (ν_e) and on lepton number conservation.

VI-2. The mass of the electron–neutrino (ν_e)

It is a well known piece of β-decay theory that should the neutrino have a finite mass then the shape of the β-spectrum near its end point is modified (e.g. Wu and Moszkowski (1966); Schopper (1966)). This is because the usual phase space factor $p_e E_e p_{\nu_e} E_{\nu_e}$ now takes the form

$$p_e E_e \{(E_0 - E_e) + m_{\nu_e} c^2\}\{[(E_0 - E_e) + m_{\nu_e} c^2] - m^2_{\nu_e} c^4\}^{1/2} \,. \qquad \text{(VI-1)}$$

This modification results in the spectrum approaching zero with a vertical tangent as against a horizontal tangent which is the situation if $m_{\nu_e} = 0$. Obviously, then, measurement of the spectrum shape near the end point can set an upper limit on m_{ν_e}.

The most accurate experiments in this field have been those on the shape of the allowed β-spectrum of ^{3}H for which $E_0 \approx 18.7$ keV. Langer and

Moffat (1952) (see also Hamilton *et al.* (1953)) for example have set an upper limit $m_{\nu_e} < 250$ eV. Daniel (1968) in reviewing the situation concludes from the same data that $m_{\nu_e} < 1$ keV. He also stresses that, to be exact, this measurement refers to the mass of the anti-neutrino and that there is really no precise value available for the neutrino (as distinct from the anti-neutrino). If the neutral particles emitted in $\beta^{\pm}$ decays are literally particle and anti-particle then there is no problem since $m_{\nu_e} = m_{\tilde{\nu}_e}$. However, albeit unlikely, they may not be. At present the best direct measurement of m_{ν_e} (as distinct from $m_{\tilde{\nu}_e}$) is (Beck and Daniel (1967)) *

$$m_{\nu_e} < 6 \text{ keV} .$$

Very recently Bergkvist (1969) has reinvestigated the ^{3}H spectrum and achieved a much lower limit for the antineutrino mass. He obtains

$$m_{\tilde{\nu}_e} < 60 \text{ eV} .$$

This result depends on the assumption that the coefficient a of an overall modifying term to the spectrum of the form

$$\left[1 + a\frac{m_e m_\nu}{E_e E_\nu}\right]$$

arising as a spinor relativistic correction (Enz (1957)) vanishes. Such is the case if the lepton current behaves like $-\mathrm{i}\,\bar{\psi}_e\gamma_\lambda(1+\gamma_5)\,\psi_{\nu_e}$ which, as we shall see, seems to be true to good accuracy.

There is therefore no evidence that the neutrino or antineutrino have a finite mass, but it is important to continue reducing the experimental upper limits.

VI-3. Possible forms of the lepton current

Assuming that the hadron current only has polar vector and axial vector parts (i.e. there is, for example, no fundamental scalar and tensor interaction) the assumption that the electron part of the lepton current has the form

* The best upper limit for the *muon* neutrino mass is due to Backenstoss *et al.* (1971) (see also Shrum and Ziock (1971)). They obtain

$$m_{\nu_\mu} < 0.6 \text{ MeV} .$$

$$\ell_\lambda^* = -\mathrm{i}\,\overline{\psi}_e \gamma_\lambda (1+\gamma_5)\,\psi_{\nu_e} \tag{VI-2}$$

is somewhat restrictive. In this section we take the viewpoint that eq. (VI-2) is certainly approximately correct, as seems to be indicated by experiments, but that there may be small deviations from it. To this end consider a more general form for ℓ_λ^* as follows

$$\ell_\lambda^* = -\mathrm{i}N\overline{\psi}_e\gamma_\lambda\{[(1+\gamma_5)+\eta(1-\gamma_5)]\,\psi_{\nu_e} + [\xi(1+\gamma_5)+\delta(1-\gamma_5)]\,\psi_{\tilde{\nu}_e}\} \tag{VI-3}$$

where $N = (1+\eta^2+\xi^2+\delta^2)^{-1/2}$ and $\psi_{\tilde{\nu}_e}$ is the (charge conjugate) field operator for the antineutrino (see App. D-4). Thus whereas ψ_{ν_e} annihilates a ν_e or creates a $\tilde{\nu}_e$, $\psi_{\tilde{\nu}_e}$ annihilates a $\tilde{\nu}_e$ or creates a ν_e. η, ξ and δ are real (assuming time reversal invariance *) parameters measuring the deviation of ℓ_λ^* from the usually accepted form. Clearly with a current of the form given in eq. (VI-3) if ξ or δ are non-zero then lepton number (see §I-5) is not conserved and if η or δ are non zero the neutral leptons emitted in β-decay will have helicities differing from the usual ± 1. It is to be noted that the particular case $\xi = 1$, $\eta = \delta = \eta'$ enables ℓ_λ^* to be written

$$\ell_\lambda^* = \frac{-\mathrm{i}\,\overline{\psi}_e}{\sqrt{1+\eta'^2}}\,[\gamma_\lambda(1+\gamma_5)+\eta'\gamma_\lambda(1-\gamma_5)]\,\psi_{\nu'_e} \tag{VI-4}$$

where

$$\psi_{\nu'_e} = \frac{1}{\sqrt{2}}\,[\psi_{\nu_e}+\psi_{\tilde{\nu}_e}] \tag{VI-5}$$

so that

$$\psi_{\tilde{\nu}'_e} = \frac{1}{\sqrt{2}}\,[\psi_{\tilde{\nu}_e}+\psi_{\nu_e}] = \psi_{\nu'_e}\,. \tag{VI-6}$$

Here we have the situation that the particle ν'_e is identical with its anti-particle $\tilde{\nu}'_e$. Such a particle is designated a Majorana neutrino (Majorana (1937)) as distinct from a Dirac neutrino for which particle and anti-particle are not identical.

* Primakoff and Sharp (1969) have suggested a *CP*-violating weak-interaction Hamiltonian for which η is imaginary and in which an equivalent term of strength η occurs in the hadron current. Such a Hamiltonian can account for *CP*-violation in K-decay and for a possible lepton non-conservation in double β-decay (see § VI-6.2).

Another way of looking at the effects of the foregoing current (eq. (VI-3)) is in terms of the handedness of, for example, the emitted neutral particle in β^- decay. The left (L) or right (R) handedness is determined by the operator $(1 \pm \gamma_5)$ and we could write eq. (VI-3) in the form

$$\ell^*_\lambda = -\mathrm{i}N\bar{\psi}_e\gamma_\lambda\{\psi_{\nu_e^L} + \eta\psi_{\nu_e^R} + \xi\psi_{\tilde{\nu}_e^R} + \delta\psi_{\tilde{\nu}_e^L}\} \tag{VI-7}$$

or, symbolically, in neutron β-decay

$$n \to p + e^- + \begin{cases} 1\tilde{\nu}_e^L & (h_\nu = +1) \\ +\eta\tilde{\nu}_e^R & (h_\nu = -1) \\ +\xi\nu_e^R & (h_\nu = +1) \\ +\delta\nu_e^L & (h_\nu = -1) \end{cases} \tag{VI-8}$$

where the amplitudes $(1, \eta, \xi, \delta)$ of the different possible states in which the neutral particle can be emitted are indicated, and where h_ν labels the helicity (handedness) of the emitted neutral lepton. Note that the labels L and R refer to the neutrino so that, for example, $\tilde{\nu}_e^L$ designates the anti-particle of a left handed neutrino. It is therefore right-handed and has helicity $h_\nu = +1$ as indicated etc.

Even more generalisation of the situation is achieved if the V–A restriction on the hadron current is removed. This fully general situation is reviewed thoroughly by Schopper (1966) and Lee and Wu (1965) on the basis of the work of a number of authors (Touschek (1957); Pauli (1957); Pursey (1957); Kahana and Pursey (1957); Enz (1957); Lüders (1958); Schopper and Müller (1959)) but we do not consider it here.

We now go on to discuss the limits which experiment impose on the values of η, ξ, and δ.

VI-4. β-decay parity experiments

Since 1959 when it was first proposed by Lee and Yang (1956) that parity is not conserved in β-decay, an extended series of experiments has been carried out aimed at detecting parity violating terms in various β-decay processes. These have mainly involved the measurement of (i) the angular distribution of electrons and positrons from polarised nuclei, (ii) the longitudinal polarisation of electrons and positrons from unpolarised nuclei and (iii) β–γ circular polarisation correlations. In addition a few experiments have been carried out involving measurements on internal bremsstrahlung and also of the nuclear recoil (see Schopper (1966)) following β-decay. Of particular note is the very elegant experiment of Goldhaber *et al.* (1958) who measured the circular polarisation of the gamma ray emitted following K-capture in $^{152}Eu^{m}$. The spin sequence is such that angular momentum conservation requires the gamma circular polarisation to be the same as that of the emitted neutrino independently of the V–A assumption. From this experiment it emerges that the neutrino is left handed with helicity of the order -1 (within about 10%). Subsequently (Palathingal (1970); see also Morita *et al.* (1958); Treiman (1958); Bincer (1958)) a direct measurement of the anti-neutrino helicity has also been made and the result is consistent with helicity $+1$.

The main body of work on parity violating experiments which was initiated by Wu *et al.* (1957) and Frauenfelder (1957a, b) is well known and part of standard β-decay lore. It has been reviewed by various authors (e.g. Frauenfelder and Steffen (1965b); Schopper (1966); Wu and Moszkowski (1966)) and will not be discussed in any detail here. We will confine ourselves to comparing experimental results with the theoretical values obtained if $\eta = \xi = \delta = 0$. Any disagreement then gives information about the values of η, ξ and δ.

Given the general current (eq. (VI-3)) it is straightforward from the standard results of β-decay theory (see Schopper (1966)) to express the asymmetry parameters in the various experiments in terms of η, ξ and δ. Taking the three types of experiment in turn, the experimentally measured quantities can be written as follows:

(i) *Angular distribution from polarised nuclei*

$$W = 1 + A_1 \frac{v}{c} \frac{\langle \boldsymbol{I} \rangle}{I} \cdot \boldsymbol{p}_e \tag{VI-9}$$

where v is the electron (or positron) speed, $\boldsymbol{p}_e$ a unit vector along its direction of momentum, $\langle \boldsymbol{I} \rangle / I$ the nuclear polarisation and A_1 the asymmetry parameter (q. v.).

(ii) *Longitudinal polarisation of electron or positron*

$$P = A_2 \frac{v}{c} \tag{VI-10}$$

where A_2 is the polarisation parameter (q.v.)

(iii) *β–γ circular polarisation correlation*

$$W = 1 + A_3 \frac{v}{c} \tau \hat{k} \cdot \hat{p}_e \tag{VI-11}$$

where $\tau = \pm 1$ corresponds to the two possible circular polarisation states of the photon, $\hat{k}$ is a unit vector in the direction of motion of the photon and A_3 is the asymmetry parameter (q.v.).

In all three cases A_1, A_2 and A_3 depend on the detailed nature of the β-decay interaction and, in the case of A_3 the nature of the γ-transition. Taking over standard results for the A_i (Lee and Yang (1956); Alder *et al.* (1957); Jackson *et al.* (1957a, b); see also Schopper (1966)) it is easy to show in all three cases that for an allowed $\Delta I = \pm 1$ (i.e. Gamow–Teller) transition

$$A_i = A_i^{(0)} \left[\frac{1 - \eta^2 + \xi^2 - \delta^2}{1 + \eta^2 + \xi^2 + \delta^2} \right] \tag{VI-12}$$

where $A_i^{(0)}$ is the value of A_i in the case $\eta = \xi = \delta = 0$. For a $\Delta J = 0$ transition the situation is slightly more complicated in the case of A_1 and A_3 due to interference between the V and A interaction and involves the values of nuclear matrix elements. But the change in the A_i is still of second order in the η, ξ and δ and this is the essential point for our purposes. In any case most experimental measurements of the A_i are for Gamow–Teller transitions.

In table VI-1 values are given of the ratio $\Gamma_i = [(A_i)_{\text{expt}}/(A_i^{(0)})]$ for a large selection of decay processes to show the accuracy with which limits can be put on the values of η, ξ and δ. The different experimental results are mainly taken from the compilations of Schopper (1966) and in the case of A_1 and A_3 only results for Gamow–Teller transitions are quoted since the interpretation of mixed transitions is more complicated.

The large selection of experimental results is given to make it clear that in spite of the tremendous effort which has been put in, the accuracy in general is not very good from the point of view of setting limits on the values of η, ξ and δ. At best an accuracy of 1.2% has been achieved (Brosi *et al.*

Table VI-1

Values of Γ_i (= $[(A_i)_{\exp}/(A_i^{(0)})]$)

Parent nucleus	$\Gamma_i = [(A_i)_{\exp}/(A_i^{(0)})]$	Reference
	(i) Γ_1	
^{8}Li	0.96 ± 0.12	Abov *et al.* (1962)
^{56}Co	1.10 ± 0.10	Ambler *et al.* (1957a, b)
^{58}Co	0.99 ± 0.15	Postma *et al.* (1957)
^{60}Co	≈ 1	Wu *et al.* (1957)
	(ii) Γ_2	
^{12}B	0.98 ± 0.06	Lipnik *et al.* (1962)
^{13}N	0.93 ± 0.20	Boehm and Wapstra (1957a, b)
^{14}O	0.73 ± 0.17	Gerhart *et al.* (1959)
	0.97 ± 0.19	Hopkins *et al.* (1961)
^{32}P	0.98 ± 0.11	Cavanagh *et al.* (1957)
	0.76 ± 0.15	De Waard and Poppema (1957)
	1.00 ± 0.13	Frauenfelder *et al.* (1957b)
	0.96 ± 0.06	Bienlein *et al.* (1958)
	0.94 ± 0.06	Geiger *et al.* (1958)
	1.05 ± 0.04	Bisi *et al.* (1960, 1963)
	1.00 ± 0.02	Ullman *et al.* (1961)
	0.981 ± 0.012	Brosi *et al.* (1962)
^{60}Co	0.98 ± 0.11	Cavanagh *et al.* (1957)
	0.74 ± 0.15	De Waard and Poppema (1957)
	0.82 ± 0.15	Frauenfelder *et al.* (1957a)
	0.982 ± 0.028	Bienlein *et al.* (1958)
	0.99 ± 0.06	Greenberg *et al.* (1960)
^{64}Cu	1.03 ± 0.15	Vishnevsky *et al.* (1957)
^{68}Ga	0.99 ± 0.09	Ullman *et al.* (1961)
^{198}Au	0.98 ± 0.03	Ullman *et al.* (1961)
	(iii) Γ_3	
^{22}Na	1.17 ± 0.24	Schopper (1957)
	0.89 ± 0.16	Appel *et al.* (1958)
	1.08 ± 0.24	Appel (1959)
	1.05 ± 0.06	Steffen (1959)
^{28}Al	0.95 ± 0.07	Mann and Bloom (1965)
^{47}Ca	1.05 ± 0.06	Behrens (1967)
	0.88 ± 0.05	Mann *et al.* (1965)
^{56}Mn	0.80 ± 0.06	Lobashov and Nazarenko (1962)
	1.02 ± 0.06	Schopper *et al.* (1964)
	1.03 ± 0.06	Behrens (1966)
^{60}Co	1.05 ± 0.15	Appel and Schopper (1957)

Table VI-1 (continued)

Parent nucleus	$\Gamma_i = [(A_i)_{\exp}/(A_i^{(0)})]$	References
	(iii) Γ_3	
^{60}Co	1.20 ±0.27	Boehm and Wapstra (1957a)
	1.02 ±0.27	Debrunner and Kundig (1957)
	0.96 ±0.21	Lundby *et al.* (1957)
	1.23 ±0.21	Schopper (1957)
	1.01 ±0.05	Appel *et al.* (1958)
	1.23 ±0.24	Boehm and Wapstra (1958)
	1.03 ±0.06	Steffen (1959)
	0.96 ±0.10	Berthier (1962)
	0.99 ±0.06	Lobashov and Nazarenko (1962)
^{95}Nb	0.98 ±0.16	Appel *et al.* (1962)
	1.02 ±0.18	Mann *et al.* (1962)
^{95}Zr	0.92 ±0.18	Appel and Schopper (1957)
^{154}Eu	0.90 ±0.30	Castle *et al.* (1964)

(1961)). If η, ξ and δ are assumed to be small, then we have from this result

$$\begin{aligned} 2|\eta^2+\delta^2| &\lesssim 0.012 \\ |\eta| &\lesssim 0.1 \\ |\delta| &\lesssim 0.1\,. \end{aligned} \tag{VI-13}$$

In terms of a Majorana neutrino ν (see eq. (VI-4)) the corresponding result is $|\eta'| \lesssim 0.1$. Thus, although the results are consistent with $\eta = \xi = \delta = 0$ there is nevertheless considerable leeway and it is important to try and improve experimental accuracy by an order of magnitude.

VI-5. **Inverse β-decay**

The famous experiment of Reines and Cowan (1959) in which the inverse β-decay process

$$\tilde{\nu}_e + p \to n + e^+ \tag{VI-14}$$

was detected is well known. It was found that the measured cross section agreed, within a sizeable experimental uncertainty (40%), with theoretical expectations assuming that $\eta = \xi = \delta = 0$. In this experiment the antineutrinos were produced in an atomic pile mainly from fission products but stemming from the basic process

$$n \to p + e^- + \tilde{\nu}_e \,. \tag{VI-15}$$

Davis (1955, 1958) using pile antineutrinos sought unsuccessfully for the process

$$\tilde{\nu}_e + {}^{37}\mathrm{Cl} \to {}^{37}\mathrm{A} + e^- \,. \tag{VI-16}$$

This is again in agreement with usual β-decay theory with $\eta = \xi = \delta = 0$. However, it is interesting to see what limits on these quantities are implied by this null result.

Referring to the symbolic equation (VI-8), it can be seen that the different species of neutrino emitted from the pile in this general situation will have proportional intensities as follows (in brackets)

$$\tilde{\nu}_e^L(1);\ \tilde{\nu}_e^R(\eta^2);\ \nu_e^R(\xi^2);\ \nu_e^L(\delta^2)\,.$$

Furthermore, using eq. (VI-7) in the usual way, the following inverse processes are now allowed

$$\left.\begin{matrix} 1\nu_e^L \\ \eta\nu_e^R \\ \xi\tilde{\nu}_e^R \\ \delta\tilde{\nu}_e^L \end{matrix}\right\} + {}^{37}\mathrm{Cl} \to {}^{37}\mathrm{A} + e^- \tag{VI-17}$$

with the amplitudes indicated. Thus given the structure of the pile antineutrino flux, the process studied by Davis (1955, 1958) should have a rate proportional to δ^2 if η, ξ and δ are taken as small quantities. From the fact that the process was unobserved an upper limit can be set on the value δ. This has been estimated, in a slightly different context by Primakoff and Rosen (1969) to be $\delta \lesssim 0.2$. Again in terms of a Majorana neutrino ν_e' (see eq. (VI-4)) the corresponding result is $\eta' \lesssim 0.2$. Here it is to be noted that although particle and antiparticle are identical for a Majorana neutrino it is still necessary for $\eta' \neq 0$ in order for a process like that in eq. (VI-16) to take place since, as well as being forbidden by lepton conservation in the conventional theory, the process is also forbidden by helicity arguments (i.e. if $\eta' = \eta = \delta = 0$ the emitted "neutrino" in neutron decay has $h_\nu = +1$ (see eq. (VI-8)) whilst under these circumstances the captured "neutrino" must be left handed (see eq. (VI-17)).

VI-6. Double β-decay

Double beta-decay is the very rare process in which a nucleus (A, Z) decays into the nucleus $(A, Z \pm 2)$ with the emission of two electrons or positrons and either (i) two neutrinos or antineutrinos or (ii) no neutrinos or antineutrinos. There are also other possible alternatives involving the capture of orbital electrons. One of the main reasons for studying double beta decay is the light the process can shed on the natures of and relation between the neutrino and antineutrino.

Basically the situation is as follows. If the neutrino and antineutrino are different (Dirac neutrino) and have opposite lepton number (i.e. $N_e(\nu_e) = +1$ and $N_e(\tilde{\nu}_e) = -1$; see §I-5) as is generally assumed and if double beta-decay is a second order effect of the ordinary β-decay interaction then the double β-decay of two neutrons will proceed as follows

$$\begin{aligned} n_1 + n_2 &\to n_1 + p_2 + e_2^- + \tilde{\nu}_{e_2} \\ &\to p_1 + e_1 + \tilde{\nu}_{e_1} + p_2 + e_2 + \tilde{\nu}_{e_2} \, . \end{aligned} \qquad \text{(VI-18)}$$

In this case two anti-neutrinos are emitted. However, if lepton number is not conserved so that the process

$$\tilde{\nu}_e + n \to p + e^- \qquad \text{(VI-19)}$$

can take place, then in the above double β-decay process $\tilde{\nu}_{e_2}$ could be absorbed by n_1 in the second stage to give overall

$$n_1 + n_2 \to p_1 + e_1^- + p_2 + e_2^- \qquad \text{(VI-20)}$$

with no neutrino or anti-neutrino emission.

Similarly, in the special case that the neutrino and anti-neutrino are identical (Majorana neutrino) so that $N_e(\nu_e) = N_e(\tilde{\nu}_e) = 0$ the neutrinoless double β-decay process can again take place.

Obviously if neutrinoless decay can take place so also can two neutrino emission as an alternative process. In general, however, neutrinoless double

β-decay (eq. (VI-20)) is expected to be much faster than double β-decay with emission of neutrinos (eq. (VI-18)) since there is much more phase space available for the virtual neutrino in the former process. But, if for some reason the matrix element for neutrinoless decay was anomalously small then, even if allowed by selection rules, the neutrinoless decay could be extremely inhibited so that decay with two-neutrino emission would dominate. There are clearly a number of alternative interpretations of different possible experimental results and they are discussed in some detail by Rosen and Primakoff (1965) (see also Greuling and Whitten (1960); Lee and Wu (1965); Konopinski (1966)). The situation can be summarised as follows:

(i) If double β-decay is shown to be a two neutrino process then the simplest conclusion is that the ν_e and $\tilde{\nu}_e$ are different and that lepton number is conserved. However, the possibility mentioned above is always there that $\nu_e \equiv \tilde{\nu}_e$ or lepton number is not conserved but that for some reason neutrinoless decay is strongly inhibited.

(ii) If double β-decay is shown to be neutrinoless then it can be concluded that lepton number is not conserved but no decision can be made as to whether the neutrino and anti-neutrino are identical or not.

(iii) Referring to the particular case of a V–A interaction and a lepton current having the form given in eq. (VI-3) then neutrinoless decay would go at a rate proportional to δ^2 (or η'^2 in the case of Majorana neutrinos).

VI-6.1. *The theory of double β-decay*

The process of double β-decay is taken to be a second order effect of the basic β-decay Hamiltonian H_β and the matrix element can be written formally as (Rosen and Primakoff (1966))

$$M_{\beta\beta} = \sum_{v} \frac{\langle \chi_f \Psi_f | H_\beta | \Psi_v \chi_v \rangle \langle \chi_v \Psi_v | H_\beta | \Psi_i \chi_i \rangle}{E_v - E_i} . \qquad \text{(VI-21)}$$

Ψ_i, Ψ_f and Ψ_v are wave functions for the initial (A, Z), final $(A, Z \pm 2)$ and intermediate $(A, Z \pm 1)$ nuclear states whilst χ_i, χ_f and χ_v refer to the initial, final and intermediate lepton states. Thus χ_i refers to a state of no leptons, χ_f to a state containing two charged leptons and either (i) two neutral leptons or (ii) no neutral leptons. E_v is an appropriate energy value for the intermediate states.

The calculation of transition rates for the two decay possibilities has been carried out by Rosen and Primakoff (1966) (see also Primakoff and Rosen (1959, 1961)) assuming a V–A interaction, that the transition is $0^+ \rightarrow 0^+$

and that the sum over intermediate states can be carried out by closure (Primakoff (1952); Rosen (1957)). For decay with emission of two-"neutrinos" they obtain for the half-life

$$T'_{1/2} \approx 2 \times 10^{20 \pm 2} \left(\frac{\langle E_v - E_i \rangle + \frac{1}{2}(\epsilon_0 + 2)}{10} \right)^2 \left(\frac{60}{Z} \right)^2$$

$$\times \left\{ 1 - \exp\left(\mp \frac{2\pi Z}{137} \right) \right\}^2 \left(\frac{8}{\epsilon_0} \right)^{10} \text{ years} \qquad \text{(VI-22)}$$

where the upper and lower signs in the square bracket refer to two electron or two positron emissoon respectively. $\langle E_v - E_i \rangle$ is the average energy difference between initial and intermediate nuclear states and ϵ_0 is the energy release in units of mc^2. The uncertainty in the exponent (20 ± 2) attempts to estimate the theoretical error stemming from the various approximations.

The neutrinoless decay process will only occur, assuming the usual V–A interaction, if the δ term in the lepton current is non-zero. Primakoff and Rosen (1969) estimate under this circumstance that the half-life is given by

$$T_{1/2} \approx \frac{1}{\delta^2} T_{1/2}^{(0)}$$

where

$$T_{1/2}^{(0)} = \frac{10^{21.9}}{f(\epsilon_0)} \left(\frac{137}{2\pi Z} \right)^2 \left[1 - \exp\left(\mp \frac{2\pi Z}{137} \right) \right]^2 \left(\frac{A}{130} \right)^{2/3} \text{ years} \qquad \text{(VI-23)}$$

and

$$f(\epsilon_0) = (\epsilon_0^5 + 13\epsilon_0^4 + 77\epsilon_0^3 + 70\epsilon_0^2)\,\epsilon_0^2 \,.$$

Given these two approximate theoretical expressions for the lifetime we can now make a comparison with experiment.

VI-6.2. *Comparison between theory and experiment in double β-decay*

It is only recently that double beta-decay has been firmly estbalished experimentally (Kirsten *et al.* (1967, 1968); see also Takaoka and Ogata (1966); Gerling *et al.* (1968). The experiments which detected it involved the mass-spectrometic analysis of tellurium ores where it was found that the relative abundance of the isotope ^{130}Xe occluded in the ores was far in excess of that in atmospheric zenon. This excess was attributed to the double beta-decay of ^{130}Te and implied a half life $10^{(21.34 \pm 0.12)}$ years (Kirsten *et al.* (1968)).

All other searches for double beta-decay (see Rosen and Primakoff (1966) and Bailin (1972) for reviews) have only been able to put a lower limit on the lifetime. In this connection some of the most effective work has been concentrated on the decay $^{48}\mathrm{Ca} \to {}^{48}\mathrm{Ti}$ (Dobrokhotov *et al.* (1959); Lazarenko and Luk'yanov (1966); Der Mateosian and Goldhaber (1966); Shapiro *et al.* (1967); Bardin *et al.* (1967, 1970)). Bardin *et al.* (1970) obtain $T_{1/2} > 2.0 \times 10^{21}$ years if the decay is neutrinoless and $T'_{1/2} > 3.6 \times 10^{19}$ years if neutrinos are emitted. The dependence of these two experimental results on the nature of the decay arises since the experimental method makes use of the assumed nature of the energy spectrum which, of course, is different in the two cases. In particular for neutrinoless decay the sum of the energies of the two electrons (or positrons) should equal the total energy released.

For the ^{130}Te decay $\epsilon_0 = 5.0\,(mc^2)$, so that for decay with neutrinos the theoretical expression (eq. (VI-22)) gives

$$T'_{1/2} \approx 10^{21.9 \pm 2} \text{ years} \tag{VI-24}$$

whilst for neutrinoless decay (eq. (VI-23))

$$T_{1/2} \approx \delta^{-2}\, 10^{15.4} \text{ years} . \tag{VI-25}$$

Comparing with the experimental value ($T_{1/2}^{\mathrm{exp}} = 10^{(21.34 \pm 0.12)}$ years) shows that experiment is consistent with the emission of two neutrinos and that an upper limit of the order 10^{-3} is imposed on δ (alternatively for a Majorana neutrino, $\eta' \lesssim 10^{-3}$). A similar limit is imposed by the lower limit on $T_{1/2}$ for the neutrinoless decay of ^{48}Ca. Primakoff and Rosen (1969) have firmed up this upper limit on δ even further by more detailed considerations in which the neutrinoless decay mode is ascribed primarily to a small ($\approx 1\%$) admixture of Δ (1236 MeV; $T = \frac{3}{2}$) in the nucleus (see Kerman and Kisslinger (1969), who first made a proposal of this kind). Since the Δ has $T = \frac{3}{2}$ the double β-decay can then proceed as follows

$$\Delta \to \mathrm{N} + \mathrm{e}^- + \mathrm{e}^- \tag{VI-26}$$

and only one nuclear particle is involved rather than the usual two (see eq. (VI-20)). This speeds up the process considerably and hence serves to set an even lower limit on δ. Primakoff and Rosen (1969) obtain $\delta \lesssim 10^{-4}$ on the basis of this model. These limits on δ are clearly far more severe than that given in eq. (VI-13) obtained by the consideration of parity non-conserving asymmetries in the single beta-decay process.

In connection with the ^{130}Te double beta-decay it is interesting to note that an associated uncertainty has arisen. In one of the mass spectrometric measurements (Takaoka and Ogata (1966)) an excess of ^{128}Xe was also found, and if this excess is similarly attributed to the double beta-decay process $^{128}\mathrm{Te} \to {}^{128}\mathrm{Xe}$ the corresponding half-life is deduced to be $10^{22.5 \pm 0.5}$ years for ^{128}Te. Accepting this value and approximate equality between the nuclear matrix elements in the ^{128}Te and ^{130}Te decays, it then turns out that the ratio of the lifetimes for these decays suggests very strongly that the decays are neutrinoless (Pontecorvo (1968)). Pontecorvo goes further and proposes that neutrinoless decay may be due to a new "superweak" interaction satisfying the seleection rule $\Delta Q = \pm 2$, $\Delta S = 0$, $\Delta N_{\mathrm{e}} = \pm 2$ which in some sense parallels the *CP*-violating "superweak" interaction proposed by Wolfenstein (1964) and discussed in Chapt. I. Primakoff and Sharp (1969) have proposed an alternative *CP*-violating theory which could account for such a neutrinoless decay (see footnote on p. 144). This is all conjecture, however, and will not be discussed further. In any case, the experimental evidence for the double beta-decay of ^{128}Te is far less strong than that for ^{130}Te.

There is obviously uncertainty here and it would be good to heighten the accuracy of experimental double-beta-decay studies – particularly in the case of ^{48}Ca which is a relatively simple nucleus from the theoretical point of view (see the calculations of Khodel (1970)). Apart from the lifetime it is most important that the energy spectrum should be investigated since for neutrinoless decay the total energy of the electrons (or positrons) should have a fixed value (the energy release) whilst for decay with emission of neutrinos a spread in the total energy is obtained.

VI-7. **Conclusions**

From the foregoing discussion it is clear that there is no evidence against the assumption that the neutrino is massless and that the lepton current has the simple form given in eq. (VI-2). However, although double β-decay phenomena seem to limit δ to be less than 10^{-3} or 10^{-4}, the limit on η is much less stringent ($|\eta| \lesssim 0.1$, (see eq. (VI-13)).

An important point to be noted is that not all of the parameters η, ξ and δ are measurable. In physical terms this stems from the arbitrariness in our definitions of $\nu_{\mathrm{e}}^{\mathrm{L}}$ and $\nu_{\mathrm{e}}^{\mathrm{R}}$. Thus, we could define "$\nu_{\mathrm{e}}^{\mathrm{L}}$" to be the left handed neutral particle emitted in β^+ decay (i.e. in the present context, define it to be "ν_e^L" $\sim \nu_e^L + \xi\tilde{\nu}_e^R$) which would be equivalent to putting $\xi = 0$ in a redefined lepton current expressed in terms of "ν_e^L". An extreme example would be the case that the emitted left handed neutral lepton was a left handed Majorana neutrino ($\nu_e^L \sim \nu_e^L + \tilde{\nu}_e^R$). More formally, the fact that η, ξ and δ are not all measurable arises because the β-decay interaction is invariant under certain unitary transformations (Pauli (1957); Lee (1957)) of the neutrino field which impose relationships between the different parameters.

The question of whether the neutrino has zero mass or not and the issues of helicity and lepton conservation are all intimately related and there has been a lot of discussion about this relationship (e.g. Lee and Wu (1965); Schopper (1966)) and its implications. One of the main points at issue is whether the neutrino lepton number can be related directly to its helicity (h_ν). Thus on the usual two-component theory (Lee and Yang (1957); Landau (1957); Salam (1957)) the neutrino lepton number is identified with $-h_\nu$. However, other alternative lepton number assignments have been suggested in which these two quantities are dissociated (e.g. Bludman (1963); Schopper (1966)) and in which the additive lepton number conservation law is replaced by a multiplicative one (Feinberg and Weinberg (1961); Cabibbo and Gatto (1960, 1961)). There is here substantial room for manoeuvre and speculation; but unless it is demonstrated experimentally that (i) $m_{\nu_e} \neq 0$ or (ii) neutrinoless double beta-decay takes place or (iii) the ratio Γ_i describing parity experiments (§VI-4) differs from unity or (iv) semi-leptonic or leptonic processes forbidden by the conventional weak interaction (e.g. $e^+e^- \to \mu^+\mu^-$) take place then such speculation need not, perhaps, be pushed very far. There is nevertheless every encouragement for nuclear experimenters to increase the accuracy of their work in the fields (i) to (iii) above.

CHAPTER VII

μ-MESON CAPTURE

VII-1. The μ-capture process

Although experimental studies of negative muon capture lie well within the province of high-energy physics, the theoretical interpretation of the experimental data is very much the concern of the nuclear structure theorist and so it is appropriate to devote a chapter to this phenomenon. The process is analogous to K-capture except that, instead of, an electron, a μ^- particle is captured from an atomic K-orbit with the resultant emission of a μ-type neutrino ν_μ. The basic process is thus

$$\mathrm{p} + \mu^- \rightarrow \mathrm{n} + \nu_\mu \qquad \text{(VII-1)}$$

and assuming the proton and muon to be initially at rest, the rest energy of the muon (≈ 106 MeV) is shared between the neutrino and the recoil neutron as follows:

$$E_{\nu_\mu} \approx 100 \text{ MeV}$$

$$E_\mathrm{n} \approx 6 \text{ MeV} .$$

In the case of capture by a complex nucleus

$$(A, Z) + \mu^- \rightarrow (A, Z-1) + \nu_\mu \qquad \text{(VII-2)}$$

two important points are to be noted. Firstly, other nucleons can take part in the process and take up some of the energy released thus leading to excited states being formed or the emission of one or more neutrons. Secondly, the capture rate increases rapidly with increasing Z. This latter effect is straightforward to understand in simple physical terms. The μ^- meson on capture by a nucleus is first of all in a highly excited Bohr orbit. However, it quickly de-excites by emission of X-rays and Auger electrons to finish up in the K orbit of radius $a_\mu = (m_e/Zm_\mu)a_0$ where a_0 is the usual Bohr radius. Thus the muon is either close to or, in the case of high Z, well inside the nucleus. Basically, the capture rate, $\Lambda_{\mu c}$, is proportional to the probability of finding the muon in the nucleus (i.e. $|\psi_\mu(0)|^2$, where $\psi_\mu(r)$ is the wave function of the muon in a K-orbit) multiplied by the number of protons in the capturing nucleus (i.e. Z). Now, for a Bohr orbit $|\psi_\mu(0)|^2$ is proportional to Z^3, so that

$$\Lambda_{\mu c} \propto Z^4 \qquad \text{(VII-3)}$$

hence accounting for the rapid increase of $\Lambda_{\mu c}$ with Z.

Actually the above formula has to be modified to take account of the fact that a muon penetrating into the nucleus does not experience the full charge Z but only the charge included within its orbit. This effect can be allowed for by replacing Z by an effective charge Z_{eff} (Wheeler (1949); Wheeler and Tiomno (1949)) so that

$$\Lambda_{\mu c} \propto (Z_{\text{eff}})^4 \,. \qquad \text{(VII-4)}$$

For light nuclei $Z_{\text{eff}} \approx Z$, whilst for heavier nuclei $Z_{\text{eff}} \approx \frac{1}{2}Z$ or less.

In addition, the rate is governed by nuclear matrix elements of the weak interaction taken with respect to the initial and different possible final nuclear states. It turns out that the rate is such that for $Z \lesssim 11$ nuclei, the muon is more likely to decay ($\mu^- \to e^- + \tilde{\nu}_e + \nu_\mu$) than to be captured whilst for $Z \gtrsim 11$ the converse is true. This means that the simplest process to analyse, namely muon capture in atomic hydrogen ($Z = 1$), is particularly difficult to study experimentally (q.v.).

We now go on to consider the nature of the basic interaction responsible for μ-capture noting that the theory and experiment of muon capture have been reviewed in many places (e.g. Primakoff (1959); Tolhoek (1963); Lee and Wu (1965); Schopper (1966); Wu and Moszkowski (1966); Rho (1967a); Weissenberg (1967)).

VII-2. The μ-capture effective Hamiltonian

As discussed in Chapt. I we assume that the same weak interaction is responsible both for beta-decay and muon capture. This means that we can take over many of the expressions obtained earlier for β-decay phenomena and use them to describe μ-capture. In particular the Hamiltonian $H_{\beta^-}(r)$ given in eq. (II-37) can lead very simply to the μ-capture Hamiltonian.

$H_{\beta^-}(r)$ refers to the process $\mathrm{n} + \nu_\mathrm{e} \to \mathrm{p} + \mathrm{e}^-$. We therefore need to (i) take its Hermitian adjoint, (ii) replace the e^- wave function by an appropriate μ^- wave function (i.e. a K-orbit wave function) and (iii) make the following changes in notation: q_λ $(= \bar{\mathrm{e}}_\lambda^- - \nu_{\mathrm{e}\lambda}) \to (\bar{\mu}_\lambda^- - \nu_{\mu\lambda})$; $\boldsymbol{q} \to \mathbf{v}_\mu = \hat{\mathbf{v}}\,\nu$ (since $\boldsymbol{\mu}^- \simeq 0$ in the capture state) where $\hat{\mathbf{v}}$ is a unit three vector in the direction of the neutrino momentum (magnitude ν).

Consider the effect of these changes on $H_{\beta^-}(r)$. The lepton part is now written (cf. eq. (II-46) for K-capture)

$$L_\lambda = -\,\mathrm{i}\,\bar{\phi}_{\nu_\mu}(\mathbf{v}, r)\gamma_\lambda(1+\gamma_5)\phi_\mu(r)\,. \tag{VII-5}$$

Appropriate normalised positive energy solutions (see App. C) for the neutrino $(m_\nu = 0)$ and muon $(E \approx m_\mu)$ functions are

$$\begin{aligned} \phi_{\nu_\mu}(\mathbf{v}, r) &= \frac{1}{\sqrt{2}}\begin{pmatrix} \chi_{\nu_\mu}(\mathbf{v}, r) \\ \boldsymbol{\sigma}_\ell \cdot \hat{\mathbf{v}}\; \chi_{\nu_\mu}(\mathbf{v}, r) \end{pmatrix} \\ \phi_\mu(r) &= \begin{pmatrix} \chi_\mu(r) \\ 0 \end{pmatrix} \end{aligned} \tag{VII-6}$$

where the suffix ℓ implies that the corresponding operator operates on lepton functions only. χ_{ν_μ} and χ_μ are non-relativistic wave functions for the neutrino and muon respectively, having the forms

$$\chi_{\nu_\mu}(\mathbf{v}, r) = \mathrm{e}^{\mathrm{i}\,\mathbf{v}\cdot\boldsymbol{r}}\,\xi_{\nu_\mu}; \qquad \chi_\mu(r) = f(r)\,\xi_\mu\,. \tag{VII-7}$$

Here the ξ are Pauli spin functions and $f(r)$ represents the space dependence of the muon in its K-orbit. $f(r)$ is normalised to unity as follows *

$$\int |f(r)|^2\,\mathrm{d}^3 r = 1\,. \tag{VII-8}$$

L_λ (eq. (VII-5)) can now be written

* Note that Primakoff (1959) uses a different normalisation.

$$L = \frac{1}{\sqrt{2}} \chi^\dagger_{\nu_\mu}(\mathbf{v}, r)(1 - \boldsymbol{\sigma}_\ell \cdot \hat{\mathbf{v}}) \boldsymbol{\sigma}_\ell \chi_\mu(r)$$

$$L_4 = -\frac{i}{\sqrt{2}} \chi^\dagger_{\nu_\mu}(\mathbf{v}, r)(1 - \boldsymbol{\sigma}_\ell \cdot \hat{\mathbf{v}}) \chi_\mu(r) . \qquad \text{(VII-9)}$$

Making use of eq. (VII-9) and the relations

$$(1 - \boldsymbol{\sigma}_\ell \cdot \hat{\mathbf{v}})^2 = 2(1 - \boldsymbol{\sigma}_\ell \cdot \hat{\mathbf{v}})$$

$$\boldsymbol{\sigma}_\ell \cdot \hat{\mathbf{v}}\,(1 - \boldsymbol{\sigma}_\ell \cdot \hat{\mathbf{v}}) = (1 - \boldsymbol{\sigma}_\ell \cdot \hat{\mathbf{v}}) \boldsymbol{\sigma}_\ell \cdot \hat{\mathbf{v}} = -(1 - \boldsymbol{\sigma}_\ell \cdot \hat{\mathbf{v}}) \qquad \text{(VII-10)}$$

$$\boldsymbol{\sigma}_\ell \cdot \hat{\mathbf{v}}\,[(\boldsymbol{\sigma} \times \hat{\mathbf{v}}) \cdot \boldsymbol{\sigma}_\ell] = \mathrm{i}\,[\boldsymbol{\sigma} - \hat{\mathbf{v}} \cdot \boldsymbol{\sigma} \hat{\mathbf{v}}] \cdot \boldsymbol{\sigma}_\ell$$

it is straightforward to manipulate H_{β^-} into a form describing μ-capture by a proton, namely

$$H_{\mu c}(r) = \chi^\dagger_{\nu_\mu}(\mathbf{v}, r) \left\{ \frac{G \cos\theta}{\sqrt{2}} \frac{1}{\sqrt{2}} (1 - \boldsymbol{\sigma}_\ell \cdot \hat{\mathbf{v}}) \left[\left(f_V^{(\mu)} \left(1 + \frac{\nu}{2M}\right) + m_\mu f_S^{(\mu)} \right) \right. \right.$$

$$+ \left(f_A^{(\mu)} - \frac{\nu}{2M} \left(f_V^{(\mu)} - 2M f_W^{(\mu)} \right) \right) \boldsymbol{\sigma} \cdot \boldsymbol{\sigma}_\ell$$

$$- \left(m_\mu f_P^{(\mu)} - f_A^{(\mu)} - \left(f_V^{(\mu)} - 2M f_W^{(\mu)} \right) + 2M f_T^{(\mu)} \right) \frac{\nu}{2M} \boldsymbol{\sigma}_\ell \cdot \hat{\mathbf{v}} \boldsymbol{\sigma} \cdot \hat{\mathbf{v}}$$

$$\left. \left. - \frac{f_V^{(\mu)}}{M} \boldsymbol{\sigma}_\ell \cdot \hat{\mathbf{v}} \boldsymbol{\sigma}_\ell \cdot \boldsymbol{P} - \frac{f_A^{(\mu)}}{M} \boldsymbol{\sigma}_\ell \cdot \hat{\mathbf{v}} \boldsymbol{\sigma} \cdot \boldsymbol{P} \right] t_- \right\} \chi_\mu(r) . \qquad \text{(VII-11)}$$

In the above expression which is accurate to order ν/M, P/M (where $\boldsymbol{P}$ is the nucleon momentum) the superfix (μ) has been added to the various form factors to indicate that they must be evaluated at the momentum transfer appropriate to μ capture. The expression has also been written in an isobaric spin notation (i.e. it includes a factor t_-).

Generalising eq. (VII-11) to the case of capture by a complex nucleus, the many-nucleon Hamiltonian takes the form

$$H_{\mu c}(r_1, ..., r_A)$$

$$= \sum_{i=1}^{A} \chi_{\nu_\mu}^{\dagger}(\mathbf{v}, r_i) \left\{ \frac{G\cos\theta}{\sqrt{2}} \frac{1}{\sqrt{2}} (1 - \boldsymbol{\sigma}_\ell \cdot \mathbf{v}) \left[\left(f_V^{(\mu)} \left(1 + \frac{\nu}{2M} \right) + m_\mu f_S^{(\mu)} \right) \right. \right.$$

$$+ \left(f_A^{(\mu)} - \frac{\nu}{2M} \left(f_V^{(\mu)} - 2M f_W^{(\mu)} \right) \right) \boldsymbol{\sigma}^{(i)} \cdot \boldsymbol{\sigma}_\ell$$

$$- \left(m_\mu f_P^{(\mu)} - f_A^{(\mu)} - \left(f_V^{(\mu)} - 2M f_W^{(\mu)} \right) + 2M f_T^{(\mu)} \right) \frac{\nu}{2M} \boldsymbol{\sigma}_\ell \cdot \hat{\mathbf{v}} \boldsymbol{\sigma}^{(i)} \cdot \hat{\mathbf{v}}$$

$$\left. \left. - \frac{f_V^{(\mu)}}{M} \boldsymbol{\sigma}_\ell \cdot \hat{\mathbf{v}} \boldsymbol{\sigma}_\ell \cdot \boldsymbol{P}^{(i)} - \frac{f_A^{(\mu)}}{M} \boldsymbol{\sigma}_\ell \cdot \hat{\mathbf{v}} \boldsymbol{\sigma}^{(i)} \cdot \boldsymbol{P}^{(i)} \right] t_-^{(i)} \right\} \chi_\mu(r_i) + H_\mu^{\mathrm{exch}} \tag{VII-12}$$

where H_μ^{exch} represents any exchange terms which may contribute to μ-capture and which, in the static approximation takes the form given in eqs. (II-49) and (II-51).

An effective μ-capture Hamiltonian of the form given in eq. (VII-12) was first obtained by Fujii and Primakoff (1959) (see also Tolhoek (1963)). The Hamiltonian has subsequently been rendered more accurate by using relativistic wave functions for the muon and neutrino (Morita and Fujii (1960); Goulard *et al.* (1964); Morita and Morita (1964); Morita *et al.* (1965)), and by taking into account second order terms in the nucleon momentum (Ohtsubo (1966); Friar (1966)). We shall not, however, concern ourselves directly with these refinements.

The next task is to consider the magnitudes of the various form factors occurring in the μ-capture Hamiltonian.

VII-3. The nucleon form factors for μ-capture

As stated in the previous section, the form factors in $H_{\mu c}$ must be evaluated at the momentum transfer k_λ appropriate to the μ-capture process. A simple application of kinematics gives in general

$$k^2 = (\mathrm{p} - \mathrm{n})^2 = m_\mu^2 + 2m_\mu \left(M_\mathrm{i} - M_f - \frac{m_\mu^2}{2M_\mathrm{f}} \right) \tag{VII-13}$$

where M_i and M_f are the masses of the initial and final nuclei.

For the case of capture by a nucleon, $M_i = M_p$, $M_f = M_n$ and $k^2 \simeq 0.9\, m_\mu^2$. Assuming CVC theory and the results of §III-1 we have

$$f_V^{(\mu)} \equiv f_V\,(0.9\, m_\mu^2) = 2F_1^{(V)}(0.9\, m_\mu^2)$$

$$f_W^{(\mu)} \equiv f_W\,(0.9\, m_\mu^2) = -\,2F_2^{(V)}(0.9\, m_\mu^2)$$

$$f_S^{(\mu)} = 0 \tag{VII-14}$$

where $F_1^{(V)}$ and $F_2^{(V)}$ are the two isovector form factors for the nucleon (see e.g. Weber (1967)). These are briefly discussed in §III-1 and their forms are such that

$$\frac{f_V(k^2)}{f_V(0)} \simeq \frac{f_W(k^2)}{f_W(0)} \simeq \left[1 + \frac{1}{12}\,k^2 R^2\right]^{-2} \tag{VII-15}$$

where R is the nucleon *RMS* radius having the value $R \triangleq 0.8$ fm. More exotic expressions have been given but for our purposes the above expressions are sufficiently accurate. They lead to the result (e.g. Fujii and Primakoff (1959); see also Frazier and Kim (1969) and Pascual (1969) for a more recent estimate)

$$f_V^{(\mu)} \simeq 0.972\, f_V^{(\beta)}$$

$$f_W^{(\mu)} \simeq 0.972\, f_W^{(\beta)}$$

where the superfix β refers to the value of the form factors in the case of β-decay ($f_V^{(\beta)} = f_V(0) = 1$; $f_W^{(\beta)} = f_W(0) = -(\kappa_p - \kappa_n)/2M$).

In the case of the axial vector current the form factor $f_A(k^2)$ can be determined from high energy neutrino experiments (Perkins (1969)) and is usually represented by a dipole fit

$$\frac{f_A(k^2)}{f_A(0)} = [1 + k^2/M_A^2]^{-2} \tag{VII-16}$$

where M_A^2 is of the order 0.99 GeV^2 (Kustom *et al.* (1969); Nambu and Yoshimura (1970)).

As far as $f_P^{(\mu)}$ is concerned appeal has to be made to the PCAC hypothesis (§IV-4). Using eq. (IV-30) and continuing to neglect the k^2 dependence of f_A and $K_{\pi NN}$ gives

$$m_\mu f_\mathrm{P}^{(\mu)} = m_\mu f_\mathrm{P}(0.9\, m_\mu^2) \simeq \frac{2Mm_\mu}{0.9\, m_\mu^2 + m_\pi^2} f_\mathrm{A}^{(\mu)} = 6.7\, f_\mathrm{A}^{(\mu)} . \tag{VII-17}$$

Alternatively using the Goldberger–Treiman relation (eq. (IV-17)) to relate f_A to F_π and $G_{\pi\mathrm{NN}}$, we obtain (cf. eq. (IV-32))

$$m_\mu f_\mathrm{P}^{(\mu)} = m_\mu f_\mathrm{P}(0.9\, m_\mu^2) = \frac{2F_\pi\, G_{\pi\mathrm{NN}}\, m_\mu}{0.9\, m_\mu^2 + m_\pi^2} \simeq -8.9 \tag{VII-18}$$

where we have taken $G_{\pi\mathrm{NN}}^2/4\pi = 14.6$ and F_π has been taken from the experimental π-decay rate ($|F_\pi|^2 m_\pi^2 \simeq 2.2 \times 10^{-14}$). Taking $f_\mathrm{A}^{(\mu)} \simeq f_\mathrm{A}^{(\beta)} =$ $= -1.24$ then gives

$$m_\mu f_\mathrm{p}^{(\mu)} \simeq 7.2\, f_\mathrm{A}^{(\mu)} \tag{VII-19}$$

which is in reasonable agreement with eq. (VII-17). There is obviously some uncertainty here stemming both from experimental input data and the theoretical uncertainty of PCAC theory. This underlines one of the major reasons for studying the phenomenon of μ-capture which is the information it can give about the size of the pseudoscalar form factor and hence the light that it throws on PCAC theory.

There is no information available about $f_\mathrm{T}^{(\mu)}$ other than that deriving from β-decay data (§V-5) which suggests that it may be non-zero and have a value of the order $1/M$. It can therefore at best be retained as a parameter in the formulation of μ-capture theory.

VII-4. The μ-capture rate $\Lambda_{\mu\mathrm{c}}$

Given the effective Hamiltonian ($H_{\mu\mathrm{c}}$) for μ-capture (eq. (VII-12)) it is straightforward to obtain an expression for the capture rate for the process

$$N_\mathrm{i}(Z) + \mu^- \to N_\mathrm{f}(Z-1) + \nu_\mu .$$

The following result is obtained (Primakoff (1959); Tolhoek (1963))

$$\Lambda_{\mu c}(\mathrm{i}\to\mathrm{f}) = \Gamma\frac{\nu^2\mathbf{v}}{2\pi}\int\frac{\mathrm{d}}{\hat{\mathbf{v}}\pi}\Big\{F_V^2|\langle \mathrm{f}|t\_|\mathrm{i}\rangle_\mu|^2 + F_A^2|\langle \mathrm{f}|t\_\boldsymbol{\sigma}|\mathrm{i}\rangle_\mu|^2$$

$$+ (F_P^2 - 2F_PF_A)|\langle \mathrm{f}|t\_\boldsymbol{\sigma}\cdot\hat{\mathbf{v}}|\mathrm{i}\rangle_\mu|^2$$

$$-\frac{F_Vf_V^{(\mu)}}{M}[\langle \mathrm{f}|t\_|\mathrm{i}\rangle_\mu^\times\langle \mathrm{f}|t\_\boldsymbol{P}|\mathrm{i}\rangle_\mu\cdot\hat{\mathbf{v}} + \mathrm{c.c.}]$$

$$-\frac{F_Af_A^{(\mu)} - F_Pf_A^{(\mu)}}{M}[(\mathbf{v}\cdot\langle \mathrm{f}|t\_\boldsymbol{\sigma}|\mathrm{i}\rangle_\mu^\times\langle \mathrm{f}|t\_\boldsymbol{\sigma}\cdot\boldsymbol{P}|\mathrm{i}\rangle_\mu + \mathrm{c.c.}]\Big\} \quad \text{(VII-20)}$$

where

$$\Gamma = \left(\frac{G\cos\theta}{\sqrt{2}}\right)^2\left(1+\frac{\nu}{[\nu^2+M_\mathrm{f}^2]^{1/2}}\right)^{-1},$$

M_f is the mass of the final nucleus and nuclear matrix elements are abbreviated as follows (see eq. (VII-7) for notation)

$$\langle \mathrm{f}|t\_O|\mathrm{i}\rangle_\mu \equiv \langle \mathrm{f}|\sum_i t_-^{(i)}\exp(-\mathrm{i}\mathbf{v}\cdot\boldsymbol{r}_\mathrm{i})\,\mathrm{f}(r_\mathrm{i})\,O_\mathrm{i}|\mathrm{i}\rangle\,. \quad \text{(VII-21)}$$

It is further assumed that in taking the square modulus an appropriate averaging and summing over the initial and final magnetic substates takes place (cf. the situation in β-decay – eq. (II-61)). The magnitude of the emitted neutrino momentum follows from the kinematics of the situation and is given by (Primakoff (1959))

$$\nu = m_\mu\left(1-\frac{\epsilon}{m_\mu}-\frac{(E_\mathrm{f}-E_\mathrm{i})}{m_\mu}\right)\left(1-\frac{m_\mu}{2(m_\mu+AM)}\right) \quad \text{(VII-22)}$$

where ϵ is the binding energy of the muon in the lowest Bohr orbit. The effective form factors F used in eq. (VII-20) are given by

$$F_V = f_V^{(\mu)}\left(1+\frac{\nu}{2M}\right) + m_\mu f_S^{(\mu)}$$

$$F_A = f_A^{(\mu)} - \frac{\nu}{2M}(f_V^{(\mu)} - 2Mf_W^{(\mu)})$$

$$F_P = (m_\mu f_P^{(\mu)} - f_A^{(\mu)} - (f_V^{(\mu)} - 2Mf_W^{(\mu)}) + 2Mf_T^{(\mu)})\,\frac{\nu}{2M}\,. \quad \text{(VII-23)}$$

The above expression (eq. (VII-20)) is for the capture rate when the nucleus makes a transition from an initial state i to a specific final state f. As mentioned in the introductory section, capture in a complex nucleus in general leads to many final states (excited states of the final nucleus and neutron emission) so that the total capture rate is given by

$$\Lambda_{\mu c} = \sum_{f} \Lambda_{\mu c}(i \rightarrow f) . \qquad \text{(VII-24)}$$

We now go on to compare theoretical and experimental values of μ-capture rates for the different processes that have been investigated.

VII-5. μ-capture in hydrogen

The simplest μ-capture process to consider is capture by protons, i.e.

$$p + \mu^- \rightarrow n + \nu_\mu .$$

Because of the low capture rate experiments on this process are best carried out in a liquid hydrogen bubble chamber. Unfortunately complications then ensue since capture mainly takes place whilst the muon is bound in a pμp muonic molecule and it becomes necessary to know the detailed form of the relevant three body wave function. However, an experiment has been performed in which muons are capture in gaseous hydrogen (Quaranta *et al.* (1969)) so that capture takes place primarily from the pμ^- atomic state. In either case, results have to be interpreted in terms of the capture rates of a μ^- from either the singlet (s) or triplet (t) state of the pμ^- system. These rates are different due to the spin dependence of $H_{\mu c}$. They can be calculated by standard procedures and are given by (Bernstein *et al.* (1958); Primakoff (1959, 1963, 1964))

$$\Lambda_{\mu c}^{(s)} = \Gamma |f(0)|^2 \frac{\nu^2}{2\pi} \{(F_V - 3F_A)^2 + F_P^2 + 2F_P(F_V - 3F_A)\}$$

$$\Lambda_{\mu c}^{(t)} = \Gamma |f(0)|^2 \frac{\nu^2}{2\pi} \{(F_V + F_A)^2 + F_P^2 - \tfrac{2}{3}F_P(F_V + F_A)\} \qquad \text{(VII-25)}$$

where Γ is given in eq. (VII-20), f(0) is the muon radial wave function for the muonic atom evaluated at $r = 0$. In eq. (VII-25) the approximation has been made of neglecting the relativistic P/M terms in $H_{\mu c}$ and the various nuclear matrix elements (eq. (VII-21)) have been given the simple values they acquire in the case of capture by a single proton (e.g. $|\langle n|t_-|p\rangle_\mu|^2 = |f(0)|^2$, $|\langle n|t_-\boldsymbol{\sigma}|p\rangle_\mu|^2 = 3|f(0)|^2$ etc.).

We are now in a position to make an evaluation of the magnitude of $\Lambda_{\mu c}^{(s)}$ and $\Lambda_{\mu c}^{(t)}$ and take the following values for the input data:

(i) $|f(0)|^2 = 1/\pi a_\mu'^3$ where a_μ' is the reduced Bohr radius for the $p\mu^-$ atom and is related to a_μ by $a_\mu' = a_\mu(1+m_\mu/M)^{-1} = 0.002847$ Å;
(ii) Using eq. (VII-22) for proton capture gives $\bar{\nu} = 0.94\, m_\mu$;
(iii) $f_V^{(\mu)} = 0.972 f_V^{(\beta)} = 0.972$, assuming CVC theory.
(iv) $f_W^{(\mu)} = 0.972 f_W^{(\beta)} = -0.972(\kappa_p - \kappa_n)/2M$, assuming CVC theory;
(v) $f_S^{(\mu)} = 0$, assuming *CVC* theory.

With these assumptions we can write (see eq. (VII-25))

$$\Lambda_{\mu c}^{(s)} = (G\cos\theta)^2 \Lambda_0 \{(F_V - 3F_A)^2 + F_P^2 + 2F_P(F_V - 3F_A)\}$$
$$\Lambda_{\mu c}^{(t)} = (G\cos\theta)^2 \Lambda_0 \{(F_V + F_A)^2 + F_P^2 - \tfrac{2}{3}F_P(F_V + F_A)\} \qquad \text{(VII-26)}$$

where $\Lambda_0 = 14.9\ \text{sec}^{-1}$ and G is measured in units of 10^{-49} erg cm^3. The form factors take the values

$$F_V = 1.023$$

$$F_A = f_A^{(\mu)} - 0.242$$

$$F_P = 0.053\,(f_{PT} - f_A^{(\mu)}) - 0.242$$

where $f_{PT} = m_\mu f_P^{(\mu)} + 2Mf_T^{(\mu)}$. In fig. VII-1 values of $\Lambda_{\mu c}^{(s)}$ and $\Lambda_{\mu c}^{(t)}$ are plotted as a function of f_{PT} for values of $f_A^{(\mu)}$ and $G\cos\theta$ which lie within the range of interest *.

It is at once obvious that there is a very considerable hyperfine effect, that the singlet capture rate is at least an order of magnitude great than the triplet capture rate and that the latter is sensitive only to the value of f_{PT}.

As far as experimental data is concerned the situation is as follows.

* A slightly more precise calculation of $\Lambda_{\mu c}^{(s)}$ and $\Lambda_{\mu c}^{(t)}$ has been carried out by Kabir (1966) and most recently by Pascual (1969) who also includes electromagnetic radiative corrections. In the approximation Pascual uses these turn out to be of the order 0.2% for $\Lambda_{\mu c}^{(s)}$. The calculations of Fujii and Yamaguchi (1964) and Ohtsubo and Fujiii (1966) on $\Lambda_{\mu c}$ in hydrogen should also be noted.

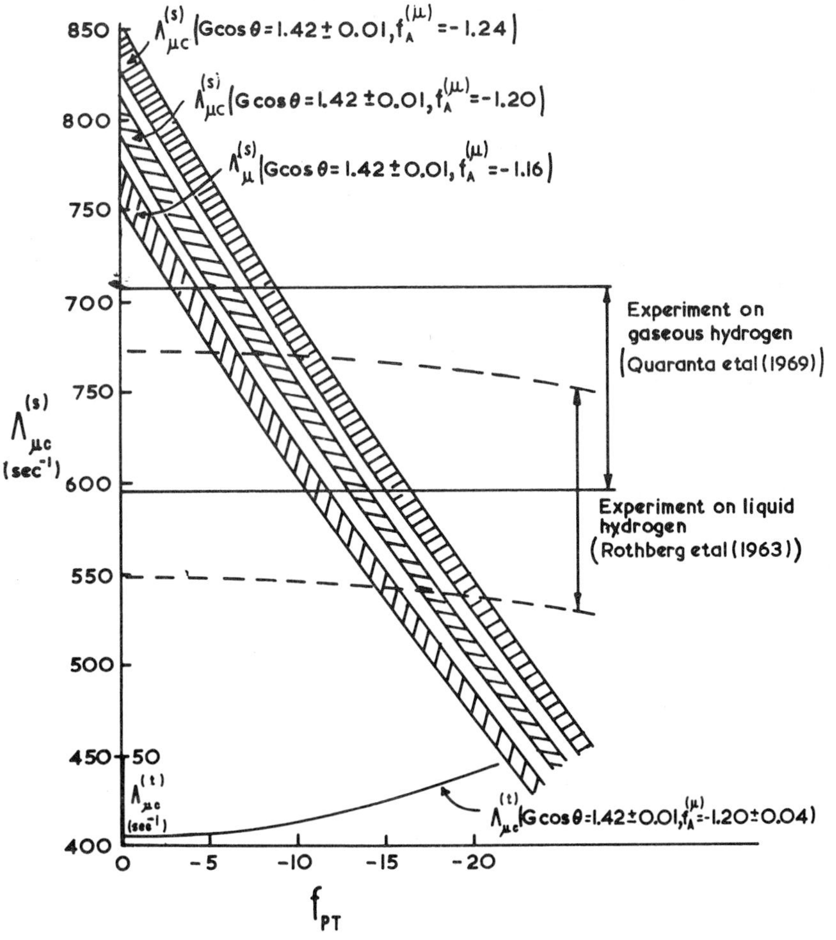

Fig. VII-1. μ-capture rate. $\Lambda_{\mu c}^{(s,t)}$, in the pμ^- singlet and triplet states as a function of f_{PT}.

Quaranta *et al.* (1969) have measured the μ-capture rate in gaseous hydrogen. Theoretical and experimental atomic and molecular analysis (Zel'dovich and Gershtein (1961); Gershtein (1958, 1959); Quaranta *et al.* (1967)) shows that for the gas density used by Quaranta *et al.* (1969), 95% of the muons decay whilst bound in μ^-p muonic atoms and are then overwhelmingly captured from the singlet state. Correcting for the 5% of muons which form pμ^-p molecules Quaranta *et al.* (1969) obtain for the experimental capture rate $\Lambda_{\text{expt}} = 651 \pm 57\ \text{sec}^{-1}$. This is compared directly with theoretical values of $\Lambda_{\mu c}^{(s)}$ in fig. VII-1. It can be seen that experiment is certainly consistent with the CVC and PCAC values of the various form factors. However, the error is too large to deduce anything more precisely – it certainly leaves room for $f_{\text{T}}^{(\mu)}$ to have a value of the order $1/M$ and therefore allows the possibility of second class currents (see §V-5).

As mentioned earlier, in the case of capture in liquid hydrogen the situation is complicated by the fact that at liquid hydrogen densities, in a time comparable to the muon lifetime (2.2×10^{-6} sec), most muons are to be found in metastable ortho $\text{p}\mu^-\text{p}$ molecules. The capture rate in this state is given by

$$\Lambda_{\mu c}(\text{ortho p}\mu^-\text{p}) = 2\gamma[\tfrac{3}{4}\xi\Lambda_{\mu c}^{(s)} + (1 - \tfrac{3}{4}\xi)\Lambda_{\mu c}^{(t)}] \tag{VII-27}$$

where $\Lambda_{\mu c}^{(s)}$ and $\Lambda_{\mu c}^{(t)}$ are given in eq. (VII-26). $\gamma = |f_{\text{M}}|^2/|f(0)|^2$, and ξ is the probability of finding the ortho ($\text{p}\mu\text{p}$) molecule in a spin $\frac{1}{2}$ state at the time of capture. In the former expression, $|f_{\text{M}}|^2$ is the probability of finding the muon at either proton and $f(0)$ is the function used in eq. (VII-25). To evaluate γ and ξ the $\text{p}\mu\text{p}$ three body problem has to be tackled. This has been done by Wessel and Phillipson (1964) and Halpern (1964a, b) who obtain $2\gamma = 1.00$ and 1.01 respectively. Halpern (1964b) obtains $\xi \approx 0.9994$. For our purposes, therefore, we take $2\gamma = 1.00 \pm 0.1$ and $\xi = 1.000$.

Since $\Lambda_{\mu c}^{(t)}$ is small and sensitive only to the value of f_{PT} an "experimental" value for $\Lambda_{\mu c}^{(s)}$ as a function of f_{PT} can be determined from the measured value of $\Lambda_{\mu c}(\text{ortho p}\mu^-\text{p})$. Thus

$$(\Lambda_{\mu c}^{(s)})_{\text{“expt”}} = \frac{1}{3}\left[\frac{2\Lambda_{\mu c}(\text{ortho p}\mu^-\text{p})}{\gamma} - \Lambda_{\mu c}^{(t)}\right]. \tag{VII-28}$$

Rothberg *et al.* have studied μ capture in liquid hydrogen and obtain $\Lambda_{\mu c}(\text{ortho p}\mu^-\text{p}) = (464 \pm 42)\ \text{sec}^{-1}$. In fig. VII-1, the corresponding value of $(\Lambda_{\mu c}^{(s)})_{\text{“expt”}}$ determined from eq. (VII-28) is plotted as a function of f_{PT}. As with the gaseous hydrogen experiment the uncertainties are large and no very precise limit can be put on f_{PT}. However, it will be noted that if $f_{\text{A}}^{(\mu)} \approx -1.24$, then the two sets of data taken together suggest that f_{PT} may lie in the region -10 to -15 which is somewhat larger (in magnitude) than the value (≈ -8 or -9) given by PCAC theory for $m_\mu f_{\text{P}}^{(\mu)}$ (see eqs. (VII-17) and (VII-18)). This in turn suggest that $f_{\text{T}}^{(\mu)}$ may indeed be non zero and of the order $-1/M$. This is opposite in sign to the value needed to account for the difference in mirror nuclear β-decays (see §V-5). It is obviously important to obtain more accurate values for the capture rates * and to tie down the value of $f_A^{(\mu)}$ more precisely.

VII-6. μ-capture in ^{3}He

There is a lot of theoretical and experimental information available about the three-nucleon nuclei ^{3}H and ^{3}He. In addition the muon capture process

$$\mu^- + {}^3\mathrm{He} \to {}^3\mathrm{H} + \nu_\mu$$

can be identified very readily in gas scintillation and diffusion cloud chamber experiments and quite accurate measurements have been made of the capture rate:

$$\begin{aligned}\Lambda_{\mu c} &= (1410 \pm 140)\ \mathrm{sec}^{-1}: \text{Falomkin } et\ al.\ (1963a, b)\\ &= (1505 \pm 46)\ \ \mathrm{sec}^{-1}: \text{Auerbach } et\ al.\ (1965)\\ &= (1465 \pm 67)\ \ \mathrm{sec}^{-1}: \text{Clay } et\ al.\ (1965)\ .\end{aligned}$$

Theoretical calculations on the capture rate have been carried out by a number of authors (Fujii and Primakoff (1959); Fujii (1960); Werntz (1960); Wolfenstein (1962); Yano (1964); Oakes (1964); Bietti (1965); Peterson (1968); Peachey (1969)) using the expression given in eq. (VII-20).

Because of the essential identity in form of the ^{3}He and ^{3}H wave functions (they differ only through small charge dependent effects) it can be shown (e.g. Peterson (1968); Peachey (1969)) that

$$\begin{aligned}&[\langle \mathrm{f}|t_-|\mathrm{i}\rangle_\mu^\times \langle \mathrm{f}|t_-\boldsymbol{P}|\mathrm{i}\rangle_\mu \cdot \hat{\boldsymbol{\nu}} + \mathrm{c.c.}] = \tfrac{2}{3}|\langle \mathrm{f}|t_-|\mathrm{i}\rangle_\mu|^2\\ &[(\hat{\boldsymbol{\nu}} \cdot \langle \mathrm{f}|t_-\boldsymbol{\sigma}|\mathrm{i}\rangle_\mu)\langle \mathrm{f}|t_-\boldsymbol{\sigma}\cdot\boldsymbol{P}|\mathrm{i}\rangle_\mu^\times + \mathrm{c.c.}] = \tfrac{2}{3}|\langle \mathrm{f}|t_-\boldsymbol{\sigma}\cdot\hat{\boldsymbol{\nu}}|\mathrm{i}\rangle_\mu|^2\ .\end{aligned} \tag{VII-29}$$

This means that in using eq. (VII-20) and integrating over neutrino angles in order to determine the capture rate, only the quantities $|\langle \mathrm{f}|t_-|\mathrm{i}\rangle_\mu|^2$ and

* Other experiments in liquid hydrogen have been carried out by Hilderbrand (1962), Bleser *et al.* (1962), Bertolini *et al.* (1962). The quoted errors are somewhat larger than those of Rotherberg *et al.* (1963) and in addition the interpretation is complicated by a sizeable contribution to the capture rate from the $p\mu^-$ atom.

$|\langle f|t_-\boldsymbol{\sigma}\cdot\hat{\mathbf{v}}|i\rangle_\mu|^2$ have to be calculated. This has been done for various assumptions about the detailed form of the ^{3}H and ^{3}He wave functions by Peterson (1968) and Peachey (1969). Unfortunately there are uncertainties here with respect to the amount of D-state and mixed symmetry S-state in the three body wave function (see §IV-5). In addition exchange effects (see §IV-5) are also expected to contribute. It is generally assumed (e.g. Peterson (1968)) that their contribution to the μ^- capture process is approximately the same as in the ^{3}H β-decay process (see §IV-5) and detailed calculations (Peachey (1969)), assuming various phenomenological forms for the exchange operators, bear out this assumption. This means that starting from the ^{3}H β-decay ft-value as input data and as a means of determining the strength of the axial vector interaction, exchange effects are automatically allowed for.

Taking the various uncertainties into account and comparing theoretical and experimental values for the capture rate Peterson (1968) and Peachey (1969) come to essentially the same conclusion that

$$4 \lesssim \frac{f_{\mathrm{PT}}}{f_{\mathrm{A}}(0)} \lesssim 33 \qquad \text{(VII-30)}$$

with a probable value lying around 11–12. This latter value is inconsistent with the Goldberger–Treiman value and suggests the presence of an induced tensor interaction. However, the theoretical and experimental uncertainties are obviously too great to come to any firm conclusion on this point. For example, Wycech (1969) has pointed out that $f_{\mathrm{F}}^{(\mu)}$ and therefore f_{PT} might be renormalised by as much as 50% due to the interaction of virtual pions (implicit in the calculation of $f_{\mathrm{P}}^{(\mu)}$ from PCAC theory) with other nucleons in the nucleus (see also Green and Rho (1969) and Baba (1970)).

An alternative approach to calculating the ^{3}He μ^- capture rate is to treat the ^{3}He and ^{3}H nuclei as elementary particles (see Chapt. IV, especially §IV-2). In this approach ^{3}He and ^{3}H are treated as $T_3 = \pm\frac{1}{2}$ substates of an $I = \frac{1}{2}$, $T = \frac{1}{2}$ elementary particle. Since these quantum numbers are identical with those of the neutron and proton the formulae relevant to the process $\mu^- + p \to n + \nu_\mu$ can be taken over at once, but with appropriate modification of the different form factors. Calculations of this sort have been carried out by Drechsler and Stech (1964); Fujii and Yamaguchi (1964); Kim and Primakoff (1965b); Kim and Ram (1967); Frazier and Kim (1969); Kim and Mintz (1971). The form factors F_{V}, F_{A} and F_{P} (cf. eq. (VII-23)) now take the form (Frazier and Kim (1969))

$$F_V = f_V^{(\mu,{}^3\mathrm{He})}\left(1 + \frac{\nu}{2M_f}\right) + m_\mu f_S^{(\mu,{}^3\mathrm{He})}$$

$$F_A = f_A^{(\mu,{}^3\mathrm{He})} - \frac{\nu}{2M_f}\left(f_V^{(\mu,{}^3\mathrm{He})} - 2M_f f_W^{(\mu,{}^3\mathrm{He})}\right) \qquad \text{(VII-31)}$$

$$F_P = \left[m_\mu f_P^{(\mu,{}^3\mathrm{He})} - f_A^{(\mu,{}^3\mathrm{He})} - \left(f_V^{(\mu,{}^3\mathrm{He})} - 2M_f f_W^{(\mu,{}^3\mathrm{He})}\right) + 2M_f f_T^{(\mu)}\right]\frac{\nu}{2M_f}$$

where the suffix (μ, ^{3}He) implies that the form factors are to be evaluated for the process $\mu^- + {}^3\mathrm{He} \to {}^3\mathrm{H} + \nu_\mu$ at the appropriate momentum transfer (see eq. (VII-13) which gives $k^2 = 0.96\ m_\mu^2$). Assuming the validity of CVC as applied to these form factors Frazier and Kim (1969) relate $f_V^{(\mu,{}^3\mathrm{He})}$ and $f_W^{(\mu,{}^3\mathrm{He})}$ to the charge and magnetic moment form factors for ^{3}H and ^{3}He. These latter have been determined from high energy electron scattering experiments (Collard *et al.* (1964)). $f_S^{(\mu,{}^3\mathrm{He})}$ is taken to be zero. As with $f_A^{(\mu)}$ there is uncertainty about the dependence of $f_A^{(\mu,{}^3\mathrm{He})}$ on k^2 and Frazier and Kim (1969), see also Kim and Primakoff (1965) assume

$$f_A^{(\mu,{}^3\mathrm{He})}(k^2) = f_A^{(\beta,{}^3\mathrm{He})}(0)\left(\frac{1 + k^2/M_V^2}{1 + k^2/M_A^2}\right)^2 \frac{f_W^{(\mu,{}^3\mathrm{He})}(k^2)}{f_W^{(\mu,{}^3\mathrm{He})}(0)} \qquad \text{(VII-32)}$$

where use has been made of the experimental dipole-fit form factors (eqs. (VII-15) and (VII-16)) with $M_V^2 = 0.71\ \mathrm{GeV}^2$, $M_A^2 = 0.99\ \mathrm{GeV}^2$. There is obviously some uncertainty here, but by relating the form factor to $f_A^{(\beta,{}^3\mathrm{He})}(0)$ which can be determined directly from the ^{3}H beta-decay data, exchange effects are automatically taken care of. Finally, assuming PCAC theory, expressions for $f_P^{(\mu,{}^3\mathrm{He})}(k^2)$ equivalent to eqs. (IV-30)–(IV-32), (VII-17) and (VII-18) can be obtained, e.g.

$$f_P^{(\mu,{}^3\mathrm{He})}(k^2) \triangleq \frac{M_i + M_f}{k^2 + m_\pi^2} f_A^{(\mu,{}^3\mathrm{He})}(k^2)\,. \qquad \text{(VII-33)}$$

As with the nucleon pseudoscalar form factor there are uncertainties according to the way in which f_P is determined and the way in which PCAC theory is formulated.

Taking $f_T^{(\mu,{}^3\mathrm{He})} = 0$, Frazier and Kim (1969) then obtain for the total capture rate

$$\Lambda_{\mu c} = (1449 \text{ to } 1525) \text{ sec}^{-1}$$

where the spread reflects the uncertainty in $f_P^{(\mu, ^3\text{He})}$. This result is certainly consistent with the experimental values quoted earlier for the capture rate.

The elementary particle approach to calculating the μ-capture rate is obviously an interesting one and, given more accurate experimental results and more information about the axial form factor, could well lead to important information about PCAC theory. Another associated development in this connection is work of Kim and Ram (1967) which, using in addition the ideas of current algebra, obtains an estimate for the *total* muon capture rate * in ^{3}He. Here a number of major approximations have to be made but, mindful of these, the agreement between theory and experiment represents some support for the validity of current commutation relations.

VII-7. μ-capture in complex nuclei

Since the muon capture rate $\Lambda_{\mu c}$ is approximately proportional to Z^4 it obviously increases rapidly as one goes to heavier nuclei so that more accurate values of $\Lambda_{\mu c}$ can be obtained (typically, accurate to a percent or so). Unfortunately, nuclear complications creep in due to (a) the possibility of many final states contributing and (b) greater uncertainty about the details of nuclear wave functions. For this reason, although consistency with theories of weak interactions can generally be demonstrated, it is difficult to unravel details of the weak interaction. Since the object of this book is to deal with issues of this latter kind we shall not here go into details about the theory and experiment of muon capture in heavier nuclei.

In some cases [e.g. ^{6}Li (Deutsch *et al.* (1968a)); ^{11}B (Deutsch *et al.* (1968b)); ^{12}C (Maier *et al.* (1964)); ^{16}O (Astbury *et al.* (1964); Cohen *et al.* (1963, 1964); Deutsch *et al.* (1969))] measurements have been made of partial capture rates to the ground or a low excited state of the final nucleus. Associated theoretical calculations have also been carried out making various assumptions about the nuclear wave functions. Typical of these calculations is the work of Ohtsubo (1966) (see also Ericson *et al.* (1964); Gillet and Jenkins (1965); Hirooka (1969)) on muon capture in ^{16}O leading to 0^-, 1^- and 2^- low lying levels in ^{16}N. Using the muon capture Hamiltonian given in

* More conventional impulse approximation calculations of the total capture rate have been carried out by Primakoff (1959), Goulard *et al.* (1964), Yano (1964) and Wang (1965) and there is general broad agreement between theory and experiment.

Table VII-1
Values of $\Lambda_{\mu c}$ for μ-capture in ^{16}O ($\Lambda_{\mu c}$ in units of 10^3 sec^{-1}).

Wavefunction	$\Lambda_{\mu c}(0^-)$	$\Lambda_{\mu c}(1^-)$	$\Lambda_{\mu c}(2^-)$
IP	2.07	4.34	24.5
EF	2.54	3.03	14.5
GM	1.61	2.41	15.3
Experiment			
C_1	0.66 ± 0.11	1.73 ± 0.10	6.76 ± 0.71
C_2	1.1 ± 0.2	1.73 ± 0.10	6.3 ± 0.7
B	1.6 ± 0.2	1.4 ± 0.2	unobserved

(IP signifies Independent Particle Model; EF signifies the wave function due to Elliott and Flowers (1957) and GM signifies the wave function due to Gillet and Vinh Mau (1964). C_1 and C_2 refer to experiments by the Columbia group (Cohen *et al.* (1963, 1964)) and B by the Berkeley group (Astbury *et al.* (1964)). $\Lambda_{\mu c}$ was evaluated theoretically by taking $f_A^{(\mu)} = -1{,}19$, $m_\mu f_P^{(\mu)} = 8 f_A^{(\mu)}$).

eq. (VII-12), and three different assumptions about the nuclear wave functions, the results for the capture rate given in table VII-1 were obtained.

It will be noted at once that the results are sensitive to the nuclear model wave function used and that the values of $\Lambda_{\mu c}$ are overestimated. Ohtsubo (1966) points out that this latter disagreement can be resolved if there is a contribution from the induced tensor form factor $f_T^{(\mu)}$ of the order $+ 1/M$. This is in agreement with some of the indications of mirror β-decay data (§V-5) but of opposite sign to the value for $f_T^{(\mu)}$ suggested by the hydrogen muon capture data. However, the conclusion about $f_T^{(\mu)}$ should not be taken seriously because of, in particular, the nuclear wave function uncertainty. As discussed by Rho (1967a, 1967b), for example, use of the Migdal model of nuclear structure (Migdal and Larkin (1964); Migdal (1964, 1966)) can lead to good agreement between theory and experiment without appealing to contributions from second class currents (see also Bunyatan (1966); Mannque (1967); Foldy and Klein (1967); Walker * (1968); Green and Rho * (1969); Green *et al.* * (1970)).

* These authors show that inclusion of 2 particle-2 hole correlations in the nuclear wave function considerably improves the agreement between theory and experiment.

An analysis of the process $\mu^- + {}^{12}C \to {}^{12}B + \nu_\mu$ leading to the ${}^{12}B$ ground state due to Foldy and Walecka (1965) is also worthy of note. By making use of inelastic electron scattering data, and the ft-value for the β-decay process ${}^{12}B \to {}^{12}C + e^- + \tilde{\nu}_e$ they are able to obtain a value for $\Lambda_{\mu c}$ rather independent of assumptions about the nuclear model wave function. From their analysis they conclude that $5 \lesssim m_\mu f_P^{(\mu)}/f_A^{(\mu)} \lesssim 28$ and that $f_A^{(\mu)}/f_A^{(\beta)} = 1.04^{+0.07}_{-0.10}$ (see also Mukhopadhyay and Macfarlane (1971)).

Total muon capture rates, which have been measured for many cases, are even more complicated to calculate since a sum has to be taken over all possible final states (see eq. (VII-24)). The general approach, which was developed by Primakoff (1959), is to remove the sum over final states by using a closure approximation. This approximation which has been elaborated and discussed by many authors (Tolhoek and Luyten (1957); Tolhoek (1959); Klein and Wolfenstein (1962); Luyten *et al.* (1963); Bell and Løvseth (1964); Luyten and Tolhoek (1965); Klein (1966); Rho (1967c); Do Dang (1972)) gives results which are in broad general agreement with experiment (e.g. Telegdi (1963); Eckhause *et al.* (1966)) but which are not reliable enough to enable fine details of the weak interaction to be determined although attempts have been made to do so (e.g. Boyssy *et al.* (1967), who find some slight evidence for an induced tensor interaction). On the contrary, assuming the correctness of the usual form of the weak interaction it has been possible to test out theories of nuclear structure both with respect to partial and total capture rates. Implicit or explicit work in this direction has been carried out by many of the authors already quoted (see also Silbar (1964); Foldy and Walecka * (1964); Lodder and Jonker (1965); Rho (1965, 1967d); Walker (1967); Raphael *et al.* (1967); Hirooka *et al.* (1968); Bogan (1968); Fujii *et al.* (1968); Kruger and Van Leuven (1969); Hrasko (1969); Bernabeu and Pascual (1969); Rho (1969)).

Muon capture in complex nuclei has also been treated theoretically from the point of view of the "elementary particle" approach initiated by Kim and Primakoff (1965a, 1965b). It is found that by making the usual CVC and PCAC assumptions and using, for example, inelastic electron scattering data to determine the weak magnetism form factor, it is possible to account for the experimental capture rates. Calculations of this sort have been carried out for the processes

* These authors make use of the relationship between the important dipole matrix elements in muon capture and the empirical value for the dipole photo-absorption matrix elements in the parent nucleus.

$\mu^- + {}^6\text{Li} \to {}^6\text{He} + \nu_\mu$ (Kim and Primakoff (1965b); Galindo and Pascual (1969); Delorme and Ericson (1970); Delorme (1970); Kim and Mintz (1970, 1971)).

$\mu^- + {}^{12}\text{C} \to {}^{12}\text{B} + \nu_\mu$ (Kim and Primakoff (1965b); Kim and Mintz (1971))

$\mu^- + {}^{16}\text{O} \to {}^{16}\text{N}(2^-) + \nu_\mu$ (Kim (1966)).

Although theory and experiment agree, it is difficult to make any more precise conclusive statement than that within experimental error and theoretical approximation experiment is consistent with an "elementary particle" treatment assuming CVC and PCAC theory. Certainly, however, the results also allow significant deviation from, for example, the Goldberger–Treiman relation for the pseudoscalar form factor deduced from PCAC theory.

VII-8. **Parity non-conserving effects in μ-capture**

Since the weak interaction responsible for μ-capture does not conserve parity it is to be expected that parity non-conserving effects should show up in muon capture experiments in which a search is made for pseudoscalar quantities. Experiments in this direction have rested on the fact that muons from a π-meson beam are 100% polarised and an appreciable fraction (15% to 20%) of this polarisation is retained until the instant of capture from the lowest Bohr orbit in a mu-mesonic atom. That this is so can be substantiated by measuring the electron asymmetry of muons which decay (rather than being captured) whilst bound in the mu-mesonic atom (e.g. Garwin *et al.* (1957); Ignatenko *et al.* (1959)). Given that the captured muon is polarised, then the recoil daughter nucleus or any emitted product particle in the final state is expected to show a directional asymmetry of the form

$$W = 1 + P_\mu \alpha_r \hat{s}_\mu \cdot \hat{\boldsymbol{p}} \tag{VII-34}$$

where P_μ is the magnitude of the muon spin polarisation, α_r is an asymmetry parameter depending on the form of the weak interaction and the nuclear details of the transition being studied, $\hat{\boldsymbol{s}}_\mu$ is a unit vector in the direction of muon spin polarisation and $\hat{\boldsymbol{p}}$ is a unit vector in the momentum direction of the recoiling nucleus or particle.

Considering, for example, muon capture by a spin zero nucleus it can be shown using the effective Hamiltonian given in eq. (VII-12) but neglecting the small momentum (***P***) dependent terms, that the nuclear recoil asymmetry parameter in a transition i → f is given by

$$\alpha_r \hat{=} \frac{F_V^2 |\langle f | t_- | i \rangle_\mu|^2 + \frac{1}{3}[-F_A^2 + F_P^2 - 2F_A F_P] |\langle f | t_- \boldsymbol{\sigma} | i \rangle_\mu|^2}{F_V^2 |\langle f | t_- | i \rangle_\mu|^2 + \frac{1}{3}[3F_A^2 + F_P^2 - 2F_A F_P] |\langle f | t_- \boldsymbol{\sigma} | i \rangle_\mu|^2} \qquad \text{(VII-35)}$$

where the form-factors F_V, F_A, F_P and the matrix elements are given in eqs. (VII-23) and (VII-21) respectively. The above result has been obtained by various authors (e.g. Ioffe (1958); Huang *et al.* (1957); Shapiro *et al.* (1957); Überall (1957); Wolfenstein (1958); Treiman (1958); Fulton (1958); Primakoff (1959)) and is subject to the frequently used assumption in muon capture (Fujii and Primakoff (1959); De Forest (1965); Rho (1965)) that

$$|\langle f | t_- \boldsymbol{\sigma} \cdot \hat{\mathbf{v}} | i \rangle|^2 \hat{=} \tfrac{1}{3} |\langle f | t_- \boldsymbol{\sigma} | i \rangle|^2 . \qquad \text{(VII-36)}$$

Using the closure approximation referred to earlier in order to calculate the recoil asymmetry coefficient appropriate to the total capture rate gives (Primakoff (1959)).

$$\alpha_r \hat{=} \frac{F_V^2 - F_A^2 + F_P^2 - 2F_A F_P}{F_V^2 + 3F_A^2 + F_P^2 - 2F_A F_P} \hat{=} -0.4 . \qquad \text{(VII-37)}$$

Similar but more complicated expressions for the asymmetry coefficient can be obtained in the case of capture by nuclei with spin (see e.g. Primakoff (1959); Galindo and Pascual (1968)). It has also been pointed out by Morita *et al.* (1965) that recoil studies can throw light on the contribution to the weak interaction of second class currents. Unfortunately at this time no experimental data is available on the size of the asymmetry coefficient.

Of particular interest, however, is the asymmetry for direct (primary) neutrons emitted in the final state of a muon capture process, i.e. a process of the form

$$\mu^- + N_i(Z) \rightarrow N_f(Z-1) + n + \nu_\mu .$$

The neutron angular distribution (the total number of neutrons emitted in unit solid angle in a given direction per capture per MeV) takes the form

$$W = \frac{N_0(E_n)}{4\pi} [1 + P_\mu \alpha_n(E_n) \hat{s}_\mu \cdot \hat{p}_n] \tag{VII-38}$$

where $\hat{p}_n$ is a unit vector in the direction of the neutron momentum and $N_0(E_n)$ is the neutron energy spectrum. The asymmetry coefficient, which is now a function of the energy E_n of the outgoing neutron, has been calculated by many authors making use of the usual effective muon capture Hamiltonian and making different nuclear structure assumptions (Ioffe (1958); Shapiro *et al.* (1957); Überall (1957); Dolinskii and Blokhintsev (1958); Primakoff (1959); Lubkin (1960); Akimova *et al.* (1961); Yovonovich and Evseev (1963); Klein *et al.* (1965); Zelvinskii (1967); Devanathan and Rose (1967); Bogan (1969a, b); Bouyssy and Vinh Mau (1972); Eramzhyan *et al.* (1972)). The simplest view is that argued by Primakoff (1959) namely that it is not unreasonable to suppose that the emitted neutron carries off most of the available recoil momentum so that after capture by a spin zero nucleus the asymmetry coefficient α_n should have approximately the same value as α_r (eq. (VII-37)) namely ≈ -0.4. Experiment, however, does not tie in with this simple result (see table VII-2, which summarises some of the experimental data). There is obviously significant disagreement among the experiments themselves as to the sign of α and obviously more work needs to be done. Further, assuming that the positive sign for α obtained in the most recent experiments is maintained, there is then violent disagreement with the value $\alpha_n \approx -0.4$ obtained by Primakoff (1959). This is also the case if the values close to $\alpha_n = -1$ obtained by Evseev *et al.* (1967a, b) are confirmed..

However, the expression for α_n given by Primakoff (1959) is obtained in too simple a fashion. For example, Klein *et al.* (1965) (see also Piketty and Procureur (1971); Eramzhyan *et al.* (1972); Longuemare and Piketty (1972)) have shown that relativistic corrections to the usual muon-capture Hamiltonian can significantly modify the value of α_n. Indeed, as shown by Bogan (1969a, b) inclusion of these effects can raise α_n from a negative to a positive value (≈ 0.1 for high energy neutrons from ^{40}Ca). Secondly, nuclear structure effects of many kinds (e.g. correlation effects, collective phenomena, final state interactions etc.) can modify the Primakoff (1959) value for α_n (e.g. Bogan (1969a, b); Fujita and Fujii (1969)). Most successful in this respect is the work of Bouyssy and Vinh Mau (1972) who are able to obtain reasonable agreement with experiment by using multi-hole-multi-particle nuclear wave functions and taking final state interactions into account. It is interesting that this agreement is improved if second class currents are allowed. This should not, however, be regarded as too significant because of the theoretical and experimental uncertainties.

Table VII-2
The neutron asymmetry parameter $\alpha_n(E_n)$ in μ-capture.

Capturing nucleus	Neutron energy E_n in MeV	$\alpha_n(E_n)$	Reference
Mg	> 5.5	0.15 ± 0.11	Baker and Rubbia (1959)
	> 10	0.15 ± 0.18	
	> 15	0.37 ± 0.28	
S	> 5	−0.23 ± 0.09	Telegdi (1960)
Mg	> 5	−0.30 ± 0.10	
S	> 5	−0.22 ± 0.07	Astbury *et al.* (1962)
Ca	6.7 ± 0.5	−0.250 ± 0.063	Evseev *et al.* (1967a)
	10.3 ± 0.8	−0.48 ± 0.09	
	13.0 ± 1.1	−0.57 ± 0.10	
	16.8 ± 1.3	−0.78 ± 0.15	
	17.8 ± 1.4	−1.05 ± 0.15	
	20.7 ± 1.7	−1.02 ± 0.21	
S	> 18	−0.85 ± 0.24	Evseev *et al.* (1967b)
Si	7.43 – 11.49	+0.030 ± 0.023	Sundelin *et al.* (1968a, b)
	19.54 – 24.55	+0.228 ± 0.104	
	35.34 – 52.33	+0.817 ± 0.597	
S	7.73 – 11.49	+0.096 ± 0.034	
	19.54 – 24.55	+0.437 ± 0.155	
	35.34 – 52.53	−0.629 ± 1.074	
Ca	7.73 – 11.49	+0.053 ± 0.033	
	19.54 – 24.55	+0.009 ± 0.151	
	35.34 – 52.53	+0.979 ± 0.816	

VII-9. **Radiative μ-capture**

Radiative μ-capture is the name given to the process *

$$N_i(Z) + \mu^- \rightarrow N_f(Z-1) + \nu_\mu + \gamma \qquad \text{(VII-39)}$$

* Note that work on the related radiative electron capture process e.g.

$$N_i(Z) + e_k^- \rightarrow N_f(Z-1) + \nu_e + \gamma$$

has not revealed significant information about the weak interaction (see e.g. Schopper (1966)).

where $N_i(Z)$ and $N_f(Z-1)$ signify initial and final nuclear states having charges Z and $Z-1$ respectively. The elemental process of this kind is, of course,

$$p + \mu^- \to n + \nu_\mu + \gamma \qquad \text{(VII-40)}$$

and the Feynman diagrams mainly contributing to this are shown in fig. VII-2.

Crude estimates and experimental data indicate that R, the ratio of the radiative capture rate to the ordinary capture rate is of the order 10^{-4}. This means that experimental measurements can at present only feasibly be carried out in nuclei with $Z \gg 1$ so that analysis is then attended by the usual nuclear complications. Detailed calculations of the radiative capture rate and associated quantities (e.g. circular polarisation of the photon, spectra etc.) have been carried out by various authors (Cantwell (1956); Huang *et al.* (1957); Bernstein (1959); Dai *et al.* (1959); Primakoff (1959); Manacher and Wolfenstein (1959); Chang and Dai (1961); Manacher (1961); Lobov and Shapiro (1962); Lobov (1963); Rood and Tolhoek (1963, 1965); Borchi and Gatto (1964); Opat (1964); Chu *et al.* (1965 (Appendix I)); Fearing (1966)).

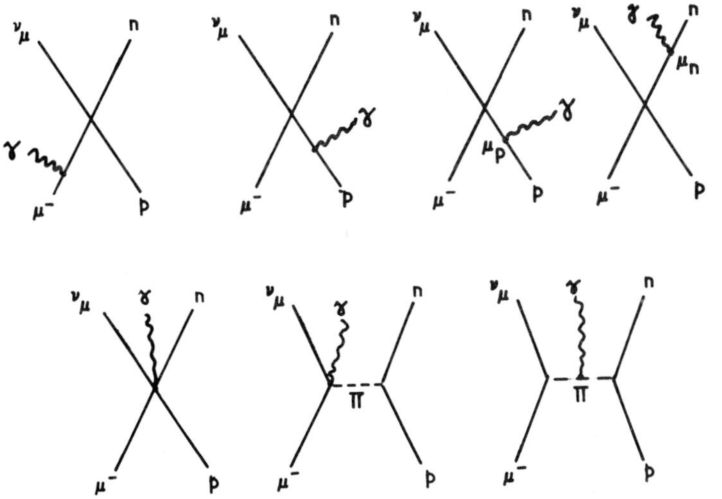

Fig. VII-2. Diagrams contributing to radiative μ-capture. Those diagrams having μ_p or μ_n at the nucleon–photon vertex refer to an interaction with the anomalous nucleon moment.

Typical of this work is that of Rood and Tolhoek (1963, 1965) who first derive an "effective" Hamiltonian for the radiative capture process similar in principle (but much more complicated in form) to that for ordinary muon capture (eq. (VII-12)). They then consider the case of capture by doubly closed shell nuclei (e.g. ^{16}O, ^{40}Ca) and, using the closure approximation, obtain an expression for the photon spectrum. It appears from their calculation that the result is not very sensitive to the nuclear model used, in particular, to the range of the nucleon–nucleon correlation function. Using harmonic oscillator wave functions for ^{16}O and ^{40}Ca they calculate forms for the photon spectrum and the radiative capture rate for different assumptions about the value of $f_P^{(\mu)}$. Their results for the energy spectrum $N(x)$ (where $x = k/k_m$ and k and k_m refer to the photon energy and its maximum value respectively) as calculated for ^{40}Ca are shown in fig. VII-3.

It will be noted that the value of R and to a lesser extent the spectrum

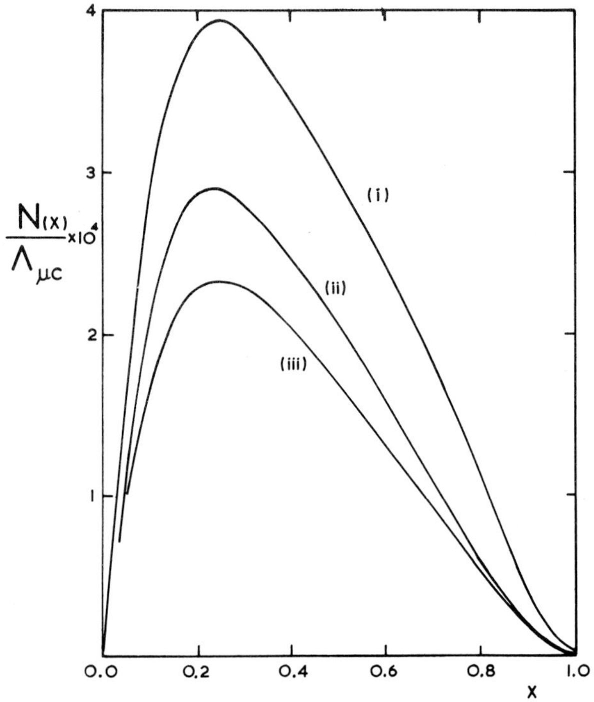

Fig. VII-3. Photon spectra for the ratio radiative μ-capture to normal capture for ^{40}Ca for the three cases (i) $m_\mu f_P^{(\mu)} = +8f_A$, (ii) $m_\mu f_P^{(\mu)} = 0$, (iii) $m_\mu f_P^{(\mu)} = -8f_A$.

shape are sensitive to the strength of the induced pseudoscalar coupling strength. On the other hand the results are not greatly dependent on the values assumed for f_A and f_W.

As far as experiment is concerned a number of measurements of R and the photon spectrum have been made (Conforto *et al.* (1962) – capture in Fe; Chu *et al.* (1963, 1965) – capture in Cu; Conversi *et al.* (1964) – capture in ^{40}Ca). Of particular interest is the work of Conversi *et al.* (1964) since this is carried out for a relatively simple nucleus for which quite detailed theoretical investigations have also been made (Rood and Tolhoek (1963); Fearing (1966); Borchi and De Gennaro (1970)). By comparing their experimental results with those of Rood and Tolhoek, Conversi *et al.* obtain $m_\mu f_P^{(\mu)} = (13.3 \pm 2.7) f_A$ which is approximately two standard deviations from the value predicted by PCAC theory (eqs. (VII-17) and (VII-18)). Fearing (1966) has reinvestigated the problem by using the giant dipole resonance model introduced by Foldy and Walecka (1964) to calculate the ordinary muon capture rate. This latter is an approach in which the important dipole parts of the nuclear matrix elements are related to integrals over the photo-absorption cross section, in the giant-dipole resonance region, of the initial nucleus. As such, it removes a large part of the nuclear uncertainty in the calculation. This has the effect of reducing the dipole contribution to the capture rate and requiring an even larger value for the induced pseudoscalar coupling strength. Thus Fearing (1966) obtains $m_\mu f_P^{(\mu)} = (16.5 \pm 3.1) f_A$ and the discrepancy with PCAC theory is worsened. Borchi and De Gennaro (1970) use Migdal's quasiparticle model (Migdal (1964, 1966)) to take account of the nucleon–nucleon residual interaction and obtain as a result of their analysis $m_\mu f_P^{(\mu)} = (12.4 \pm 2.8) f_A$.

The difference between the values of $m_\mu f_P$ obtained by Conversi *et al.* (1964), Fearing (1966) and Borchi and De Gennaro (1970) indicates that the results are by no means completely model independent. Evenso the disagreement with PCAC is worrying. One solution is to appeal to contributions from second class currents (§V-5) and this point has been looked into by Borchi and Gatto (1964), Rood and Tolhoek (1965) and by Fearing (1966). It is found that PCAC theory and experiment can be reconciled if an induced tensor interaction of strength $f_T^{(\mu)} \gtrsim +7.5/M$ (in the case of Conversi *et al.*'s use of the theory of Rood and Tolhoek (1966)) and $f_T^{(\mu)} \gtrsim 17.5/M$ (in the case of Fearing's analysis). Such values are very large indeed compared with the value $\approx 1/M$ possibly suggested by mirror β-decay data. They are, however, of the same sign. Another possibility in this direction pointed out by Conversi *et al.* and Fearing is that if CVC theory is abandoned and $f_S^{(\mu)} \approx 1.5/M$ then agreement between theory and experiment can be obtained. Neither of these solutions is at all palatable and one can only hope that nuclear and experimental uncertainties (particularly the method used for comparing theory and experiment) are such that in due course the discrepqncy will be removed. Every incentive is here to pursue both the theory and experiment much further since the radiative capture process is clearly very sensitive to the value of $f_P^{(\mu)}$. Indeed, as pointed out by Rood and Tolhoek (1965) (see also Oziewicz and Pikulski (1967)) additional information might be obtained by measuring the angular distribution of photons after the capture of polarised muons or by measuring the circular polarisation of the photons. Here, it has been shown (Cutkosky (1957); Huang *et al.* (1967)) that for the basic process $\mu^- + p \to n + \nu_\mu + \gamma$ the asymmetry parameter α_γ takes the value + 1 for the standard bare V–A interaction. Of course, as with the neutron asymmetry (see §VII-7), taking into account strong interaction effects (e.g. induced pseudoscalar interaction) and calculating α_γ for capture in a nucleus, its value is modified. For example Rood and Tolhoek (1965) obtain $\alpha_\gamma \approx 0.75$ for capture in ^{40}Ca averaging over the γ-ray interval from 57 to 75 MeV. However, a recent measurement of this same quantity by DiLella *et al.* (1971) gives $\alpha_\gamma \lesssim -0.32 \pm 0.48$ in significant disagreement. As usual there are wave function uncertainties, but it is noteworthy that Rood and Tolhoek (1965) judge that their results are rather independent of nuclear structure assumptions. More work is clearly needed here. (See Rood and Yano (1971) and Yano *et al.* (1971) who have undertaken a programme to improve on the approximations in earlier calculations).

VII-10. **Conclusions**

Muon capture is an interesting field in that it allows a study of the semileptonic weak interaction to be made at an energy-momentum transfer significantly higher than that for nuclear beta-decay. Under this circumstance terms such as the induced pseudoscalar and tensor interactions which are proportional to the energy-momentum transfer become important and their magnitude can in principle be determined.

In the case of normal muon capture, nuclear complexities cloud the issue except in the case of capture by, for example, hydrogen or very simple nuclei like ^3He. Even in these cases experimental and, to a lesser extent, theoretical uncertainties do not give very precise answers so that although experiment and standard theory agree, there is still leeway for inclusion of, for example, an induced tensor interaction contribution. For more complex nuclei it seems that μ-capture is more useful as a study of nuclear structure than vice-versa. This is also the case with respect to the investigation of the angular distribution of direct neutrons emitted after muon capture.

Radiative muon capture, on the other hand, even though only observed in complex nuclei, is particularly sensitive to the value of the induced pseudoscalar interaction ($m_\mu f_P^{(\mu)}$) whilst at the same time appearing to be rather independent of the details of nuclear structure. At the present time analysis of experiment gives values of $m_\mu f_P^{(\mu)}$ in conflict with those determined from muon capture in hydrogen. The way in which this conflict will be resolved is not clear. It may be due to the contribution of second class currents (i.e. of an induced tensor interaction) or represent some modification of the basic interaction due to the presence of nuclear matter (e.g. exchange effects, modification of the pion propagator) or it may be due to too great a faith being placed in nuclear models and wave functions. Finally, it may be that the experimental results of which there are few, are unreliable. Whatever the cause, it is at present unknown, and much more theoretical and experimental work is required in this field.

CHAPTER VIII

THE WEAK INTERNUCLEON POTENTIAL

VIII-1. Introduction

In §I-5.3 it was pointed out that part of the usual weak interaction Hamiltonian density can lead to a weak potential between nucleons. This part conserves strangeness and has the form

$$\mathcal{H}_W^{(\text{n.l.}:\Delta S=0)} = \frac{-G}{2\sqrt{2}}[\cos^2\theta\,\{\mathcal{G}_\lambda, \mathcal{G}_\lambda^*\}_+ + \sin^2\theta\,\{\mathcal{S}_\lambda, \mathcal{S}_\lambda^*\}_+] \qquad \text{(VIII-1)}$$

where $\mathcal{G}_\lambda$ and $\mathcal{S}_\lambda$ are hadron currents satisfying the strangeness selction rules $\Delta S = 0$ and $\Delta S = \pm 1$ respectively. Both $\mathcal{G}_\lambda$ and $\mathcal{S}_\lambda$ can be decomposed into polar vector and axial vector components and, according to the ideas of SU3, these components are member of an octet of polar vector currents $(\mathcal{F}_\lambda^{(i)})$ and an octet of axial vector currents $(\mathcal{F}_\lambda^{5(i)})$, thus

$$\begin{aligned} \mathcal{G}_\lambda &= (\mathcal{F}_\lambda^{(1)} + \mathrm{i}\mathcal{F}_\lambda^{(2)}) + (\mathcal{F}_\lambda^{5(1)} + \mathrm{i}\mathcal{F}_\lambda^{5(2)}) \\ \mathcal{S}_\lambda &= (\mathcal{F}_\lambda^{(4)} + \mathrm{i}\mathcal{F}_\lambda^{(5)}) + (\mathcal{F}_\lambda^{5(4)} + \mathrm{i}\mathcal{F}_\lambda^{5(5)})\,. \end{aligned} \qquad \text{(VIII-2)}$$

The Hamiltonian density $\mathcal{H}_W^{(\text{n.l.}:\Delta S=0)}$ will obviously have matrix elements between two nucleon states thereby generating a weak potential between the nucleons of strength G and such a potential concept should be valid for low energy nuclear physics processes. On dimensional grounds we expect the relative strengths of the weak and strong potentials in a nucleus to be of the order $G/\hbar c\bar{r}^2 \approx 10^{-7}$ where $\bar{r}$ is the average separation of nucleons in the nucleus *. This is a very small effect and if it were just a straightforward

* It should be noted that even if the current × current description of the weak interaction were incorrect a weak potential would still exist between nucleons due to, for example, the exchange of an electron–neutrino pair. Such a potential, however, would have a strength of the order $(G/\hbar c\bar{r}^2)^2 \approx 10^{-14}$ and with present techniques would be quite undetectable.

correction to the strong interaction would be impossible to detect. However, the weak interaction contains both parity conserving (p.c.) and parity violating (p.v.) parts so that the weak potential also contains a part which does not conserve parity (to be denoted by $V_{\text{p.v.}}$). As will be seen it is possible, though still experimentally difficult, to detect this part and hence to obtain information about $\mathcal{H}_{\text{W}}^{(\text{n.l.}:\Delta S=0)}$.

The different possible forms for a parity violating potential allowed by symmetry considerations have been investigated by Blin-Stoyle (1960a), Gammel and Thaler (1960) and Herczeg (1963) using the methods initiated by Eisenbud and Wigner (1941) for parity conserving potentials. Suffice it to say here that there are many possible forms, both static and velocity dependent, not all of which arise naturally in the calculations which follow.

This present chapter will be concerned with determining the form of $V_{\text{p.v.}}$ on the basis of current theories of the weak interaction (see Fischbach and Tadić (1972) for a review) whilst the following chapter will deal with the detection of $V_{\text{p.v.}}$ and the interpretation of experimental results.

VIII-2. Early calculations of the weak parity violating potential $V_{\text{p.v.}}$

The earliest calculations (Blin-Stoyle (1960a); Michel (1964)) of $V_{\text{p.v.}}$ concerned themselves only with the $\cos^2\theta$ part of $\mathcal{H}_{\text{W}}^{(\text{n.l.}:\Delta S=0)}$ since $\cos^2\theta/\sin^2\theta \approx 20$, and consisted of straightforward perturbation evaluations of diagrams (a) and (b) in fig. VIII-1. Diagram VIII-1 (a) was evaluated by making the "factorisation" approximation in which two-nucleon amplitudes of the form $\langle \text{N}_1\text{N}_2 | \mathcal{J}_\lambda \mathcal{J}_\lambda^* | \text{N}_1'\text{N}_2' \rangle$ are replaced by $\langle \text{N}_1 | \mathcal{J}_\lambda | \text{N}_1' \rangle \langle \text{N}_2 | \mathcal{J}_\lambda^* | \text{N}_2' \rangle$. Neglecting the momentum dependence of the nucleon weak interaction form factors (§II-2.1) which immediately arise in evaluating the two brackets, assuming CVC theory (§III-1) and restricting consideration to first-class currents (§V-3) only, leads to a contact parity-violating two nucleon potential having the form

$$V_{\text{p.v.}}^{(\text{factorisation})} = \frac{G f_{\text{A}}(0)\cos^2\theta}{2\sqrt{2}M} [(\boldsymbol{\sigma}^{(1)} - \boldsymbol{\sigma}^{(2)}) \cdot \{\boldsymbol{p}, \delta(\boldsymbol{r})\}_+$$

$$+ (1 + \kappa_{\text{p}} - \kappa_{\text{n}})(\text{i}\boldsymbol{\sigma}^{(1)} \times \boldsymbol{\sigma}^{(2)}) \cdot [\boldsymbol{p}, \delta(\boldsymbol{r})]_-] T_+^{(1,2)} \qquad \text{(VIII-1)}$$

where $\boldsymbol{r} = \boldsymbol{r}_1 - \boldsymbol{r}_2$, $\boldsymbol{p} = \boldsymbol{p}_1 - \boldsymbol{p}_2$ and $T_+^{(1,2)} = (t_+^{(1)} t_-^{(2)} + t_-^{(1)} t_+^{(2)})$. In the above expression the axial vector form factor * $f_{\text{A}}(0) \approx -1.2$ (§IV-3) and κ_{p} and

* In the original calculation of Blin-Stoyle (1960a) and Michel (1964), $f_{\text{A}}(0)$ was taken equal to -1.

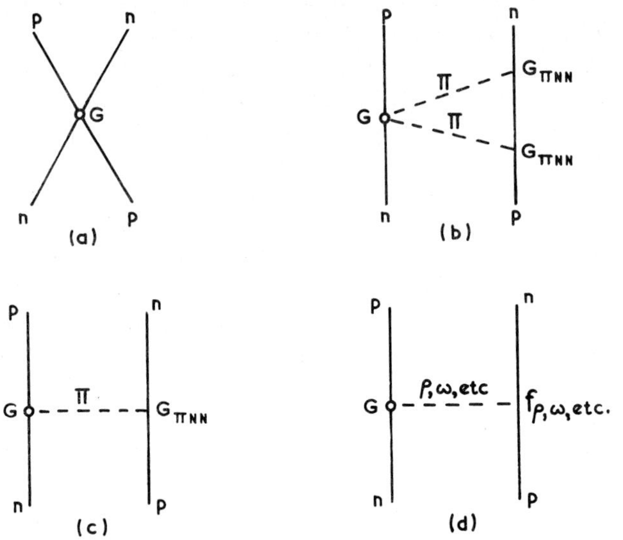

Fig. VIII-1. Diagrams contributing to the weak internucleon potential.

κ_n are the anomalous magnetic moments of the proton and neutron respectively.

The potential stemming from diagram VIII-1 (b) can also be evaluated straightforwardly and can be written

$$V_{\text{p.v.}} = \frac{-GG_{\pi NN}^2 \cos^2\theta \mu^6}{16\pi^2 M^2}\left[\frac{1}{(\mu r)^4} + \frac{2}{(\mu r)^5} + \frac{1}{(\mu r)^6}\right]$$

$$\times\, e^{-2\mu r}\, \mathbf{r}\cdot(\boldsymbol{\sigma}^{(1)} \times \boldsymbol{\sigma}^{(2)})\, T_+^{(1,2)} \qquad \text{(VIII-2)}$$

where μ is the pion Compton wavelength and $G_{\pi NN}$ the renormalised pion–nucleon pseudoscalar coupling constant ($G_{\pi NN}^2/4\pi \approx 14.6$).

Although both potentials have been used in calculations of parity non-conserving phenomena in nuclei they involve too many approximations to be useful in any other respect than indicating the order of magnitude of effect to be expected. The contact potential, for example, has negligible effect in the presence of the strong internucleon potential because of the repulsive core nature of the latter. Here an improvement can be effected by taking account of the momentum dependence of the nucleon form factors which, in turn, corresponds to allowing for the finite extension of the meson cloud around a nucleon which extends beyond the strong repulsive core. Equivalently, the weak internucleon potential can be calculated by evaluating directly diagrams such as VIII-1 (c) and VIII-1 (d) in which a single meson coupled weakly at one vertex (G) and strongly at the other (f) is exchanged between nucleons.

The simple evaluation of the two pion exchange potential given in eq. (VIII-2) is also not satisfactory insofar as no account is taken of $\pi-\pi$ and π–N rescattering effects which are known to be of importance in calculations of the two pion exchange contribution to the strong internucleon potential. Lacaze (1968) has looked at this point in some detail and concludes that the resulting potential has a longer range than that given in eq. (VIII-2). In addition he also obtains a velocity dependent part. Again, however, it is probably safe to assume that the dominant part of the two-pion exchange contribution to $V_{\text{p.v.}}$ can be calculated by evaluating diagrams such as VIII-1 (d) – (ρ-meson exchange).

In view of these comments and the fact that successful calculations of the strong internucleon potential in terms of the exchange of single mesons (O(ne) B(oson) E(xchange) P(otentials)) have been performed (see e.g. Signell (1969)) we shall here carry out further discussion of $V_{\text{p.v.}}$ in terms of the OBEP model as first suggested by Dashen *et al.* (1964). These latter authors also pointed out that because of the repulsive core due to the strong nucleon–nucleon potential exchange of the lightest bosons will be most important for $V_{\text{p.v.}}$. This immediately suggests that the most important contribution to $V_{\text{p.v.}}$ will arise from the one pion exchange diagram. However, as will be seen in the next section, *CP* and charge symmetry severely limit the circumstances in which a one pion exchange can contribute to $V_{\text{p.v.}}$. Our next task is therefore to consider these symmetries with respect to $\mathcal{H}_{\text{weak}}^{(\text{n.l.}:\Delta S=0)}$ in some detail.

VIII-3. Limitations imposed by *CP* invariance and isospin symmetry on $V_{\text{p.v.}}$

Consider first the form of the weak parity violating vertex for the coupling of the neutral pion to a nucleon. One possible form for this vertex is $[\overline{\psi}\psi - \widetilde{\psi}\widetilde{\overline{\psi}}]\phi_0$ where ψ and ϕ_0 represent the nucleon and pion field functions respectively. The interaction is written in a symmetrised form involving a commutator bracket in order to avoid singularities (e.g. Schweber (1964); p. 269).

In Appı D the transformation properties of field functions like ψ and ϕ_0

under the C and P operations are discussed. By successive application of these two transformations it follows that

$$
\begin{aligned}
\psi^{CP}(x) &= \eta_C \eta_P \gamma_4 C \tilde{\bar{\psi}}(x') \\
\bar{\psi}^{CP}(x) &= -\eta_C^{\times} \eta_P^{\times} \tilde{\psi}(x') C^{-1} \gamma_4 \\
\phi_0^{CP}(x) &= -\phi_0^{\dagger}(x') = -\phi_0(x') \text{ (pseudoscalar, neutral field)}
\end{aligned}
\tag{VIII-3}
$$

where $x' \equiv (-x_1, -x_2, -x_3, x_4)$, η_C and η_P are phase factors of modulus unity and the matrix C satisfies the relations

$$
C^{-1} \gamma_\lambda C = -\tilde{\gamma}_\lambda; \quad C^{\dagger} = C^{-1} \text{ (see App. C)} . \tag{VIII-4}
$$

Thus under the CP operation, the parity violating vertex function transforms as follows

$$
\begin{aligned}
[\bar{\psi}\psi - \tilde{\psi}\tilde{\bar{\psi}}]\phi_0 \to {} & [\bar{\psi}^{CP}\psi^{CP} - \tilde{\psi}^{CP}\tilde{\bar{\psi}}^{CP}]\phi_0^{CP} \\
&= [-(\tilde{\psi} C^{-1}\gamma_4)(\gamma_4 C \tilde{\bar{\psi}}) + (\overline{\psi \tilde{C}} \tilde{\gamma}_4)(\tilde{\gamma}_4 \tilde{C}^{-1}\psi)](-\phi_0) \\
&= -[\bar{\psi}\psi - \tilde{\psi}\tilde{\bar{\psi}}]\phi_0 \, .
\end{aligned}
\tag{VIII-5}
$$

It is therefore not invariant under the CP operation and must vanish if the weak interaction is CP invariant. The above argument can be extended to include all possible P violating pion–nucleon vertices (e.g. Barton (1961); Henley (1969a)) and applies generally to the coupling of any neutral (C-even) pseudoscalar meson (e.g. π°, η° etc.) to a nucleon.

In the case of charged pions the situation is more complicated. Under the CP operation $\phi_{\pm} \to -\phi_{\mp}$ so that a possible P-violating pion–nucleon coupling transforms as follows:

$$
[\bar{\psi}_p \psi_n - \tilde{\psi}_n \tilde{\bar{\psi}}_p]\phi_+ \to -[\bar{\psi}_n \psi_p - \tilde{\psi}_p \tilde{\bar{\psi}}_n]\phi_- \, . \tag{VIII-6}
$$

Thus a CP invariant coupling can be constructed having the form

$$
[\bar{\psi}_p \psi_n - \tilde{\psi}_n \tilde{\bar{\psi}}_p]\phi_+ - [\bar{\psi}_n \psi_p - \tilde{\psi}_p \tilde{\bar{\psi}}_n]\phi_- \, . \tag{VIII-7}
$$

This vertex interaction has the isospin transformation properties of the third component of an isovector ($\sim(\boldsymbol{\tau} \times \boldsymbol{\phi})_3$) and is governed by the selection rule $\Delta T = \pm 1$. Further, under the charge symmetry operation U_S (see App. E) which leads to the transformations $\psi_p \to \psi_n$; $\psi_n \to -\psi_p$; $\phi_{+,0,-} \to \phi_{-,0,+}$; it is clear that the interaction (eq. (VIII-7)) is *odd*.

However, if the current $\mathcal{J}_\lambda$ (see eq. (VIII-1)) is completely regular or irregular (see §V-3) i.e.

$$\mathcal{J}_\lambda^* = \mp U_S \mathcal{J}_\lambda U_S^{-1} \tag{VIII-8}$$

then it follows that

$$U_S \{\mathcal{J}_\lambda, \mathcal{J}_\lambda^*\}_+ U_S^{-1} = \{\mathcal{J}_\lambda, \mathcal{J}_\lambda^*\}_+ . \tag{VIII-9}$$

The $\cos^2\theta$ part of the $\mathcal{H}_W^{(\text{n.l.}:\Delta S=0)}$ is thus *even* under the charge symmetry operation and cannot, therefore, lead to an *odd* vertex interaction, i.e. it can only lead to an isoscalar ($\Delta T = 0$) or isotensor ($\Delta T = \pm 2$) interaction and *not* an isovector. This means that a weak one-pion exchange potential cannot be generated by the $\cos^2\theta$ part of $\mathcal{H}_W^{\Delta S=0}$ if only regular or only irregular currents occur. On the other hand, as pointed out and exploited by Blin-Stoyle and Herczeg (1966, 1968) (see also Albright and Oakes (1970); Herczeg (1971)), the presence of *both* regular and irregular currents in $\mathcal{J}_\lambda$ can lead to an isovector vertex and hence a one-pion exchange potential proportional to $\cos^2\theta$. Further, should $\mathcal{J}_\lambda$ be enlarged so as to include neutral currents then an isovector component is again introduced. Inclusion of neutral currents has been suggested on two grounds (i) as a means of accounting for the $\Delta T = \frac{1}{2}$ rule which holds well for nonleptonic decays of strange particles (see e.g. d'Espagnat (1963)) and (ii) as a mechanism for *CP*-violation (Oakes (1968)).

As far as the $\sin^2\theta$ part of $\mathcal{H}_W^{(\text{n.l.}:\Delta S=0)}$ is concerned, this has no definite symmetry under the charge symmetry operation and can therefore lead to a one-pion exchange potential.

The foregoing discussion has been carried out assuming that the initial and final nucleons are free and on their mass shell. However, as pointed out by Henley (1971), nucleons bound in the nucleus are off their mass shell by $\sim V/M$ where V is the nuclear potential and M the nucleon mass. Under these circumstances, even with irregular and neutral currents excluded a *P*-violating one-pion exchange potential proportional to $\approx (V/M)\cos^2\theta$ can arise. In a more detailed model calculation Henley (1971) finds that this $\cos^2\theta$ potential has much the same strength as the $\sin^2\theta$ potential.

In the case of vector meson exchange the only restriction imposed by *CP* invariance and charge symmetry is that the vertex interaction must be isospin *even* in the case of charged mesons (Feuer (1969)). This means that both

$\sin^2\theta$ and $\cos^2\theta$ parts can appear in the parity violating potential.

In summary then, if there are no irregular or neutral currents present, the $\cos^2\theta$ part of $\mathcal{H}_W^{(\mathrm{n.l.}:\Delta S=0)}$ can only lead to a $\Delta T = 0, \pm 2$ potential * between free nucleons deriving from vector meson exchange (e.g. ρ, ω, etc.). However, for nucleons off the mass shell, as they are in the nucleus, a one-pion exchange potential proportional to $\cos^2\theta$ can arise. The $\sin^2\theta$ part of $\mathcal{H}_W^{(\mathrm{n.l.}:\Delta S=0)}$ can lead to a $\Delta T = 0 \pm 1$ potential and, in particular, allows charged pseudoscalar meson exchange (e.g. $\pi^\pm$) as well as vector meson exchange. If irregular or neutral currents are present, then a $\Delta T = \pm 1$ pion exchange potential can also arise from the $\cos^2\theta$ part. We now consider the form of these potentials in detail.

VIII-4. The one pion exchange parity violating potential $V_{\mathrm{p.v.}}^{(\pi)}$

This potential derives from the diagram given in fig. VIII-2 in which a π^- is exchanged between neutron and proton. The only unknown is the amplitude n_-^0 for the vertex $\mathrm{n} \to \mathrm{p} + \pi^-$ and, in terms of this amplitude, it is a straightforward matter to calculate the corresponding one-pion exchange potential (e.g. McKellar (1967); Fischbach (1968); Fischbach and Trabert (1968)) which has the form

$$V_{\mathrm{p.v.}}^{(\pi)} = \frac{G_{\pi\mathrm{NN}} m_\pi^{-1/2} n_-^0}{8\pi\sqrt{2M}} (\boldsymbol{\sigma}^{(1)} + \boldsymbol{\sigma}^{(2)}) \cdot [\boldsymbol{p}, \exp(-m_\pi r)/r]_- T_-^{(1,2)} \qquad \text{(VIII-10)}$$

where $\boldsymbol{r} = \boldsymbol{r}_1 - \boldsymbol{r}_2$, $\boldsymbol{p} = \boldsymbol{p}_1 - \boldsymbol{p}_2$, $T_-^{(1,2)} = (t_+^{(1)} t_-^{(2)} - t_-^{(1)} t_+^{(2)}) = 2\mathrm{i}(\boldsymbol{t}^{(1)} \times \boldsymbol{t}^{(2)})_3$ and $G_{\pi\mathrm{NN}}$ is the strong nucleon-pion coupling constant ($G_{\pi\mathrm{NN}}^2/4\pi \approx 14.6$). Note that this potential has the expected $\Delta T = \pm 1$ form.

Assuming that no neutral or second class currents contribute to the weak interaction Hamiltonian, then the value of n_-^0 can be related via current algebra and SU3 to the amplitudes for the non-leptonic decays of hyperons (this was first attempted by McKellar (1967)) **. Earlier, Suzuki (1965) and Sugawara (1965 a, b) by using PCAC theory and current algebra had shown how to express the parity violating (i.e. S-wave) non-leptonic hyperon decay

* It will be noted that the isospin operator $T_+^{(1,2)}$ occurring in the potential proportional to $\cos^2\theta$ given in eq. (VIII-1) is a composite of an isoscalar and an isotensor.

** Recently Schulke (1972) has suggested that this approach may not be reliable and goes on to argue that n_-^0 and therefore $V_{\mathrm{p.v.}}^{(\pi)}$ are negligible.

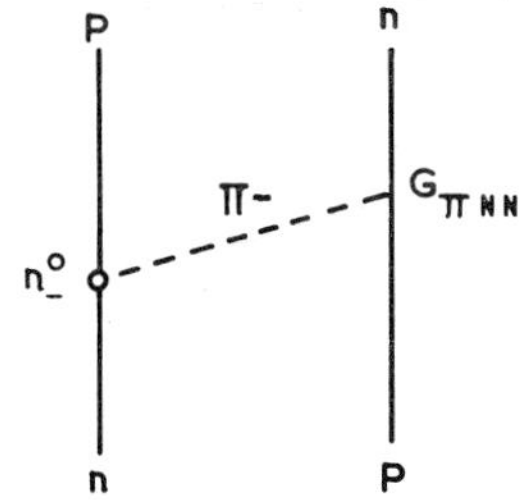

Fig. VIII-2. One-pion exchange contribution to the parity-violating potential.

amplitudes in terms of three reduced matrix elements. These matrix elements are proportional to the 27-plet, symmetrical octet (8_{ss}) and antisymmetrical octet (8_{as}) parts of the $8 \otimes 8$ product occurring in $\mathcal{H}_W^{(n.l.)}$ of the two current octets $\mathcal{F}_\lambda^{(i)}$ and $\mathcal{F}_\lambda^{5(i)}$ (see §I-5). The matrix elements are also proportional to $\sin\theta\,\cos\theta$ (since they refer to strangeness changing transitions and involve the product $(\mathcal{J}_\lambda \mathcal{S}_\lambda^* + \mathcal{J}_\lambda^* \mathcal{S}_\lambda)$) and are denoted by $\alpha(27)\sin\theta\,\cos\theta$, $\alpha(8_{ss})\sin\theta\,\cos\theta$ and $\alpha(8_{as})\sin\theta\,\cos\theta$ respectively.

In terms of the parameters α, the parity violating amplitudes for the processes $n \to p + \pi^-$ and $p \to n + \pi^+$ which, from our previous discussion (§VIII-3) are proportional to $\sin^2\theta$, can be written (McKellar (1967))

$$n_-^0 = -p_+^+ = \sin^2\theta\left[\frac{1}{5}\alpha(27) - \frac{3}{10}\alpha(8_{ss}) + \frac{1}{\sqrt{20}}\alpha(8_{as})\right]. \qquad \text{(VIII-11)}$$

A related way of looking at the situation is as follows (Fischbach *et al.* (1970); Bailin (1970)). Using only SU3 symmetry and *CP*-invariance it can be shown that

$$n_-^0 = -p_+^+ = \frac{\tan\theta}{\sqrt{6}}\,[4\Lambda_-^0 - 2\Xi_-^- + \delta_\Lambda - \sqrt{\tfrac{3}{2}}\,\delta_\Sigma] \qquad \text{(VIII-12)}$$

where

$$\delta_\Lambda = \Lambda_-^0 + \sqrt{2}\,\Lambda_0^0$$

$$\delta_\Sigma = \Sigma_+^+ - \Sigma_-^- - \sqrt{2}\,\Sigma_0^+$$

and the other symbols represent the decay amplitudes for the particles concerned. (Note that Λ_-^0 etc. are proportional to $\sin\theta\,\cos\theta$ so that n_-^0 is proportional to $\sin^2\theta$ as in eq. (VIII-11)).

Use of current algebra and soft pion techniques, give further, $\delta_\Lambda = 0$ and $\delta_\Sigma = 2\Sigma_+^+ \doteq 0$ (from experiment). Thus n_-^0 can be determined directly using the experimental values for Λ_-^0 and Ξ_-^-.

Taking $\theta = \theta_V = \theta_A \approx 0.22$ and the experimental data of Berge (1967) on hyperon decays, Fischbach and Trabert (1968) obtain $n_-^0 = -0.20 \times 10^5$ sec$^{-1/2}$.

This result is obtained using conventional Cabibbo theory and, in particular, assuming that no neutral or second class currents occur in the weak interaction. If this restriction is relaxed then, as pointed out in the previous section, the $\cos^2\theta$ part of $\mathcal{H}_W^{\Delta S=0}$ may contribute to the one-pion exchange potential whose strength will correspondingly be increased by a factor of the order $\cos^2\theta/\sin^2\theta \approx 20$.

Various authors (e.g. McKellar (1968a,b, 1969a,b, 1970); Fischbach and Trabert (1968); Tadić (1968); Herczeg (1972)) have investigated the dependence of n_-^0 on the different models of the weak interaction in detail. For example Fischbach and Trabert (1968) obtain the values of n_-^0 given in table VIII-1 on the basis of the models indicated. Note, however, that the analysis cannot determine the relative sign of $G_{\pi NN}$ and n_-^0 so that there is an overall sign ambiguity for $V_{\text{p.v.}}^{(\pi)}$.

There is considerable variation in the values of n_-^0 and obviously if reliable experimental information could be obtained about the magnitude of $V_{\text{p.v.}}^{(\pi)}$ and hence of n_-^0 it would be possible to eliminate at least some theories of weak interactions. This point will be returned to in Chapt. IX when we discuss the effects of $V_{\text{p.v.}}^{(\pi)}$ in nuclear physics processes. Further, as mentioned in §VIII-3, the additional contribution to $V_{\text{p.v.}}^{(\pi)}$ from the $\cos^2\theta$ part of the Hamiltonian when account is taken of the fact that nucleons in the nucleus

Table VIII-1
Values of n_-^0 for different weak interaction models.

Value of n_-^0 (units of 10^5 sec$^{-1/2}$)	Model	Reference
−0.20	Conventional Cabibbo	Cabibbo (1963)
3.9	Neutral current	d'Espagnat (1963)
0	Segré, γ_5 invariant, schizon	Segré (1968)
−5.0	Segré, γ_5 non-invariant, schizon	Segré (1968)
2.0	Lee, schizon	Lee (1968)

are not on their mass shell (Henley (1971)) must not be forgotten. Unfortunately no details have been worked out of the magnitude of this effect. In addition, no account has been taken of higher order strong interaction effects. On this point Pignon (1971) suggests that they enhance the magnitude of $V^{(\pi)}_{\text{p.v.}}$. (see also Desplanques (1972)).

VIII-5. The vector meson exchange parity violating potential $V^{(\rho)}_{\text{p.v.}}$

Vector meson contributions to the parity violating potential can derive from the exchange of the ρ, ω, ϕ mesons and we here concern ourselves primarily with ρ exchange. The relevant diagram is shown in fig. VIII-1(d) and the main problem is to calculate the parity violating amplitude for the process "nucleon → nucleon + ρ". Consider for example the vertex $\text{n} \to \text{p} + \rho^-$. This amplitude has the form

$$A(\text{n} \to \text{p} + \rho^-) = \langle \text{p}\,\rho\,|\mathcal{H}^{(\text{n.l.}:\Delta S=0)}_{\text{W(p.v.)}}|\text{n}\rangle = \epsilon_\lambda M_\lambda(k) \qquad \text{(VIII-12)}$$

where $\mathcal{H}^{(\text{n.l.}:\Delta S=0)}_{\text{W(p.v.)}}$ is the parity-violating part of $\mathcal{H}^{(\text{n.l.}:\Delta S=0)}_{\text{W}}$ and k and ϵ are the four momentum and polarisation vector respectively for the emitted ρ. Using covariance arguments and noting that $M_\lambda(k)$ transforms as an axial vector we can write (cf. eq. (II-17)), apart from a multiplicative factor,

$$M_\lambda(k) = \mathrm{i}(\bar{u}(p)|[-h_{\text{A}}(k^2)\gamma_\lambda\gamma_5 + \mathrm{i}h_{\text{P}}(k^2)k_\lambda\gamma_5 + h_{\text{T}}(k^2)\sigma_{\lambda\nu}k_\nu\gamma_5]|u(n)) \qquad \text{(VIII-13)}$$

where the h's are form factors and $u(n)$ and $u(p)$ are nucleon spinors. The terms proportional to h_{P} and h_{T} both have small values compared with that proportional to h_{A} at low momentum transfer, which is the region of interest here. Neglecting them it is then a standard calculation to evaluate the resulting parity violating potential. It has the form

$$V^{(\rho)}_{\text{p.v.}} = \frac{h_{\text{A}}(0) f_\rho}{8\pi\sqrt{2}M} [(\boldsymbol{\sigma}^{(1)} - \boldsymbol{\sigma}^{(2)}) \cdot \{\boldsymbol{p}, \exp(-m_\rho r)/r\}_+$$

$$+ (1 + \kappa_{\text{p}} - \kappa_{\text{n}})(\mathrm{i}\boldsymbol{\sigma}^{(1)} \times \boldsymbol{\sigma}^{(2)}) \cdot [\boldsymbol{p}, \exp(-m_\rho r)/r]_-] T^{(1,2)}_+ \qquad \text{(VIII-14)}$$

where the strong ρNN vertex has been described by the phenomenological Lagrangian density

$$\mathcal{L}_{\rho NN} = f_\rho \boldsymbol{\rho}_\lambda \cdot \mathrm{i}\bar{N}\left(\gamma_\lambda + \frac{\mathrm{i}(\kappa_\mathrm{p}-\kappa_\mathrm{n})}{2M}\sigma_{\lambda\nu}\partial_\nu\right)\frac{\boldsymbol{\tau}}{2}N \tag{VIII-15}$$

f_ρ is the ρNN coupling constant ($f_\rho^2/4\pi \approx 2.4$) and $\boldsymbol{\rho}_\lambda$ is an isovector field operator for the ρ-meson.

Three points are to be noted about the expression for $V_{\text{p.v.}}^{(\rho)}$ given in eq. (VIII-14). Firstly, it has the expected $\Delta T = 0, \pm 2$ form. Secondly, its overall strength is determined by the (as yet) unknown quantity $h_\mathrm{A}(0)$. Thirdly, it has the same general form as the contact potential given in eq. (VIII-1) resulting from the "factorisation" approximation. Indeed the latter can be obtained from the former by the replacements

$$h_\mathrm{A}(0) \to \frac{G}{\sqrt{2}}\cos^2\theta\left(\frac{\sqrt{2}m_\rho^2 f_\mathrm{A}(0)}{f_\rho}\right) \tag{VIII-16}$$

$$\frac{m_\rho^2}{4\pi}\frac{\mathrm{e}^{-m_\rho r}}{r} \to \delta(r)\,. \tag{VIII-17}$$

The relation given in eq. (VIII-17) is a natural development of the factorisation approach if account is taken of the nucleon form factors. This approach has been exploited by Blin-Stoyle and Herczeg (1968) (who also take account of second class current contributions which lead to isovector contributions) and by Tadić (1968). In this approach the δ-functions in the contact potential (eq. (VIII-1)) are replaced by integrals over the nucleon form factors and the potential can be written

$$V_{\text{p.v.}}^{(\text{factorisation})} = \frac{G\cos^2\theta}{2\sqrt{2}M}\left[(\boldsymbol{\sigma}^{(1)} - \boldsymbol{\sigma}^{(2)})\cdot\{\boldsymbol{p}, v(r)\}_+ \right.$$
$$\left. + (1+\kappa_\mathrm{p}-\kappa_\mathrm{n})(\mathrm{i}\boldsymbol{\sigma}^{(1)}\times\boldsymbol{\sigma}^{(2)})\cdot[\boldsymbol{p}, v'(r)]_-\right]T_+^{(1,2)} \tag{VIII-18}$$

where

$$v(r) = \frac{1}{(2\pi)^3}\int \mathrm{e}^{\mathrm{i}\boldsymbol{k}\cdot\boldsymbol{r}} f_\mathrm{V}(k^2) f_\mathrm{A}(k^2)\,\mathrm{d}^3k \tag{VIII-19}$$

$$(1+\kappa_\mathrm{p}-\kappa_\mathrm{n})v'(r) = \frac{1}{(2\pi)^3}\int \mathrm{e}^{\mathrm{i}\boldsymbol{k}\cdot\boldsymbol{r}}[f_\mathrm{V}(k^2) - 2Mf_\mathrm{W}(k^2)]f_\mathrm{A}(k^2)\,\mathrm{d}^3k\,. \tag{VIII-20}$$

In the above, second class currents have been neglected and the form factors f_V, f_A and f_W (see Chapt. II) have been taken to be real.

The main problem connected with $V^{(\rho)}_{\text{p.v.}}$, however, is not so much its detailed form but more its overall strength as determined by the value of $h_A(0)$. The present situation in this respect is not completely clear. In order to evaluate $h_A(0)$, Fischbach *et al.* (1969) used the current-field identity (Gell-Mann and Zachariasen (1961); Kroll *et al.* (1967)). This identity relates the polar-vector current $\mathcal{V}^{(\alpha)}_\lambda$ to the ρ-meson field operator $\rho^{(\alpha)}_\lambda$ as follows

$$\rho^{(\alpha)}_\lambda(x) = \frac{f_\rho}{m^2_\rho}\,\mathcal{V}^{(\alpha)}_\lambda(x) \tag{VIII-21}$$

where α is an isospin index. As a result of their calculations they obtained the result quoted in eq. (VIII-16) so justifying the factorisation approximation (Fischbach and Tadić (1972)). However, Olesen and Rao (1969) and Feuer (1969) using a similar approach found a much smaller (as to be negligible) value for $h_A(0)$. The discrepancy between these different calculations lies in the assumptions made about the so-called "seagull" and Schwinger terms which arise in the calculation and the extent to which they cancel. An alternative approach (McKellar and Pick (1971, 1972)) calculates the ρNN vertex from the observed parity violating hyperon decays using SU(6) symmetry and octet dominance and leads to a result close to the factorisation approximation. But basically there is an unresolved arbitrariness here (see e.g. Danilov (1970)) so that to determine the presence or otherwise of $V^{(\rho)}_{\text{p.v.}}$ and hence the value of $h_A(0)$ would help to resolve a complex problem in the area of current and field algebra.

The discussion so far has centred around the exchange of charged ρ mesons. There can also, in principle, be a contribution from the exchange of the neutral ρ and also the neutral ω, ϕ etc. Since the weak interaction currents are charged contributions from neutral meson exchange can only arise from strong interaction renormalisation effects and are expected to be much weaker. It is usual, therefore, to neglect them. However, should neutral currents contribute to the weak Hamiltonian then neutral meson exchange can take place. The resulting potentials (which include isovector terms) have been evaluated by Fischbach *et al.* (1969) (see also Fischbach and Tadić (1972)).

VIII-6. The single-particle parity violating potential $W_{\text{p.v.}}$

For the purposes of nuclear structure calculations it is clearly useful to be able to work as a first approximation with an effective single-particle parity violating potential. Such a potential can be regarded as the average potential experienced by a single nucleon outside a core of nucleons. Formally it can be defined as follows (Michel (1964); McKellar (1968a))

$$\langle\alpha|W_{\text{p.v.}}|\beta\rangle = \langle\Psi_\alpha|\sum_{i<j} V_{\text{p.v.}}(i,j)|\Psi_\beta\rangle \qquad \text{(VIII-22)}$$

where $|\alpha\rangle$ and $|\beta\rangle$ are one-particle states and Ψ_α and Ψ_β are many particle states with the last particle in the states α and β and the other particles in a "core" state. $V_{\text{p.v.}}(i,j)$ is a two-particle parity violating potential.

Writing Ψ_α and Ψ_β as Slater determinants of core states γ_i and α or β respectively eq. (VIII-22) reduces to

$$\langle\alpha|W_{\text{p.v.}}|\beta\rangle = \sum_i \langle\alpha\gamma_i|V_{\text{p.v.}}|\beta\gamma_i - \gamma_i\beta\rangle\,. \qquad \text{(VIII-23)}$$

Michel (1964) originally evaluated the foregoing expression for the potential $V_{\text{p.v.}}^{(\text{factorisation})}$ (eq. (VIII-1)) using plane wave states for the γ_i. He obtained the following effective single particle potential (denoted by W_{M})

$$W_{\text{M}} = G'_{\text{M}}\left[1+\frac{N-Z}{A}\tau_3\right]\boldsymbol{\sigma}\cdot\boldsymbol{p} \qquad \text{(VIII-24)}$$

where *

$$G'_{\text{M}} = +\frac{G\cos^2\theta\, f_{\text{A}}(0)}{2\sqrt{2}}\,\frac{(1+\kappa_{\text{p}}-\kappa_{\text{n}})}{M}\,\rho_0$$

and ρ_0 is the nucleon density in the nucleus. The momentum $\boldsymbol{p}$ now refers to a fixed coordinate system at the centre of mass of the core nucleons.

McKellar (1968a) has carried this approach further and averaged the potentials $V_{\text{p.v.}}^{(\pi)}$ and $V_{\text{p.v.}}^{(\rho)}$ as well as taking hard core correlation effects among the nucleons into account (the importance of these latter effects was first pointed out by Adams (1967); see also Gari *et al.* (1971a)). For this latter

* Note that there is a sign error in Michels (1964) paper (Coutinho (1970); Eman and Tadić (1971)).

purpose the two particle plane wave state was replaced by the corresponding Bethe–Goldstone state. Using his results the following expression can be written down for the effective single particle potential

$$W = G' \left[1 + \frac{N-Z}{A}\tau_3\right] \boldsymbol{\sigma} \cdot \boldsymbol{p} \tag{VIII-25}$$

where

$$G' = -\left[\frac{h_A(0) f_\rho (1+\kappa_p - \kappa_n)}{m_\rho^2}\,\omega_\rho + \frac{n_-^0 G_{\pi NN}}{m_\pi^{5/2}}\right] \frac{\rho_0}{2\sqrt{2M}}$$

$$= -\left[\frac{h_A(0) f_\rho}{m_\rho^2 G \cos^2\theta\, f_A(0)}\,\omega_\rho + \frac{n_-^0 G_{\pi NN}}{G \cos^2\theta\, f_A(0)(1+\kappa_p - \kappa_n)}\,\omega_\pi\right] G'_M \,. \tag{VIII-26}$$

G'_M is defined in eq. (VIII-24) and ω_ρ and ω_π are momentum dependent reduction factors related to hard core effects. McKellar (1968) estimates their average values to be $\overline{\omega}_\rho \approx 0.4$, $\overline{\omega}_\pi \approx 0.14$. Writing

$$h_A(0) = \alpha \frac{G}{\sqrt{2}} \cos^2\theta \left(\frac{\sqrt{2} m_\rho^2 f_A(0)}{f_\rho}\right)$$

(see eq. (VIII-16)) where α is a parameter whose value is 1 if the factorisation approximation is correct (but see the discussion in §VIII-5) and inserting the values of the different constants gives finally

$$G' \approx [-0.4\,\alpha + 3.3 \times 10^{-6} n_-^0]\, G'_M \,. \tag{VIII-27}$$

This latter relation has been exploited by McKellar (see §IX-4.1) in order to use the various calculations (Chapt. IX) which have been carried out using the simple Michel potential characterised by a strength G'_M.

CHAPTER IX

THE EFFECTS OF A PARITY VIOLATING POTENTIAL IN LOW ENERGY NUCLEAR PHYSICS PROCESSES

IX-1. General consideration of parity violating effects

In the last chapter an account was given of the attempts which have been made to determine the strength and form of the parity violating potential $V_{\text{p.v.}}$ from our knowledge of the fundamental weak interaction $\mathcal{H}_{\text{W}}$. As is to be expected the nature of $V_{\text{p.v.}}$ is dependent on the form of this interaction. What is remarkable is the extreme sensitivity of $V_{\text{p.v.}}^{(\pi)}$ in this respect. There is, thus considerable incentive to study the effect of $V_{\text{p.v.}}$ in nuclear physics processes in order to obtain information about its form and strength and, in turn, about $\mathcal{H}_{\text{W}}$. This chapter is concerned with the theory of these effects and with the comparison between theory and experiment.

Including a parity violating potential, the total Hamiltonian for a nuclear system can be written

$$H = H_0 + V_{\text{p.v.}} \tag{IX-1}$$

where H_0 is a spatial scalar and is the usual nuclear Hamiltonian including the strong internucleon potential. Let the eigenstates and eigenvalues of H_0 be denoted by ψ_i and E_i respectively, so that

$$H_0 \psi_i = E_i \psi_i \,. \tag{IX-2}$$

Since $V_{\text{p.v.}}$ is very small ($\sim 10^{-6}$ or 10^{-7}) compared with H_0 it is perfectly proper to treat it by perturbation theory so that to a very good approximation the eigenfunctions of H can be written

$$\Psi_i = \psi_i + \sum_{j \neq i} \frac{\langle j | V_{\text{p.v.}} | i \rangle}{E_i - E_j} \psi_j \tag{IX-3}$$

where, of course, the ψ_j have opposite parity to ψ_i.

Symbolically, eq. (IX-3) can be written

$$\Psi_i = \psi_i + \mathcal{F}\phi_i \tag{IX-4}$$

where ϕ_i has opposite parity to ψ_i and $\mathcal{F}$ is a measure of the parity impurity in the state Ψ_i; it is, in general, expected to have a value of the order $\mathcal{F} \approx 10^{-6}$ to 10^{-7}. ψ_i is frequently referred to as the "regular" part of the wave function and ϕ_i as the "irregular" part. For preliminary purposes we shall discuss parity violating effects in terms of the wave function given in eq. (IX-4).

Wilkinson (1958a) was the first author to discuss in detail the types of experiment in nuclear physics which could lead to the observation of parity violating effects. These experiments can be divided into two classes

(i) those in which a search is made for the violation of an absolute parity selection rule.

(ii) those in which a search is made for an observable pseudoscalar quantity.

Experiments of class (i) are typified by considering the process of α-decay. Consider the situation that α-decay is energetically possible between a 2^- state and a 0^+ state. Since the orbital angular momentum carried away by the α-particle must be $L = 2$ and the corresponding parity even, it follows that the decay cannot take place since a change of parity is required. However, if there is admixed an amplitude $\mathcal{F}$ of 2^+ state into the 2^- state then the decay can take place through this component of the wave function with an intensity proportional to $\mathcal{F}^2$. Although $\mathcal{F}^2$ is in the region $10^{-14} - 10^{-12}$ we shall see that sensitivities of this order can be achieved in practice.

Experiments of the class (ii) type are familiar from β-decay (see §VI-4) where, for example, measurements of the angular distribution of electrons from polarized nuclei or of the longitudinal polarisation of electrons emitted from unpolarised nuclei, indicate that parity is not conserved in the process. Effects of this kind arise from an interference between the scalar and pseudoscalar parts of the Hamiltonian for the system. In our case, similar effects are to be expected in electromagnetic transitions resulting from the interference between electric and magnetic multipoles of the same order. One multipole would derive from the regular part of the wave function and the other, of amplitude $\mathcal{F}$, from the irregular part. The resulting interference term can then manifest itself, for example, as an asymmetry of the form $\boldsymbol{k} \cdot \boldsymbol{I}$ (where $\boldsymbol{k}$ is the gamma ray momentum and $\boldsymbol{I}$ is the nuclear spin) in the angular distribution of the emitted radiation. This leads to a $\cos\theta$ (i.e. $P_1(\theta)$) term in the angular distribution function in addition to the usual even Legendre polynomials which may occur. It can also lead to higher odd Legendre polynomials appearing. Similarly a circular polarisation of the radiation emitted from unpolarised nuclei can also occur. Here it is important to note that since we are dealing with an interference phenomenon, the effect is proportional to $\mathcal{F}$ rather than to $\mathcal{F}^2$.

To be rather more precise, consider a regular electric (or magnetic) multipole transition (operator O_L) with corresponding irregular magnetic (or electric) multipole transition (operator O'_L) between two states Ψ_i and Ψ_f where (see eq. (IX-4))

$$\begin{aligned} \Psi_i &= \psi_i + \mathcal{F}\phi_i \\ \Psi_f &= \psi_f + \mathcal{F}\phi_f . \end{aligned} \tag{IX-5}$$

The regular electric (or magnetic) multipole matrix element will have the form

$$\langle \Psi_f | O_L | \Psi_i \rangle = \langle \psi_f | O_L | \psi_i \rangle + O(\mathcal{F}^2) \tag{IX-6}$$

whilst the corresponding irregular magnetic (or electric) multipole matrix element to first order in $\mathcal{F}$ will be

$$\langle \Psi_f | O'_L | \Psi_i \rangle = \mathcal{F}[\langle \phi_f | O'_L | \psi_i \rangle + \langle \psi_f | O'_L | \phi_i \rangle] . \tag{IX-7}$$

Such a transition will be referred to symbolically as an $O_L(O'_L)$ transition. To order $\mathcal{F}$, the resulting interference term Q_γ then has the order of magnitude

$$\begin{aligned} Q_\gamma \approx \frac{\langle \Psi_f | O'_L | \Psi_i \rangle}{\langle \Psi_f | O_L | \Psi_i \rangle} &\approx \frac{\mathcal{F}[\langle \phi_f | O'_L | \psi_i \rangle + \langle \psi_f | O'_L | \phi_i \rangle]}{\langle \psi_f | O_L | \psi_i \rangle} \\ &= \mathcal{F}R . \end{aligned} \tag{IX-8}$$

Clearly to get a large interference term Q_γ and hence a significant experimental effect it is important for the matrix element ratio R to be as large as possible. This means that a nuclear transition must be chosen in which, for one reason or another, the regular matrix element $\langle \psi_f | O_L | \psi_i \rangle$ is inhibited. In this way an amplification effect is obtained which can be as large as two or three orders of magnitude. Even so experimental accuracies (q.v.) of at least 1 part in 10^3 must be achieved.

From the detailed theoretical point of view any calculation of parity violating effects contains two parts. Firstly the amplitudes of parity impurities in a given nuclear state have to be calculated taking some specific form for $V_{\text{p.v.}}$ and using some form of nuclear model. Secondly the effect of these parity impurities on some nuclear process (e.g. forbidden α-decay, asymmetry or circular polarisation of γ-rays etc.) has to be determined. In the next sections we review the calculations of this kind that have been carried out and compare them with experimental results.

IX.2. Parity violating α-decay

In the previous section it was pointed out that the experimental detection of a parity forbidden alpha-decay would demonstrate the presence of parity impurities in nuclear states and therefore the presence of a parity violating potential. Theoretical and experimental work in this direction has centred around the possible α-decay of the 2^- state in ^{16}O at 8.87 MeV to the 0^+ ground state of ^{12}C and discussion in this section will be concerned primarily with this decay.

The theory of the process has been studied by Michel (1964), Gari and Kümmel (1969), Henley *et al.* (1969) and Gari (1969, 1970) and in the following discussion we use the approach of Gari and Kümmel.

IX-2.1. *Theory of parity violating α-decay*

Parity impurities in the 2^- (8.87 MeV) state in ^{16}O will arise primarily from the nearby 2^+ states at 6.9 MeV, 9.8 MeV and 11.5 MeV. Admixtures of more distant states and parity admixtures into the ground state of ^{12}C and the α-particle itself are expected to be much smaller because of the large energy denominators involved in the perturbation theory (see eq. (IX-3)). Further, the 9.8 MeV state has a very small reduced width for α-decay and it seems reasonable to neglect its contribution. Thus the parity impurity admixture in the 8.87 MeV (2^-) state can be written

$$|8.87, 2^+\rangle_{\text{adm}} = F_{11.5}\,|11.5, 2^+\rangle + F_{6.9}\,|6.9, 2^+\rangle \tag{IX-9}$$

where, in an obvious notation, the admixture amplitudes F_i are given by

$$F_i = \frac{\langle i, 2^+|V_{\text{p.v.}}|8.87, 2^-\rangle}{E_i - E\,(8.87)}\,. \tag{IX-10}$$

Here it should be noted that since all the nuclear states involved have $T = 0$, only $V^{(\rho)}_{\text{p.v.}}$ can contribute since only this part of $V_{\text{p.v.}}$ contains a $T = 0$ component and can therefore lead to non-vanishing values for F_i.

A suitable expression for calculating α-decay widths has been given by Mang (1964), namely

$$\Gamma = \frac{P_L(\epsilon, R)}{2MR} |\langle \text{daughter} + \alpha | \text{parent} \rangle_R|^2 \qquad \text{(IX-11)}$$

where R is the sum of daughter and α-particle radii, and $P_L(\epsilon, R)$ is the penetrability for angular momentum L and energy ϵ.

The irregular α-decay width for the case under consideration (for which $L = 2$) is thus given by the following expression

$$\Gamma_{\text{irr}}(8.87) \triangleq \frac{P_2(8.87, R)}{2MR} \Big| \sum_{i=6.9,11.5} F_i \langle \text{daughter} + \alpha | i, 2^+ \rangle_R \Big|^2 . \qquad \text{(IX-12)}$$

Unfortunately α-decay theory is not very satisfactory for the purposes of calculating absolute values of allowed widths and so little confidence could be put in a direct calculation of Γ_{irr} using eq. (IX-12). However, Gari and Kümmel (1969) have pointed out that in the simple case considered here it is possible to relate Γ_{irr} to known measured α-decay widths. Comparing the α-widths of the 11.5 MeV and 6.9 MeV levels it is clear, as might intuitively be expected, that the dominant contribution is from 4 particle – 4 hole (4p-4h) components of the nuclear wave functions. This is supported by the fact already pointed out that the nearly pure 2p-2h state at 9.8 MeV has a very small reduced width. Since the 6.9 MeV and 11.5 MeV states are believed to be primarily (2p-2h) + (4p-4h) states (e.g. Celenza *et al.* (1966)) it then follows that we can write for their regular decay widths

$$\Gamma_{\text{reg}}(i) = \frac{P_2(i, R)}{2MR} |C_4^{(i)} \langle \text{daughter} + \alpha | (i, 2^+)_{\text{4p-4h}} \rangle_R |^2 \qquad \text{(IX-13)}$$

where $C_4^{(i)}$ is the 4p-4h amplitude (wavefunction $|(i, 2^+)_{\text{4p-4h}}\rangle$) in the level i.

Correspondingly

$$\Gamma_{\text{irr}}(8.87) = \frac{P_2(8.87, R)}{2MR} \Big| \sum_{i=6.9,11.5} F_i C_4^{(i)} \langle \text{daughter} + \alpha | (i, 2^+)_{\text{4p-4h}} \rangle_R \Big|^2 .$$

$$\text{(IX-14)}$$

Using eqs. (IX-13) and (IX-14), it finally follows that

$$\Gamma_{\text{irr}}(8.87) = \frac{P_2(8.87,R)}{P_2(11.5,R)} \left| \frac{F_{6.9}C_4^{(6.9)} + F_{11.5}C_4^{(11.5)}}{C_4^{(11.5)}} \right|^2 \Gamma_{\text{reg}}(11.5)\,. \quad \text{(IX-15)}$$

Gari and Kümmel (1969) and Gari (1969, 1970) have evaluated $\Gamma_{\text{irr}}(8.87)$ using eqs. (IX-10) and (IX-15). They use the wavefunctions of Celenza *et al.* (1966), Camiz and Vinh Mau (1964) and Gillet and Vinh Mau (1964) and take account of hard core effects by following the method of Kallio and Day (1969). Although absolute penetrabilities and α-decay rates are sensitively dependent on the value chosen for the interaction radius R, sensitivity is considerably diminished here since $\Gamma_{\text{irr}}(8.87)$ depends on a ratio of penetrabilities. The calculations were carried out using the potential $V_{\text{p.v.}}^{(\rho)}$ given in eq. (VIII-14) taking $h_{\text{A}}(0) = (G/\sqrt{2}) \cos^2\theta\,(\sqrt{2}m_\rho^2 f_{\text{A}}(0)/f_\rho)$ and allowing for the modification stemming from different assumptions about the form of the basic weak interaction which were calculated by Fischbach *et al.* (1969). Here it should be remembered as mentioned earlier that since all the states concerned have $T = 0$, there can be no contribution from $V_{\text{p.v.}}^{(\pi)}$. The overall results are dependent on the value r_{c} taken for the hard core radius and this variation is shown in fig. IX-1 for two different assumptions about the values of m_ρ (the mass of the ρ-meson).

The values obtained for $\Gamma_{\text{irr}}(8.87)$ are similar to those obtained by Henley *et al.* (1969) although the details of the two calculations are somewhat different. The values are also close to the value 0.8×10^{-10} eV obtained by Michel (1964) using the single particle parity violating potential W_M (see eq. (VIII-24)) in a sophisticated order of magnitude estimate.

IX-2.2. *Experimental results parity violating α-decay*

A fair number of experiments aimed at detecting parity violation in α-decay have been carried out (Tanner (1957); Segel *et al.* (1958); Bromley *et al.* (1959); Alburger *et al.* (1961); Segel *et al.* (1961a, b); Kaufmann and Wäffler (1961); Donovan *et al.* (1961); Bassi *et al.* (1962, 1963); Boyd *et al.* (1968); Hättig *et al.* (1969); Wäffler (1970); Sprenkel-Segel *et al.* (1970); Jaus *et al.* (1970)). Of these experiments the most recent (in ^{16}O) are on the way to achieving accuracies which might enable revealing comparisons with theory to be made. The results obtained are given in table IX-1. These results are in essential agreement with the theoretical values given in fig. IX-1. However, they are not precise enough, as yet, to distinguish between the different possible forms of weak interaction. Nevertheless, if positive results continue to be achieved then this is clear evidence that $h_{\text{A}}(0)$ is non-zero.

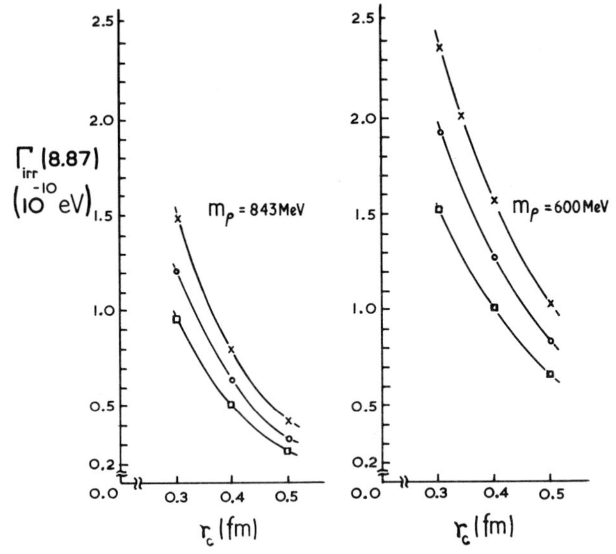

Fig. IX-1. The irregular α-decay width $\Gamma_{irr}(8.87)$ for the 8.87 MeV 2^- state in ^{16}O as a function of the hard core radius r_c and for different models of the weak interaction (□ = conventional Cabibbo (1963) and Segré (1968) models; ○ = Lee (1968) model; X = neutral current (d'Espagnat (1963) model).

Table IX-1
Recent experimental results on parity violating α-decay in ^{16}O.

$\Gamma_{irr}(8.87)$	Reference
$\leq 0.94 \times 10^{-10}$ eV	Hättig *et al.* (1969)
$(1.8 \pm 0.8) \times 10^{-10}$ eV	Hättig *et al.* (1970)
$(2 \pm 1) \times 10^{-10}$ eV	Wäffler (1970)
1×10^{-9} eV	Sprenkel-Segel *et al.* (1970)
$\leq 2.1 \times 10^{-9}$ eV	Jones *et al.* (1970)

This in turn, as was pointed out in §VIII-5, can throw some light on various problems in the area of current and field algebra (e.g. Cheng *et al.* (1970) conclude that just to establish experimentally that $\Gamma_{irr}(8.87) \approx 10^{-10}$ eV demands the presence of *non-cancelling* seagull and Schwinger terms). Clearly it would be of great value if the experimental results could be significantly improved and experiments carried out on other parity violating α-decays, e.g. from the 3^+, $T = 0$, 11.06 MeV state in ^{16}O (see Fink *et al.* (1972) for a related theoretical calculation).

IX-3. Parity violation in electromagnetic transitions

We now consider in some detail the form and magnitude of pseudoscalars occurring in electromagnetic transitions arising from parity impurities in the nuclear states concerned. The following physical situations will be considered:

(i) The angular distribution of gamma-rays from polarised nuclei.
(ii) Beta–gamma angular correlations.
(iii) The angular distribution of gamma-rays following the capture of polarised thermal neutrons by unpolarised medium weight nuclei.
(iv) The emission of circularly polarised gamma-rays from unpolarised nuclei.
(v) Emission and absorption of electromagnetic radiation by few-nucleon nuclei.
(vi) The longitudinal polarisation of internal conversion electrons emitted by unpolarised nuclei.
(vii) The intensity ratio of the components of the hyperfine spectrum as shown by the Mössbauer effect.

In order to develop the theory of processes of this kind it is first necessary to consider some of the general features of the angular distribution of electromagnetic radiation in the case of a transition between two nuclear magnetic sub-states.

IX-3.1. *Asymmetry effects in γ-transitions. General results*

The general theory of gamma emission by nuclei and associated angular distributions, angular correlations, polarisations etc. is discussed, for example, in detail in various articles in Siegbahn (1965) (e.g. Moszkowski (1965); Frauenfelder *et al.* (1965a); and de Groot *et al.* (1965) – see also Biedenharn and Rose (1953); Rose (1955)) and in the following some familiarity with the methods involved is assumed. The development follows closely that of Blin-Stoyle (1960b).

Consider a gamma transition between two nuclear substates $|IM)$ and $|I'M')$, both of which may contain parity impurities. The appropriate matrix element of the electromagnetic interaction (denoted by $\mathcal{H}(A)$) for the transition can be written (e.g. Rose, 1955)

$$\langle I'M'|\mathcal{H}(A)|IM\rangle$$

$$= \pi \sum_{L\mu P} (2L+1)^{1/2} f(P) D^{(L)}_{\mu P}(\boldsymbol{k})[\langle I'M'|\mathcal{H}(A^{\mu}_{L}(m))|IM\rangle$$

$$+ (-P)\langle I'M'|\mathcal{H}(A^{\mu}_{L}(e))|IM\rangle] \qquad \text{(IX-16)}$$

where $f(P)$ ($P=+1$ or -1 corresponding to left or right circular polarisation) defines the polarisation of the gamma-ray emitted in the direction $\boldsymbol{k}$. $\mathcal{H}(A^{\mu}_{L}(m))$ and $\mathcal{H}(A^{\mu}_{L}(e))$ are the usual magnetic and electric multipole operators (e.g. Moszkowski (1966)) with choice of phase factors such that the corresponding reduced matrix elements are relative real (see Biedenharn and Rose (1953) and also Chapt. X). The $D^{(L)}_{\mu P}(\boldsymbol{k})$ are elements of the (unitary) rotation matrix whose argument $\boldsymbol{k}$ implies a rotation from the coordinate system defining the radiation to the quantisation coordinate system.

In such a transition interference effects arise between multipoles of opposite parity because of the parity impurities in the states $|IM\rangle$ and $|I'M'\rangle$ and will manifest themselves as pseudoscalar quantities in the angular and polarisation distributions of the emitted radiation. The angular distribution $W_{M,M-\mu}$ of the radiation emitted in the transition $|IM\rangle \to |I'M-\mu\rangle$ taking into account the contribution from all multipoles and for arbitrary polarisation of the radiation is obtained by taking the square modulus of the matrix element given in eq. (IX-16). The calculation is straightforward and, using the techniques of Biedenharn and Rose (1953), the following expression is obtained:

$$W_{M,M-\mu} = |\langle I'M-\mu|\mathcal{H}(A)|IM\rangle|^2$$

$$= \pi^2 \cdot \sum_{LL'} [(2L+1)(2L'+1)]^{1/2} C(I'LI; M-\mu\mu)$$

$$\times C(I'L'I; M-\mu\mu) F^{\mu}_{LL'}(\boldsymbol{k}) \qquad \text{(IX-17)}$$

where the C's are Clebsch–Gordan coefficients and

$$F^{\mu}_{LL'}(k) = (-)^{\mu+1} \sum_{\nu} \{C(LL'\nu;-11)\, C(LL'\nu;-\mu\mu)[G^{\nu}_{LL'}(m^{\times}_{L} m_{L'} + e^{\times}_{L} e_{L'})$$

$$+ H^{\nu}_{LL'}(e^{\times}_{L} m_{L'} + m^{\times}_{L} e_{L'})]\, P_{\nu}(\cos\theta)$$

$$+ C(LL'\nu;11)\, C(LL'\nu;-\mu\mu)[(\nu-2)!/(\nu+2)!]^{\frac{1}{2}} [I^{\nu}_{LL'}(m^{\times}_{L} m_{L'} - e^{\times}_{L} e_{L'})$$

$$+ J^{\nu}_{LL'}(e^{\times}_{L} m_{L'} - m^{\times}_{L} e_{L'})]\, P^{(2)}_{\nu}(\cos\theta)\} \,. \qquad \text{(IX-18)}$$

In the above expression P_ν and $P_\nu^{(2)}$ are Legendre functions,

$$G^{\nu}_{LL'} = |f(1)|^2 + (-)^{L+L'-\nu} |f(-1)|^2$$

$$H^{\nu}_{LL'} = |f(1)|^2 - (-)^{L+L'-\nu} |f(-1)|^2$$

$$I^{\nu}_{LL'} = f^{\times}(-1)\, f(1) + (-)^{L+L'-\nu} f^{\times}(1)\, f(-1)$$

$$J^{\nu}_{LL'} = f^{\times}(-1)\, f(1) - (-)^{L+L'-\nu} f^{\times}(1)\, f(-1) \qquad \text{(IX-19)}$$

and m_L and e_L are the reduced matrix elements

$$m_L = \langle I' \| \mathcal{H}(A_L(m)) \| I \rangle\,; \qquad e_L = \langle I' \| \mathcal{H}(A_L(e)) \| I \rangle\,. \qquad \text{(IX-20)}$$

To proceed further it is necessary to specify both the orientation of the initial nuclear system, that is, the distribution of the nuclear substates $|IM)$, and also the polarisation of the emitted radiation.

IX-3.2. *Angular distribution of unpolarised radiation from polarised nuclei*

Suppose that the initial nuclear state has been polarised by some mechanism and that the probability * of finding the nucleus in the substate $|IM)$ is $p(M)$. The resulting angular distribution of the emitted radiation is then given by

$$W = \sum_{M\mu} p(M)\, W_{M,M-\mu}\,. \qquad \text{(IX-21)}$$

Now for a plane wave with the electric field polarised at an angle χ to the direction defined by the intersection of the plane normal to $\boldsymbol{k}$ and the plane

* The assumed noncoherence of the substates $|IM)$ is valid for the situations considered in this chapter. See Biedenharn and Rose (1953) for a discussion of this point.

of $\boldsymbol{k}$ and the axis of quantisation, the polarisation function $f(P)$ is given by (Rose (1955))

$$f(P) = (1/\sqrt{2}) \exp(-\mathrm{i} P\chi) . \tag{IX-22}$$

Thus

$$G^{\nu}_{LL'} = \delta_{L+L'+\nu,\text{even}}$$

$$H^{\nu}_{LL'} = \delta_{L+L'+\nu,\text{odd}}$$

$$I^{\nu}_{LL'} = (\cos 2\chi)\,\delta_{L+L'+\nu,\text{even}} - \mathrm{i}\,(\sin 2\chi)\,\delta_{L+L'+\nu,\text{odd}}$$

$$J^{\nu}_{LL'} = (\cos 2\chi)\,\delta_{L+L'+\nu,\text{odd}} - \mathrm{i}\,(\sin 2\chi)\,\delta_{L+L'+\nu,\text{even}} \, . \tag{IX-23}$$

However, if polarisation insensitive detectors are used for the detection of the radiation, then an average has to be taken over χ, giving

$$(I^{\nu}_{LL'})_{\text{average}} = (J^{\nu}_{LL'})_{\text{average}} = 0 \, . \tag{IX-24}$$

The resulting angular distribution then simplifies to the following form

$$W(\theta) = \pi^2 \sum_{LL'\nu} B_\nu(I) F_\nu(LL'I'I)[\delta_{L+L'+\nu,\text{even}}(m^{\times}_L m_{L'} + e^{\times}_L e_{L'})$$

$$+ \delta_{L+L'+\nu,\text{odd}}(m^{\times}_L e_{L'} + e^{\times}_L m_{L'})] P_\nu(\cos\theta) \tag{IX-25}$$

where $B_\nu(I)$ is given by

$$B_\nu(I) = \sum_M (2\nu+1)^{1/2} C(I\nu I; M0)\, p(M) \tag{IX-26}$$

and is a parameter introduced by various authors (e.g. de Groot *et al.* (1965)) to describe the nature of the nuclear orientation. For $\nu = 0, 1$ it has the following explicit forms

$$B_0(I) = 1 \; ; \quad B_1(I) = \frac{\sqrt{3} \sum_M M p(M)}{[I(I+1)]^{1/2}} \, . \tag{IX-27}$$

Here it will be noted that $B_1(I)$ gives a direct measure of the nuclear polarisation.

The factor $F_\nu(LL'I'I)$ is defined by

$$F_\nu(LL'I'I) = (-)^{I'+3I-1}[(2I+1)(2L+1)(2L'+1)]^{\frac{1}{2}}C(LL'\nu;1-1)W(LL'II;\nu I'). \tag{IX-28}$$

The F-coefficients are frequently used in the representation of angular distributions and have been tabulated in various places (e.g. Alder *et al.* (1957)); Frauenfelder and Steffen (1965a)).

Inspection of eq. (IX-25) shows that the terms signifying parity violation, namely those with ν odd, can occur in two ways. Firstly there can be interference between multipoles of like character (i.e. electric or magnetic) but whose orders differ by an odd number (e.g. E1–E2 interference). Secondly there can be interference between multipoles of unlike character but which have the same multipolarity (e.g. E1–M1) or multipolarity differing by an even number (e.g. E2–M4). In practice the situation most likely to lead to maximum effect is the case of interference between electric and magnetic multipoles of the same order.

The actual magnitude of the asymmetry depends not only on the actual fact of parity violation, but also on the non-vanishing of the $B_\nu(I)$ with ν odd and obviously these terms should be as large as possible. Three ways for producing $B_\nu(I)$ with ν odd are discussed in the following sections.

IX-3.3. *Polarisation by non-nuclear methods*

Two groups of methods for orienting nuclei can be distinguished which depend essentially on the interaction of the nuclear electromagnetic moments with electromagnetic fields. On the one hand are the methods which depend on the separation in energy of nuclear magnetic substates and then ensure that these are unequally populated by reducing the temperature of the system so far that the low-lying states become preferentially populated, i.e. the $p(M)$ of eqs. (IX-21) and (IX-27) are not equal to one another. On the other hand, there are the optical and microwave methods which depend on the atomic absorption and emission of radiation; in these preferential nuclear magnetic substate populations may be obtained by emission of atomic radiation through different channels from those by which it was absorbed.

These methods have been reviewed and the orientation parameters discussion by de Groot *et al.* (1965). So far little accurate experimental work involving nuclear gamma transitions has been carried out using these methods although with the advent of continuously operating helium-dilution refrigerators this state of affairs will not continue (see for example the recent measurements in ^{159}Tb by Pratt *et al.* (1970) and in ^{192}Pt by Holmes *et al.* (1971)).

IX-3.4. *β–γ angular correlations*

Since it is now well established that parity is not conserved in nuclear beta-decay (see §VI-4) it follows that measurement of the direction of the emitted β-ray leads to a polarisation of the final nuclear state ($I \neq 0$) following the transition (corresponding to a pseudoscalar term of the form $\boldsymbol{I} \cdot \boldsymbol{p}$). Thus, if parity is not conserved in a succeeding gamma-transition a forward-backward asymmetry is to be expected in the β–γ angular correlation.

We now restrict ourselves to the case of an allowed β-decay $I_\beta \to I$ where I_β is the spin of the β-emitting state and I that of the daughter (γ-emitting) state. In such a decay the angular momentum restrictions are such that only the orientation parameters $B_0(I)$ and $B_1(I)$ are non-vanishing. From eq. (IX-27), $B_0(I) = 1$. Further $B_1(I)$ can be determined from the standard expression for the angular distribution of β-particles from polarised nuclei (e.g. Jackson *et al.* (1957b); see also the discussion for the particular case $I_\beta = I$ in §III-4.2.1) using time reversal invariance arguments which can relate asymmetries and polarisation (e.g. Satchler (1958); see also §X-6)). The following result is obtained in the case of an allowed β-decay. for the polarisation P of the daughter nucleus along the direction of emission of the β-particles

$$P = \frac{1}{I} \sum_M M p(M) = -A \frac{I+1}{3I} \frac{p}{E} \tag{IX-29}$$

where p and E are the momentum and energy of the emitted β-particle and $-A$ is given by

$$-A = \left[\pm \lambda_{I_\beta I} + 2\delta_{I_\beta I} \left(\frac{I}{I+1} \right)^{\frac{1}{2}} \right] [1 + y^2]^{-1} . \tag{IX-30}$$

Here,

$$\begin{aligned} \lambda_{I_\beta I} &= 1 && \text{for} \quad I_\beta = I - 1 \\ &= 1/(I+1) && \text{for} \quad I_\beta = I \\ &= -I/(I+1) && \text{for} \quad I_\beta = I + 1 \end{aligned}$$

and $y = (G_\mathrm{V} \langle t_\pm \rangle)/(G_\mathrm{A} \langle \boldsymbol{\sigma} t_\pm \rangle)$ where $\langle t_\pm \rangle$ and $\langle \boldsymbol{\sigma} t_\pm \rangle$ are the usual allowed β-decay matrix elements. The upper sign in the foregoing expression refers to

β^--decay and the lower to β^+-decay. In the above it has been assumed that the β-decay interaction is time reversal invariant and has a V–A form.

From eqs. (IX-27), (IX-29) and (IX-30) it therefore follows that

$$B_1(I) = \frac{p}{E}\left(\frac{I+1}{3I}\right)^{\frac{1}{2}}\left[\pm\lambda_{I_\beta I} + 2\delta_{I_\beta I}|y|\left(\frac{I}{I+1}\right)^{\frac{1}{2}}\right][1+|y|^2]^{-1}. \qquad \text{(IX-31)}$$

Substitution of the above expression for $B_1(I)$ into the general formula for the angular distribution $W(\theta)$, given in eq. (IX-25) for the case of *ML–EL* interference gives

$$W(\theta) = \pi^2(|m_L|^2 + |e_L|^2)[1+\alpha\cos\theta] \qquad \text{(IX-32)}$$

where

$$\alpha = \frac{2p}{E}\left\{\left[\pm\lambda_{I_\beta I} + 2\delta_{I_\beta I}|y|\left(\frac{I}{I+1}\right)^{\frac{1}{2}}\right][1+y^2]^{-1}\right\} F_1(LLI'I)\frac{m_L^\times e_L}{|m_L|^2+|e_L|^2}.$$

In writing α in the above form the fact that m_L and e_L are relatively real has been used. It is now a straightforward matter to use eq. (IX-32) for any particular β–γ angular correlation.

Krüger (1959) has obtained a similar expression to eq. (IX-32) and has also investigated the dependence of α on the value of y. His results indicate that for certain values of y α can be vanishingly small so that it is important that the decays used for studying parity violating effects by this method are suitably chosen.

IX-3.5. *Capture of polarised thermal neutrons*

A method of producing polarised nuclei particularly suitable for investigating parity violating effects and involving the capture of polarised thermal neutrons was first suggested by Haas *et al.* (1959). Capture of polarised S-wave neutrons leads to a compound nuclear state which retains most (all in the case of capture by a spin 0 nucleus) of the polarisation of the incoming neutron. Any forward-backward asymmetry of subsequent gamma rays emitted by the compound nucleus relative to the direction of polarisation is then indicative of parity non-conservation. The detailed form of this asymmetry is again obtained by calculating the forms of the $B_\nu(I)$ for this case and inserting them into the general expression (eq. (IX-25)) for the angular distribution of radiation from a polarised nucleus.

Let the spin of the initial nuclear state (before capture) be I_i and let the spin of the compound (γ-emitting) state be I. If $p_n(\mu)$ is the probability of the neutron having a z-component μ ($\mu = \pm\frac{1}{2}$), the orientation weighting factor $p(M)$ for the compound state is given by

$$p(M) = \frac{2}{2I+1} \sum_{\mu} p_{\mathrm{n}}(\mu) |C(I_i \tfrac{1}{2} I; M-\mu\mu)|^2 . \tag{IX-33}$$

Inserting this expression into eq. (IX-27), then gives *:

$$B_0(I) = 1; \qquad B_1(I) = \frac{\frac{3}{4} + I(I+1) - I_i(I_i+1)}{[3I(I+1)]^{\frac{1}{2}}} P_{\mathrm{n}} \tag{IX-34}$$

with all other $B_\nu(I)$ zero. In the above expression P_{n} is the polarisation of the captured neutrons, given by

$$P_{\mathrm{n}} = \frac{p_{\mathrm{n}}(\frac{1}{2}) - p_{\mathrm{n}}(-\frac{1}{2})}{p_{\mathrm{n}}(\frac{1}{2}) + p_{\mathrm{n}}(-\frac{1}{2})} . \tag{IX-35}$$

Substitution of $B_0(I)$ and $B_1(I)$ in the general formula (eq. (IX-25)) for the angular distribution $W(\theta)$ in the case of ML–EL interference, gives

$$W(\theta) \propto 1 + \alpha \cos\theta \tag{IX-36}$$

where the asymmetry factor α is given by:

$$\alpha = 2P_{\mathrm{n}} \frac{\frac{3}{4} + I(I+1) - I_i(I_i+1)}{[3I(I+1)]^{\frac{1}{2}}} F_1(LLI'I) \frac{m_L e_L^{\times}}{|m_L|^2 + |e_L|^2} . \tag{IX-37}$$

IX-3.6. *Emission of circularly polarised γ-radiation*

In the case of a γ-emitting nuclear system which is unpolarised, the radiation will be circularly polarised if parity violating effects are present. This corresponds, for example, to the longitudinal polarisation of electrons observed in the β-decay of unpolarised nuclei. For generality we consider the case of an arbitrarily oriented nuclear system and then specialise to the case of unpolarised nuclei. The former case might, in fact, be of interest since an orientation may be produced in forming a suitable γ-emitting state (Wilkinson (1958a)).

The angular distribution of left circularly polarised radiation $W_L(\theta)$ from

* Note that there is a sign error in eq. (24) of Blin-Stoyle (1960b).

oriented nuclei is given by taking (see eq. (IX-16) *et seq.*) $f(1) = 1$ and $f(-1) = 0$ in eqs. (IX-17) to (IX-19), and then using eq. (IX-21). Similarly the distribution of right circularly polarised radiation $W_R(\theta)$ is given by following the same procedure but with $f(1) = 0$ and $f(-1) = 1$. Using the methods already described, the following expression is obtained for the degree δ of circular polarisation:

$$\delta = \frac{W_L(\theta) - W_R(\theta)}{W_L(\theta) + W_R(\theta)}$$

$$= \frac{\sum_{LL'\nu} B_\nu(I) F_\nu(LL'I'I)[\delta_{L+L'+\nu,\mathrm{odd}}(m_L^\times m_{L'} + e_L^\times e_{L'}) + \delta_{L+L'+\nu,\mathrm{even}}(m_L^\times e_{L'} + e_L^\times m_{L'})] P_\nu(\cos\theta)}{\sum_{LL'\nu} B_\nu(I) F_\nu(LL'I'I)[\delta_{L+L'+\nu,\mathrm{even}}(m_L^\times m_{L'} + e_L^\times e_{L'}) + \delta_{L+L'+\nu,\mathrm{odd}}(m_L^\times e_{L'} + e_L^\times m_{L'})] P_\nu(\cos\theta)} .$$

(IX-38)

In the particularly simple case that the initial nucleus is unpolarised, $B_0(I) = 1$ and $B_\nu(I) = 0$ for $\nu \neq 0$. Further, $F_0(LL'I'I) = \delta_{LL'}$. The foregoing expression for δ then reduces to

$$\delta = 2 \sum_L m_L^\times e_L \Big/ \sum_L (|m_L|^2 + |e_L|^2) . \qquad \text{(IX-39)}$$

An effect is therefore only obtained if there is interference between electric and magnetic multipoles of the same order.

This approach to the investigation of parity violating effects is particularly important from the experimental point of view since it does not require an angular correlation measurement – only a detector sensitive to circular polarisation.

IX-3.7. *Longitudinal polarisation of internal comversion electrons*

In the same way that the longitudinal polarisation of electrons in β-decay signifies parity violation, a longitudinal polarisation of internal conversion electrons will result from parity impurities in the nuclear states concerned. This was first pointed out by Michel (1964) who stressed that the effect could be significant because of the large difference (as large as 10^3 or even greater) in the internal conversion coefficients for *EL* and *ML* radiation. We define the longitudinal polarisation P_e as follows:

$$P_e = \frac{W_e(+\frac{1}{2}) - W_e(-\frac{1}{2})}{W_e(+\frac{1}{2}) + W_e(-\frac{1}{2})} \tag{IX-40}$$

where $W_e(m_s)$ is the intensity of internal conversion electrons with spin component m_s along their direction of momentum. For the case of a regular EL transition and an irregular ML transition the magnitude of P_e for K-conversion electrons is then expected to be of the order

$$P_e \approx \left(\frac{\beta_L}{\alpha_L}\right)^{1/2} \frac{m_L}{e_L} \tag{IX-41}$$

where β_L and α_L are the K-internal conversion coefficients for the ML and EL conversion respectively.

Carhart (1967) has studied the situation in more detail and has obtained a more exact expression for P_e. He has also shown that in the limit of large transition energies the value of P_e approaches the value of the corresponding γ-ray circular polarisation.

IX-3.8. *Intensity ratio of the components of the hyperfine spectrum*

An alternative experimental way of studying parity-violating effects in gamma transitions is to use the Mössbauer effect * to compare the intensities of two transitions of the type $|IM\rangle \to |I'M-\mu\rangle$ and $|I-M\rangle \to |I'-M+\mu\rangle$ between nuclear substates. In these two cases the 3-component of angular momentum carried away by the emitted radiation has the values $+\mu$ and $-\mu$ respectively. Inspection of eqs. (IX-17) and (IX-18) shows that the parity violating terms (ν odd) have opposite signs in these two cases so that the intensities of the two lines (I_+ and I_- respectively) in the same direction are different if there are parity impurities in the nuclear states concerned and provided that the odd ν Legendre functions do not vanish in that direction. The magnitude of the effect ($R = (I_+ - I_-)/(I_+ + I_-)$) is directly related to the magnitude of the circular polarisation of the emitted radiation. Unfortunately, this method is limited to nuclei for which the Mössbauer effect can be observed.

* The Mössbauer effect, but with a polarisation condition, has also been used to study T-violation (see §X-7.4).

IX-4. Parity violation in electromagnetic transitions – detailed results and comparison with experiment

In the foregoing sections different methods have been described for investigating parity violating effects in nuclear electromagnetic transitions. These effects are all dependent on the relative size of the relevant irregular and regular multipole matrix elements (e_L and m_L). The regular matrix element can be calculated by standard procedures given the appropriate nuclear wave functions and in any case, is generally known from the experimental data. It will not be further described here. In order to calculate the irregular matrix element use must be made of an expression for the nuclear wave function of the type given in eq. (IX-3) namely

$$\Psi_i = \psi_i + \sum_{j \neq i} \frac{\langle j | V_{\text{p.v.}} | i \rangle}{E_i - E_j} \psi_j \qquad \text{(IX-42)}$$

where ψ_i and ψ_j have opposite parities.

To proceed further it is now necessary to be more specific. There are essentially three cases to consider. Firstly there is the situation in which the electromagnetic transition is between low-lying nuclear states of a complex nucleus, as is usually the case in situations where the γ-ray follows a beta-transition. Secondly, there is the special case of transitions in "few-nucleon" nuclei. Thirdly, there is the situation in which the transition is between a state of relatively high energy ($\approx$ 8 MeV) in a complex nucleus and a low-lying state as in radiative neutron capture. In the first case the calculation can be carried out using wavefunctions generated by some appropriate nuclear model. In the second, a more direct approach is possible. In the third, the capture state and the total situation are much more complicated and only crude theoretical calculations are possible. We consider these three cases separately.

IX-4.1. *Parity violating γ-transitions between low-lying states in complex nuclei*

Consider an electromagnetic transition between the nuclear states Ψ_1 and Ψ_2 where, allowing for parity violating effects,

$$\Psi_1 = \psi_1 + \sum_{i \neq 1} \frac{\langle i | V_{\text{p.v.}} | 1 \rangle}{E_1 - E_i} \psi_i \,; \qquad \Psi_2 = \psi_2 + \sum_{j \neq 2} \frac{\langle j | V_{\text{p.v.}} | 2 \rangle}{E_2 - E_j} \psi_j \,. \qquad \text{(IX-43)}$$

To first order in the parity violating potential the regular multipole matrix element can be written

$$\langle \Psi_2 | O_L | \Psi_1 \rangle \triangleq \langle \psi_2 | O_L | \psi_1 \rangle \equiv \langle 2 | O_L | 1 \rangle \tag{IX-44}$$

where the multipole operator O_L signifies $\mathcal{H}(A_L(e))$ or $\mathcal{H}(A_L(m))$ as the case may be.

Correspondingly, the irregular multipole matrix element has the form

$$\langle \Psi_2 | O_L' | \Psi_1 \rangle \triangleq \sum_{i \neq 1} \frac{\langle 2 | O_L' | i \rangle \langle i | V_{\text{p.v.}} | 1 \rangle}{E_1 - E_i} + \sum_{j \neq 2} \frac{\langle 2 | V_{\text{p.v.}} | j \rangle \langle j | O_L' | 1 \rangle}{E_2 - E_j} \tag{IX-45}$$

where O_L' is the irregular multipole operator and has the opposite parity to O_L.

To calculate the irregular multiple matrix element then requires some assumed form for $V_{\text{p.v.}}$, appropriate model wave functions for the states 1, 2, i and j and some method of dealing with the sum over the states i and j which contribute to the parity impurity.

The first detailed calculations of irregular multipole matrix elements and the magnitude of the related parity violating effects were by Blin-Stoyle and his collaborators (e.g. Blin-Stoyle (1960b) – general calculations; Blin-Stoyle and Spector (1961) – calculation in ^{7}Li; Blin-Stoyle and Maqueda (1966) – calculation in ^{181}Ta; Maqueda (1966) – calculation in ^{19}F.) These calculations all used the static parity violating potential given in eq. (VIII-2) and because of its restricted form and the more recent developments in the calculation of $V_{\text{p.v.}}$ (see Chapt. VIII) the detailed results given in these papers should only be taken to indicate the order of magnitude of the effect.

An important step forward in calculations of this sort was taken by Michel (1964) who considered the effects of a single particle parity violating potential of the form given in eq. (VIII-24), namely,

$$V_{\text{p.v.}} = W_M = G_M' \left[1 + \frac{N-Z}{A} \tau_3 \right] \boldsymbol{\sigma} \cdot \boldsymbol{p} \,. \tag{IX-46}$$

Taking a simple model Hamiltonian of the form

$$H = p^2/2M + V(r) + G_M'' \boldsymbol{\sigma} \cdot \boldsymbol{p} = H_0 + G_M'' \boldsymbol{\sigma} \cdot \boldsymbol{p} \tag{IX-47}$$

it is straightforward to show that it can be solved to order $G_M''^2$ by the transformation

$$H = \mathrm{e}^{\mathrm{i}S} H_0 \mathrm{e}^{-\mathrm{i}S}\ ; \qquad \Psi_k = \mathrm{e}^{\mathrm{i}S} \psi_k \tag{IX-48}$$

where $S = MG_M'' \boldsymbol{\sigma} \cdot \boldsymbol{r}$ and the ψ_k are eigenstates of H_0. Correspondingly the matrix element of an irregular multipole operator O_L' between two states Ψ_1 and Ψ_2 can be written

$$\langle \Psi_2 | O_L' | \Psi_1 \rangle = \langle 2 | \mathrm{e}^{-\mathrm{i}S} O_L' \mathrm{e}^{\mathrm{i}S} | 1 \rangle \triangleq \langle 2 | -\mathrm{i}[S, O_L'] | 1 \rangle \text{ to order } G_M''^2 \tag{IX-49}$$

since $\langle 2 | O_L' | 1 \rangle = 0$ because of the irregular nature of O_L'. Thus the irregular multipole matrix element reduces to that of an "effective" multipole operator $(-\mathrm{i}\,[S, O_L'])$, taken between the two zero-order states concerned. In this approximation, therefore, the sum over states in eq. (IX-45) is avoided and the calculation much simplified.

To first order in G_M'' it follows from eq. (IX-48) that the wave function Ψ_k has the form

$$\Psi_k = \psi_k + \mathrm{i} MG_M'' \boldsymbol{\sigma} \cdot \boldsymbol{r} \psi_k \tag{IX-50}$$

so that, referring to eq. (IX-4), a convenient measure of the parity impurity $\mathcal{F}$ is

$$\mathcal{F} = MG_M'' R \tag{IX-51}$$

where R is the nuclear radius. Typical values are

$A = 16;\ \mathcal{F} = 3.1 \times 10^{-7}$

$A = 160;\ \mathcal{F} = 8.0 \times 10^{-7}$ for an extra proton

$\mathcal{F} = 5.5 \times 10^{-7}$ for an extra neutron .

A further consequence is that irregular electric multipole elements are vanishingly small since, apart from a very small component, the electric multipole operator commutes with S. This particular result only holds if H_0 is spin independent, however, and is no longer valid if, for example, the usual shell-model spin-orbit potential is introduced into H_0. It should also be noted that the result embodied in eq. (IX-49) does not depend on the potential $V(\boldsymbol{r})$ being spherically symmetrical. The result can therefore be applied to both spherical and deformed nuclei.

Michel (1964) used this method to calculate the general form of asymmetry effects in varying circumstances. In particular, using Nilsson (1955) wave-functions he evaluated the magnitude of these effects for a number of distorted nuclei (^{173,175,177}Lu, ^{181}Ta, ^{177,179}Hf) in which the regular transition is E1 and the irregular M1 (an E1(M1) transition). Wahlborn (1965) (see also Grabowski (1970)) developed and firmed up Michel's work using the same effective single-particle potential and allowing for a spin dependent Hamiltonian and produced numerical results for a number of transitions (in ^{181}Ta, ^{175}Lu and ^{57}Fe) in which the regular transition is M1 and the irregular E1 (an M1(E1) transition). Similar calculations to those of Wahlborn (1965) were carried out by Szymanski (1966) for an M1(E1) transition in ^{203}Tl. These calculations all lead to values for asymmetries (e.g. circular polarisation) proportional to the strength G'_M of the single particle parity violating potential W_M. Because of this it is straightforward to take over the results obtained by Michel (1964), Wahlborn (1965) and Szymanski (1966) and adjust them for other assumed strengths for the single particle potential. This has been done by McKellar (1968a, b; 1969a, b; 1970) and McKellar and Rajaraman (1968) for electromagnetic transitions in ^{175}Lu, ^{181}Ta and ^{203}Tl using the more general single particle potential given in eq. (VIII-25) which allows for short range correlation effects and enables different theories of the fundamental weak interaction to be allowed for. The results obtained by McKellar are included in table IX-3 where a comparison is made with experimental data.

Of course a significant approximation is made in representing the effects of a parity violating two-body potential by an effective one-body potential. Recently, a number of calculations have been made using two-body potentials directly. Here the difficulty arises that the sum over states in eq. (IX-45) has to be carried out explicitly and inevitably has to be limited to a sum over states close in energy, to the perturbed state. Calculations of this type using $V^{(\pi)}_{\text{p.v.}}$ and $V^{(\rho)}_{\text{p.v.}}$ (see eqs. (VIII-10) and (VIII-14)) have been carried out by Henley (1968) – for ^{10}B, ^{14}N, ^{18}F; Vinh Mau and Bruneau (1969) – for ^{181}Ta; Wambach *et al.* (1970) – for ^{18}F; Eman and Tadić (1971) – for ^{175}Lu and ^{181}Ta; Desplanques and Vinh Mau (1971) – for ^{175}Lu and ^{181}Ta; Gari *et al.* (1971b) – for ^{181}Ta; Vogel (1971) – for ^{159}Tb and ^{180}Hf; Dumitrescu *et al.* (1971a) – for ^{175,177}Lu; Dumitrescu *et al.* (1971b) – for odd mass deformed nuclei in general. A calculation using a potential of the type given in eq. (VIII-2) has been carried out by Blin-Stoyle and Maqueda (1966) – for ^{181}Ta. The results obtained by some of these authors are referred to in §IX-4.1.1.

IX-4.1.1. *Comparison between theory and experiment*

During the last few years a large number of experiments have been carried out which measure the different asymmetries discussed in this chapter and which are indicative of parity violation in the electromagnetic transition concerned. The more recent and accurate of these experimental results (which mainly consist of γ-ray circular polarisation measurements) are gathered together in table IX-2 together with a description of the transitions involved. The details of and techniques used in these experiments have been extensively reviewed by Hamilton (1968) (see also Henley (1969); Hamilton (1969)).

The various transitions are chosen for investigation bearing in mind three factors. Firstly, they should be suitable from a technical viewpoint so that it is possible to carry the experiment out and obtain accuracies of the required order of magnitude. Secondly, there should, if possible, be nearby states to those between which the transition takes place having opposite parity so leading to small energy denominators in eq. (IX-45). Thirdly, the regular transition should be hindered as much as possible with respect to the irregular transition so that the effect being measured is amplified significantly (see the end of §IX-1). For example, in the much used 482 keV transition in ^{181}Ta, the regular M1 transition is hindered by about 3×10^6. Even so, experimental accuracies of at least 10^{-3} and, in general, of the order 10^{-5} have to be achieved. Inevitably many spurious effects (e.g. electronic instability, bremsstrahlung, Compton scattering etc.) have to be allowed for and the final statistical analysis has to be carried out very carefully (see e.g. Hamilton (1969)). The disagreements and inconsistencies between some of the results given in table IX-2 clearly indicate the difficulties of this sort of work. Nevertheless, there seems to be no question now but that parity violating effects have been observed in different nuclei.

The most fully investigated decay, both experimentally and theoretically, is the 482 keV transition in ^{181}Ta and it is of interest to compare theory and experiment in this case. Consider first those calculations which have been carried out using the effective single particle parity violating potential discussed in §VIII-6, namely

$$W = G' \left[1 + \frac{N-Z}{A} \tau_3\right] \boldsymbol{\sigma} \cdot \boldsymbol{p} \tag{IX-50}$$

where $G' \approx [-0.4\,\alpha + 3.3 \times 10^{-6} n_-^0]\, G'_M$. The quantities α, n_-^0 (which refer to the strengths of $V^{(\rho)}_{\text{p.v.}}$ and $V^{(\pi)}_{\text{p.v.}}$ respectively) and G'_M are defined in §§VIII-4 and VIII-6. McKellar (1970 – see also 1968a, b; 1969a) has evaluated W as a multiple of the Michel (1964) potential W_M

Table IX-2
Experimental results for parity-violating γ-transitions.

Nucleus	Regular transition	Irregular transition	Energy (keV)	Measured asymmetry	Reference
^{19}F	$\frac{1}{2}^- \xrightarrow{E1} \frac{1}{2}^+$	M1	110	$\alpha = +(4.3 \pm 5.2) \times 10^{-4}$	Moline *et al.* (1970)
^{41}K	$\frac{7}{2}^- \xrightarrow{M2} \frac{3}{2}^+$	E2	1290	$\delta = +(1.9 \pm 0.3) \times 10^{-5}$	Lobashov *et al.* (1969)
^{57}Fe	$\frac{3}{2}^- \xrightarrow{M1} \frac{1}{2}^-$	E1	14.4	$R = +(2 \pm 6) \times 10^{-5}$	Kankeleit (1964)
^{159}Tb	$\frac{5}{2}^- \xrightarrow{E1} \frac{3}{2}^+$	M1	363	$\alpha = +(4.5 \pm 3.0) \times 10^{-3}$ (for 100% Tb polarisation)	Pratt *et al.* (1970)
				$\delta = -(1 \pm 5) \times 10^{-4}$	Lipson *et al.* (1971a, b)
^{175}Lu	$\frac{5}{2}^+ \xrightarrow{M1+E2} \frac{7}{2}^+$	E1	343	$\delta = +(2 \pm 3) \times 10^{-5}$	Boehm and Kankeleit (1968)
				$= -(0.15 \pm 0.60) \times 10^{-5}$	Bock and Jenschke (1970)
	$\frac{9}{2}^- \xrightarrow{E1+M2} \frac{7}{2}^+$	M1	396	$\delta = +(4 \pm 1) \times 10^{-5}$	Lobashov *et al.* (1966, 1967)
				$= +(6.3 \pm 1.0) \times 10^{-5}$	Vanderleeden and Boehm (1969, 1970)
^{180}Hf	$8^- \xrightarrow{M2} 6^+$	E2	501	$\delta = -(1.4 \pm 0.7) \times 10^{-3}$	Povel (1968); Bock (1969)
				$= -(2.80 \pm 0.45) \times 10^{-3}$	Jenschke and Bock (1970)
				$= -(2.3 \pm 0.6) \times 10^{-3}$	Lipson *et al.* (1971a, b)
				$= -(2.8 \pm 0.45) \times 10^{-3}$	Krane *et al.* (1971a, b)
				$= -(2.0 \pm 0.4) \times 10^{-3}$	Kuphal (1972)
^{181}Ta	$\frac{5}{2}^+ \xrightarrow{M1+E2} \frac{7}{2}^+$	E1	482	$\delta = -(0.6 \pm 0.1) \times 10^{-5}$	Lobashov *et al.* (1966, 1967)
				$= -(1.0 \pm 4.0) \times 10^{-5}$	Boehm and Kankeleit (1968)
				$= -(2.0 \pm 4.0) \times 10^{-5}$	Bock (1969)
				$= -(9.0 \pm 6.0) \times 10^{-5}$	Cruse and Hamilton (1969)
				$= -(2.8 \pm 0.6) \times 10^{-5}$	Bodenstedt *et al.* (1969a, b)
				$= -(1.3 \pm 0.7) \times 10^{-5}$	Diehl *et al.* (1969)

Table IX-2 (continued)

Nucleus	Regular transition	Irregular transition	Energy (keV)	Measured asymmetry	Reference
				$\delta = -(0.39 \pm 0.12) \times 10^{-5}$	Vanderleeden and Boehm (1969, 1970)
				$= +(7.0 \pm 7.0) \times 10^{-5}$	van Rooijen *et al.* (1967)
				$= -(0.41 \pm 0.13) \times 10^{-5}$	Bock and Jenschke (1970)
				$= -(2.1 \pm 1.1) \times 10^{-5}$	De Saintignon *et al.* (1970)
				$= -(0.31 \pm 0.25) \times 10^{-5}$	Lipson *et al.* (1971a, b)
				$= +(0.1 \pm 0.4) \times 10^{-5}$	Kuphal (1972)
^{182}W	$2^- \xrightarrow{E1+M2} 2^+$	M1 + E2	1189	$\delta = -(0.25 \pm 0.40) \times 10^{-4}$	Lipson *et al.* (1971b)
				$\alpha = -(2.8 \pm 1.7) \times 10^{-4}$	Krane *et al.* (1972)
^{203}Tl	$\frac{3}{2}^+ \xrightarrow{M1+E2} \frac{1}{2}^+$	E1	273	$\delta = -(2 \pm 3) \times 10^{-5}$	Boehm and Kankeleit (1968)
				$= -(5.6 \pm 7.6) \times 10^{-5}$	Baker and Hamilton (1970, 1971) *
				$= -(3.0 \pm 1.8) \times 10^{-5}$	de Saintignon and Chabre (1970)
				$= -(0.4 \pm 1.0) \times 10^{-5}$	Lipson *et al.* (1971b)
				$= -(0.2 \pm 0.5) \times 10^{-5}$	VanderLeeden *et al.* (1971)
				$= +(0.1 \pm 0.7) \times 10^{-5}$	Kuphal (1972)
				$= -(25.2 \pm 6.5) \times 10^{-5}$	Dydak *et al.* (1971) *

* This experiment involved the measurement of a $\beta-\gamma$ angular correlation (i.e. a measurement of α in eq; (IX-32) and the value of δ quoted here was obtained by using the expression for α in eq. (IX-32) in combination with that for δ in eq. (IX-39);

$$W_M = G'_M \left[1 + \frac{N-Z}{A}\tau_3\right] \boldsymbol{\sigma} \cdot \boldsymbol{p} \tag{IX-51}$$

for the various weak interaction models (and their corresponding values of α and n_-^0) listed in table IX-3 together with a CP-violating theory due to Oakes (1968) and assuming that α takes the value appropriate to the factorisation approach (see §VIII-5). These results are given in table IX-3 together with the corresponding values of δ, the latter being derived from the work of Wahlborn (1965) by direct proportional adjustment by the factor W/W_M.

Inspection of the results and comparison with the experimental values listed in table IX-2 shows that certain theories (notably those of d'Espagnat (1963), Segré (1968) – γ_5 non-invariant, and Oakes (1968)) gives values for δ approaching an order of magnitude larger than experiment. Further, the Cabibbo (1963) theory, for example, gives the wrong sign for δ. At face value, therefore, these theories are ruled out. However, there are many uncertainties both of a fundamental kind (e.g. those concerning the strength of $V_{\text{p.v.}}^{(\rho)}$ and therefore $W^{(\rho)}$ discussed in §VIII-5) and a nuclear physics kind (e.g. use of an effective single particle potential, nuclear wavefunctions etc.). These uncertainties mean that no firm conclusions should be drawn at this point. On the other hand table IX-3 does illustrate the potential power of this approach as a means of studying the weak interaction.

As mentioned in the previous section some calculations on the ^{181}Ta decay have also been carried out using a two body potential. Vinh Mau and Bruneau (1969) using $V_{\text{p.v.}}^{(\pi)}$ alone obtained $|\delta| \approx 0.1 \times 10^{-5}$ on the basic Cabibbo theory to be compared with a value $|\delta| \approx 0.6 \times 10^{-5}$ using only $|W_\pi|$. There is, therefore, a significant difference in the one-body and two-body calculation of the effect of $V_{\text{p.v.}}^{(\pi)}$. Eman and Tadić (1971) use both $V_{\text{p.v.}}^{(\pi)}$ and $V_{\text{p.v.}}^{(\rho)}$ and also allow approximately for the effects of short range correlations. For the conventional Cabibbo theory they obtain $\delta = 8.97^{+8.25}_{-4.07} \times 10^{-6}$ or $6.12^{+5.63}_{-1.87} \times 10^{-6}$ according to the relative signs of $V_{\text{p.v.}}^{(\pi)}$ and $V_{\text{p.v.}}^{(\rho)}$. These values are significantly smaller than experiment and of the opposite sign. They also investigate the values of δ for other weak interaction models. In addition they consider the effects of making the parity violating potentials gauge invariant (see also Gari and Huffman (1971) and Fischbach and Tadić (1972)), a point which was first discussed by Michel (1964). The situation is not straightforward, but has the effect of introducing additional parity violating electromagnetic interactions into the nuclear Hamiltonian which do not, however, contribute to EL transitions (Fischbach and Tadić (1972); McKellar

Table IX-3
Theoretical values of the circular polarisation δ for ^{181}Ta on various weak interaction theories.
(W is written in the form $W = W_\rho \pm |W_\pi|$ allowing for the sign uncertainty in $V^{(\pi)}_{\text{p.v.}}$)

Model	Reference	W_ρ (units of W_M)	$\lvert W_\pi \rvert$ (units of W_M)	$W = W_\rho \pm \lvert W_\pi \rvert$ (units of W_M)		δ (units of 10^{-5})	
Conventional Cabibbo	Cabibbo (1963)	−0.4	0.1	−0.3	−0.5	+2	+3
Neutral current	d'Espagnat (1963)	+0.1	1.5	+1.6	−1.4	−11	+8
Segré, γ_5 invariant, schizon	Segré (1968)	−0.4	0	−0.4	−0.4	+2	+2
Segré, γ_5 non-invariant, schizon	Segré (1968)	−0.4	1.9	+1.5	−2.3	−9	+14
Lee, schizon	Lee (1968)	−0.2	0.7	+0.5	−0.9	−3	+5
Oakes, *CP* violating	Oakes (1968)	−0.2	3.2	+3.0	−3.4	−18	+22

(1972)) in agreement with Siegert's theorem (Siegert (1937); see also §XI-2). Further calculations by Gari *et al.* (1971a, b) and Desplanques and Vinh Mau (1971) which take much more careful account of pairing effects and short range correlations in the nuclear structure calculations lead to theoretical values one to three orders of magnitude smaller (according to the weak interaction model) than the experimental values for δ and, again, of the opposite sign. There is thus a considerable measure of ambiguity and uncertainty here and further work is necessary paying particular attention to gauge invariance, the effect of short range correlations and to off-mass shell effects (Henley (1971)) on $V^{(\pi)}_{\text{p.v.}}$.

We do not give here any further details of calculations relevant to other transitions listed in table IX-2 which have been measured experimentally since the foregoing discussion about ^{181}Ta amply illustrates the situation. There are, however, one or two transitions on which theoretical calculations have been made and which are particularly sensitive to certain features of the weak interaction.

Henley (1968) (– see also Wambach *et al.* (1970)) for example, takes up a point originally touched upon by Dashen *et al.* (1964) and suggests that a search should be made for parity-violating effects in E1 or M1 γ-transitions between levels having $T = 0$ in self conjugate light nuclei. Such transitions are expected to be particularly sensitive to the presence of $V^{(\pi)}_{\text{p.v.}}$. The argument goes simply as follows. If the initial and final states have $T = 0$, then only $T = 0, 1$ admixtures are important because of the $\Delta T \neq 2$ selection rule in electromagnetic transitions (see §XI-6). Both E1 and M1 transitions are inhibited in $T = 0 \rightarrow T = 0$ transitions (Trainor (1952); Radicati (1952); Morpurgo (1958)) and so an irregular transition resulting from a $T = 1$ admixture in either the initial or final states should be relatively enhanced. The $T = 1$ admixtures are, of course, induced by $V^{(\pi)}_{\text{p.v.}}$. Henley (1968) makes some detailed estimates of these effects and finds that the 2^- (5.105 MeV) $\xrightarrow{\text{E1}}$ 1^+ (0.71 MeV) transition in ^{10}B and 0^- (1.080 MeV) $\xrightarrow{\text{E1}}$ 1^+ (0.0 MeV) transition in ^{18}F are good candidates for detecting $V^{(\pi)}_{\text{p.v.}}$. In both cases the $V^{(\pi)}_{\text{p.v.}}$ contribution is estimated to be at least an order of magnitude larger than the $V^{(\rho)}_{\text{p.v.}}$ contribution.

IX-4.2. *Parity violating γ-transitions in "few-nucleon" systems*

One of the major theoretical handicaps which arises in the study of parity-violating electromagnetic transitions in complex nuclei is lack of detailed knowledge about nuclear wave functions. This handicap would be significantly reduced if measurements could be made on transitions in few-nucleon nuclei for which much more reliable information about wave functions is available.

For this reason a number of theoretical calculations have been carried out on the form and magnitude of the effects to be expected (Blin-Stoyle and Feshbach (1961); Partovi (1964); Danilov (1965, 1971); Dal'karov (1965); Tadić (1968); Bonar *et al.* (1968); Moskalev (1968, 1969); Feuer (1969); Hadjimichael and Fischbach (1971); Hadjimichael *et al.* (1971)). These calculations are concerned with processes such as $p + n \to d + \gamma, \gamma + d \to p + n$, $n + d \to {}^3H + \gamma$. A particularly interesting point in connection with the (n, p) capture process was noted by Danilov (1965). He showed that for the capture of unpolarised neutrons, the circular polarisation of the emitted photons depends only on the $T = 0, 2$ components of $V_{\text{p.v.}}$ (i.e. $V_{\text{p.v.}}^{(\rho)}$) whilst the asymmetry of the photons with respect to the direction of polarisation of captured neutrons depends only on the $T = 1$ components of $V_{\text{p.v.}}$ (i.e. $V_{\text{p.v.}}^{(\pi)}$). Hence, in principle, there is a means here of determining the relative size of these two parts of the weak potential.

Unfortunately, the magnitudes of the predicted effects turn out to be out of reach of experiment at the present time. For example, Hadjimichael and Fischbach (1971) have calculated (a) the asymmetry of the emergent proton about the plane of polarisation of linearly polarised incident photons for the case of deuteron photodisintegration ($\gamma + d \to n + p$) and (b) the degree of circular polarisation of emergent photons in the process $n + p \to d + \gamma$. In these calculations, the two parity violating potentials $V_{\text{p.v.}}^{(\rho)}$ and $V_{\text{p.v.}}^{(\pi)}$ are used based on the different weak interaction models mentioned in §VIII-4. They find that the parity violating effects are mainly attributable to $V_{\text{p.v.}}^{(\pi)}$ and lie in the range 10^{-9} to 10^{-7} according to the weak interaction model used. Their results are in general agreement with the estimates of Blin-Stoyle and Feshbach (1961) and Partovi (1964) but do not agree with those of Danilov (1965) and Tadić (1968). This disagreement can probably be attributed to the use of approximate wave functions for the initial and final states by the latter authors. Hadjimichael and Fischbach conclude that when experimental measurements on these processes of the required accuracy become available then it should be possible to discriminate between the different weak interaction models. The only measurement made so far is by Lobashov *et al.* (1970, 1972) who obtained for the circular polarisation of (n, p) capture gamma rays, the value $\delta = -(1.30 \pm 0.45) \times 10^{-6}$.

In the (n, p) capture process at very low energies the transitions which contribute to the parity violating effect are a regular M1(${}^1S_0$ ($T=1$) $\to$ ${}^3S_1$ ($T=0$)) transition and irregular E1(${}^3S_1$ ($T=0$) $\to$ ${}^3P_1$ ($T=1$) and ${}^1S_0$ ($T=1$) $\to$ $\to$ ${}^1P_1$ ($T=0$)) transitions. The magnitude of the effect is then proportional to the ratio of the E1 to M1 matrix elements. Blin-Stoyle and Feshbach (1961) pointed out that in the corresponding (n, d) capture process ($n + d \to {}^3H + \gamma$), where again the magnitude of the effect is proportional to the ratio of an irregular E1 to a regular M1 matrix element, an amplification might be expected since the M1 matrix element is inhibited. This was first demonstrated by Verde (1950) who showed that for a completely space symmetric ^{3}H wave function the M1 matrix element actually vanishes. Preliminary calculations of this effect by Moskalev (1969) and Hadjimichael *et al.* (1971) show that parity violating effects of the order 10^{-6} are to be expected. These effects, however, do not appear to be very sensitive to the weak interaction model used (Hadjimichael (1972)). An experimental investigation of this reaction would be welcome.

IX-4.3. *Parity violating γ-transitions in nuclei following neutron capture*

As described in §IX-3.5 parity violating effects can be detected by measuring the up–down asymmetry of gamma-rays following the capture of polarised thermal neutrons. This approach was first suggested by Haas *et al.* (1959) who carried out experiments on the process $n + {}^{113}Cd \rightarrow {}^{114}Cd + \gamma$. In this reaction a polarised 1^+ state at 9.05 MeV in ^{114}Cd is formed which has a small branch M1 decay to the 0^+ ground state. Interference with the corresponding irregular E1 transition can then lead to a parity violating symmetry. The form of this asymmetry is given by eqs. (IX-36) and (IX-37) and its precise magnitude depends on the value of the quantity $m_1 e_1^{\times}/(|m_1|^2 + |e_1|^2)$ $(\approx e_1/m_1)$. Unfortunately, it is extremely difficult to do more than estimate the size of this expression because of the complexity and overlapping of the nuclear states involved. Such an estimate has been made by Blin-Stoyle, (1960b) and Abov *et al.* (1965).

In this estimate, because of the high level density at the nuclear excitation concerned, it is assumed that very near the capture state ψ_c in ^{114}Cd is another state ψ_c' having the same spin but opposite parity. The admixture of ψ_c' into ψ_c is then taken to be the dominant parity impurity. The amplitude of this impurity is estimated using a harmonic oscillator single particle model to be of the order of magnitude $50\mathcal{F}$ where $\mathcal{F}$ is a measure of the relative strength of the parity violating potential. Thus the quantity e_1/m_1 is given approximately by $50\mathcal{F} R$ where R is the ratio of typical single particle e_1/m_1 reduced matrix elements. Taking $\mathcal{F} \approx 6 \times 10^{-7}$ (see §IX-4.1 and eq. (IX-51)) and $R \approx 10$ gives $e_1/m_1 \approx 3 \times 10^{-4}$ and, using eq. (IX-37) an asymmetry of the order $\alpha \approx \pm(6 \times 10^{-4})P_n$ where P_n is the neutron polarisation.

This value for α should not be taken too seriously since it depends on a number of crude approximations. McKellar (1968a, 1969a) has modified the foregoing value of α in order to take account of different theories of the weak interaction and of short range internucleon correlations. He obtains values for

Table IX-4
γ-ray asymmetry of the 9.05 MeV transition in ^{114}Cd.

Measured asymmetry (α/P_n) ($\times 10^4$)	Corrected asymmetry * ($\times 10^4$)	Reference
-3.7 ± 0.9	-6.1 ± 1.5	Abov *et al.* (1964, 1965)
-2.5 ± 2.2	-2.5 ± 2.2	Warming *et al.* (1967)
-3.5 ± 1.2	-4.4 ± 1.5	Abov *et al.* (1968)
-0.6 ± 1.7	-0.6 ± 1.8	Warming (1969)
-3.6 ± 0.7	-4.5 ± 1.0	Abov *et al.* (1970)

* Corrected for the presence of the unresolved 8.48 MeV γ-ray in the decay spectrum (Warming (1969)).

α in the range $(\pm 2$ to $\pm 10) \times 10^{-4}$ according to the theory. Because of the crudity of the original estimate for α, however, these different values are not significant.

A summary of the most recent experimental results is given in table IX-4 where it can be seen that the values are consistent with the value predicted theoretically. Alberi *et al.* (1972) have also measured the circular polarisation δ for the combined 9.05 MeV and 8.48 MeV transitions following the capture of unpolarised thermal neutrons and obtain $\delta = (6.0 \pm 1.5) \times 10^{-4}$. However, because of the nuclear complexities it seems unlikely that significant information about the weak interaction will derive from experiments of this kind.

IX-5. **Parity violation in other nuclear processes**

It is obviously of little value to search for parity violating effects in β-decay processes since such processes in any case violate parity conservation maximally and the effects being searched for would therefore be completely masked.

As far as other nuclear reaction and scattering processes are concerned, they are generally dominated by electromagnetic or nuclear forces which would again virtually obliterate parity non-conserving effects. Michel (1964) has suggested that experiments involving the transmission of thermal neutrons through a crystal might be an exception since "the Coulomb force is absent and the nuclear scatterings add coherently to appear simply as an "index of refraction" for neutron waves in the crystal". He makes a theoretical estimate of the order of magnitude of the effect but so far no experimental investigation has been made. Berovic (1971) has suggested studying polarisation phenomena in reactions having the structure $1^+ + 0^+ \to 0^+ + 0^+$ in spin space. He makes a theoretical study of a direct reaction (^{12}C (d, α) ^{10}B) particularly sensitive to the presence of a $\Delta T = 1$ component in the parity violating potential but the size of the predicted effect ($\sim 5 \times 10^{-5}$) is too small to be detected by present experimental techniques.

IX-6. Conclusions

This chapter has been concerned with studies of parity impurities in nuclear states induced by the weak parity violating potential $V_{\text{p.v.}}$. It is clear that by these studies the first steps have been taken in clarifying the structure of this potential and, in turn, determining the nature of the strangeness conserving hadronic weak interaction $\mathcal{H}_W^{(\text{n.l.}:\Delta S=0)}$. At the present time work of this kind in the field of low energy nuclear physics seems to be the only practical way of obtaining information about this part of the weak interaction. In general, since it conserves strangeness, the strong interaction completely masks the effects of $\mathcal{H}_W^{(\text{n.l.}:\Delta S=0)}$ in an elementary particle process and it is only in the field of low energy nuclear physics where experimental accuracies of the order 1 in 10^4, or better, can be achieved that its effects can be detected. Even here only the parity violating part of $\mathcal{H}_W^{(\text{n.l.}:\Delta S=0)}$ shows itself, the parity non-violating part being much smaller than the uncertainties in the strong internucleon potential.

It is interesting to note that in the different asymmetry measurements, since the phenomena being studied result from interference effects, it is possible to determine the sign of the weak interaction coupling constant G. Thus, referring to table IX-3, the sign difference between the circular polarisation, δ, for ^{181}Ta as determined from the Cabibbo theory and the experimental value (table IX-2) could be accounted for if G was taken to be negative. This would be contrary to the predictions of an Intermediate Vector Boson theory unless the coupling constant f (see eq. (I-41)) was imaginary as in indefinite metric theories (e.g. Lee (1971)).

This, together with other information that might be deduced about the form of $\mathcal{H}_W^{(\text{n.l.}:\Delta S=0)}$, is good reason for pushing theory and experiment in this field a good deal further forward.

CHAPTER X

TIME REVERSAL NON-INVARIANCE IN NUCLEAR REACTIONS AND ELECTROMAGNETIC TRANSITIONS

X-1. Introduction

In Chapt. I possible microscopic T (or CP)-violating theories able to account for the observed CP-violation in the decay of the K_L^0 were briefly described. The theories can be grouped as follows

(i) C- and T-violating milli-strong interaction (e.g. Prentki and Veltman (1965); Cabibbo (1965); Lee and Wolfenstein (1965)).

(ii) C- and T-violating electromagnetic interaction (e.g. Bernstein *et al.* (1965); Barshay (1965); Lee (1965a); Arzubov and Filippov (1968)).

(iii) C-, P- and T-violating milli-weak interaction (e.g. Cabibbo (1964, 1965); Glashow (1965); Zachariasen and Zweig (1965); Nishijima and Swank (1966, 1967a, b); Oakes (1968); Das (1968); Maiani (1968); Okubo (1968a, b)).

(iv) CP-violating superweak theories (e.g. Wolfenstein (1964); Lee and Wolfenstein (1965)).

For theories of type (i) nuclear effects (e.g. T-non-invariant asymmetries) of the order of the observed CP-violating effect in K_L^0 decay (i.e. $\approx 10^{-3}$) are to be expected. Similarly for theories of type (ii) effects of the order 10^{-3} are again expected. Here it should be noted that effects of order unity which might at first sight be anticipated in nuclear electromagnetic transitions cannot occur because of restrictions imposed by the conservation of the electromagnetic current (see §X-3). With theories of type (iii) in which the weak interaction is made CP-violating then the expected effects are very small ($\lesssim 10^{-6}$ to 10^{-9}) except in the case of a weak interaction process itself in the nucleus. The only possibility at the present time is β-decay and a discussion of the situation in this respect has been given in §V-4. Finally, with superweak theories (type (iv)) in which T-non-invariant effects show up effectively only through impurities in the K_S and K_L states and for which the coupling constant F of the CP-violating Hamiltonian is of the order $10^{-8} \times$ (weak interaction coupling constant) then no measurable effects are expected in nuclear physics processes.

The concern of this chapter will therefore be with effects stemming from *P*-conserving *C*- and *T*-violating milli-strong and electromagnetic interactions although brief mention will be made in §X-8 of effects arising from simultaneous *P*- and *T*-violation.

X-2. *T*-violating internucleon potentials

The existence of a *C*- and *T*-violating elementary particle interaction leads to the generation of a *T*-violating internucleon potential in much the same way that the weak interaction leads to a *P*-violating potential (see Chapt. VIII).

The general form that such a *T*-violating potential can take has been considered by Herczeg (1966) allowing for the possibility that the potential might also be *P*-violating. Unfortunately the general form has considerable complexity (the parity conserving part alone involves eighteen arbitrary functions of the internucleon distance) and so is not reproduced here. It is of interest, however, to write down the most general single particle *T*-violating potentials. These have the forms:

$$U_{\text{t.v.}}(r) = \tfrac{1}{2}[(\alpha_1(r) + \alpha_2(r)\tau_3)(\boldsymbol{r}\cdot\boldsymbol{p} + \boldsymbol{p}\cdot\boldsymbol{r}) + \text{h.c.}] \tag{X-1}$$

$$U_{\text{p.t.v.}}(r) = \tfrac{1}{2}[(\beta_1(r) + \beta_2(r)\tau_3)\,\boldsymbol{\sigma}\cdot\boldsymbol{r} + \text{h.c.}] \tag{X-2}$$

where $\alpha_1, \alpha_2, \beta_1$ and β_2 are arbitrary functions of the position coordinate r. $U_{\text{t.v.}}$ is a *T*-violating scalar potential (i.e. parity conserving), whilst $U_{\text{p.t.v.}}$ is a *T*-violating pseudoscalar potential (i.e. parity violating). This latter potential, being *P*-violating will not concern us further. An equivalent two-body potential of the form given in eq. (X-1) has also been used in the literature (Henley and Huffman (1968); Huffman (1970b)).

Attempts to calculate the detailed form of *T*-violating potentials on the basis of some assumed form for a *C*- and *T*-violating theory of the electromagnetic interaction have been made by Huffman (1970a) and Clement and Heller (1971) and their calculations are briefly discussed in the following section. Suffice it to say here that their results depend on restricted and specific assumptions about the form of a *T*-violating electromagnetic interactions and can only be used to indicate the possible magnitude of a *T*-violating internucleon potential. So far no other calculations of *T*-violating potentials have been carried out.

X-3. *T*-violating electromagnetic interaction

The usual electromagnetic current J_λ is taken to be Hermitian, conserved, a member of an octet of polar vector currents (see §§III-1, XI-1) and odd under the charge conjugation operator * U_C (see App. D), i.e.

$$J_\lambda^* = J_\lambda = -U_C J_\lambda U_C^{-1} . \tag{X-3}$$

J_λ is also odd under the time reversal operation as follows (cf. eq. (V-5))

$$J_\lambda^\dagger = -U_T J_\lambda U_T^{-1} . \tag{X-4}$$

In order to introduce *T*-violation (i.e. *C*-violation assuming parity conservation) into the electromagnetic interaction, Bernstein *et al.* (1965) suggested that the electromagnetic current might contain another (Hermitian and conserved) component, K_λ, *even* under the charge conjugation operation, i.e.

$$K_\lambda^* = K_\lambda = +U_C K_\lambda U_C^{-1} \tag{X-5}$$

and *even* under the time reversal operation, i.e.

$$K_\lambda^\dagger = +U_T K_\lambda U_T^{-1} . \tag{X-6}$$

The total electromagnetic current $J_\lambda^{\text{e.m.}}$ is then given by

$$J_\lambda^{\text{e.m.}} = J_\lambda + K_\lambda . \tag{X-7}$$

Both currents are assumed to transform as polar vectors under the parity operation (neglecting the effects of weak interactions) and *C* or *T* invariance requires that $K_\lambda = 0$. The total charge operator Q corresponding to the current $J_\lambda^{\text{e.m.}}$ is given by (cf. eq. (I-23)

$$Q = Q_J + Q_K \tag{X-8}$$

where

$$Q_J = -\mathrm{i} \int J_4(x)\, \mathrm{d}^3x \tag{X-9}$$

* The result in eq. (X-3) can be demonstrated explicitly, for example, using the expression for the current given in App. D eq. (D 11) and the relations given in eq. (D 12).

$$Q_K = -\mathrm{i} \int K_4(x)\, \mathrm{d}^3x \tag{X-10}$$

and theories of T-violation in the electromagnetic interaction can be divided into those for which Q_K is zero for all physical states (e.g. Lee (1965b)) and those for which $Q_K \neq 0$ (e.g. Lee (1965a)). We shall not be concerned with the details of these theories but note, in passing, that the latter type requires the existence of particles designated by Lee (1965a) as $a^{\pm}$. Furthermore the isospin nature of K_λ is unspecified.

The T-violation generated by the additional current K_λ would be expected to manifest itself in nuclear physics processes by a T-violating contribution to the nuclear electromagnetic current. Specifically if the initial and final nucleon states are on the mass shell, the matrix element of $J_\lambda^{\text{e.m.}}$ can be written (see §I-4 and App. F)

$$\langle N' | J_\lambda^{\text{e.m.}}(0) | N \rangle = \mathrm{i}(\bar{u}(N') | F_1(k^2)\gamma_\lambda - F_2(k^2)\sigma_{\lambda\nu}k_\nu + \mathrm{i}\, F_3(k^2)\, k_\lambda \,| u(N)) \tag{X-11}$$

where $k_\lambda = N'_\lambda - N_\lambda$ and N_λ, N'_λ designate the four-momenta of the initial and final nucleon states respectively. In general, each form factor F_1, F_2 and F_3 will contain both isoscalar and isovector parts. F_1 and F_2 are the usual charge and anomalous magnetic moment form factors whilst F_3 is normally taken to be zero.

We now have to consider briefly the implications of the presence of K_λ for the different form factors. Since $J_\lambda^{\text{e.m.}}$ is a Hermitian operator ($J_\lambda^{\text{e.m.}} = J_\lambda^{\text{e.m.}*}$) this requires

$$\langle N' | J_\lambda^{\text{e.m.}} | N \rangle = \epsilon_\lambda \langle N | J_\lambda^{\text{e.m.}} | N' \rangle^{\times} \tag{X-12}$$

where $\epsilon_\lambda = (1, 1, 1, -1)$ and *no* sum is implied by the repeated index λ. This in turn, using eq. (X-11) requires F_1, F_2 real and F_3 imaginary. However, if K_λ is zero, i.e. if the theory is invariant under time reversal invariance, then the relation given in eq. (X-4) requires that F_1, F_2 and F_3 are real (see §V-3 where similar arguments are used for the weak interaction currents). Thus F_3 is identically zero. On the other hand, if the current K_λ contributes and time reversal invariance does not hold, then the relation given in eq. (X-6) allows F_1, F_2 and F_3 to have imaginary components. Hermiticity, however, forbids any imaginary contribution to F_1 and F_2. The situation is then that J_λ contributes only to F_1 and F_2 which are real, whilst the T-violating current K_λ contributes only to F_3 which is imaginary.

Unfortunately, however, for nucleons on the mass shell the fact that $J_\lambda^{\text{e.m.}}$ is conserved is alone sufficient to require that F_3 is identically zero (see App. F) and so under these circumstances no T-violating effects can be observed. This was first pointed out by Bernstein *et al.* (1965). Specifically this means that there can be no T-violating nucleon–photon vertex interaction and, correspondingly, no long range T-violating force resulting from one photon exchange (fig. X-1(a)).

However, should the photon line couple to nucleon states off the mass shell (as in fig. X-1(b, c)) then the previous arguments do not hold. F_3 (and other additional T-violating form-factors) is non-vanishing and both J_λ and K_λ contribute thus leading to a T-violating internucleon potential. The magnitude of this potential compared with the strong interaction potential can be estimated crudely as follows. A factor $\approx (m_\pi/M)^2$ has to be introduced to account for the off-mass-shell character of the force, a factor $k/M \approx m_\pi/M$ to account for the F_3 form factor term and a factor α for the electromagnetic vertex. Thus

$$V_{\text{t.v.}} \approx \alpha \left(\frac{m_\pi}{M}\right)^3 V_{\text{strong}} \approx 0.002\%\; V_{\text{strong}}\,. \tag{X-13}$$

Detailed calculations of the form and strength $V_{\text{t.v.}}$ arising from diagrams of the type X-1(b) have been carried out by Huffman (1970a) who makes a number of simplifying assumptions. He finds the following characteristics for the potential: (i) it is rather singular and may be stronger than indicated by the estimate given in eq. (X-13), (ii) it has an exchange character and only the isovector component of K_λ contributes, (iii) it is spin and momentum depen-

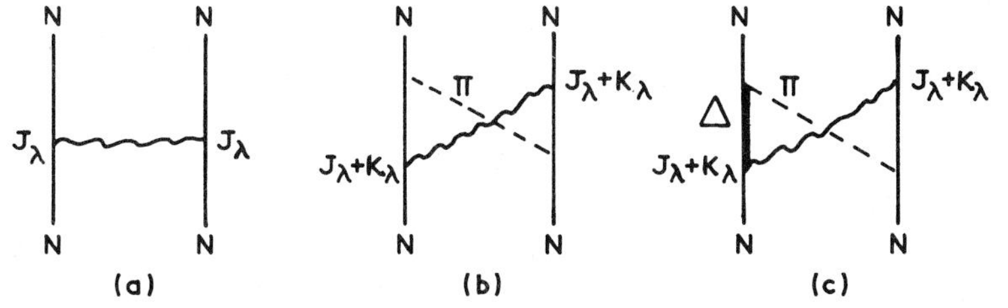

Fig. X-1. Typical one-photon exchange diagrams which might contribute to a T-violating internucleon potential. (a) J_λ only contributes – T-invariant, (b) and (c) J_λ and K_λ contribute – T-violating.

dent. The possibility of T-violation in the vertex for

$$\gamma + \mathrm{N} \rightleftharpoons \Delta \tag{X-14}$$

which is included in the diagram X-1(c) and where Δ refers to the first nucleon isobar ($I = \frac{3}{2}$, $T = \frac{3}{2}$) was first suggested by Barshay (1966) who considered T-violating effects in the reactions $\gamma + \mathrm{d} \rightleftharpoons \mathrm{n} + \mathrm{p}$. There have, however, been no estimate of the form of the related internucleon potential.

In addition to the T-violating internucleon potential generated by the presence of K_λ or of a T-violating $\Delta\mathrm{N}\gamma$ vertex, there will also be T-violating contributions to the effective electromagnetic interaction in nuclei. These will arise from diagrams of the type given in fig. X-2 and will take the form of two-body T-violating operators, $\mathcal{H}_{\mathrm{t.v.}}(A)$. Such terms will clearly be important, for example, in dealing with T-violating effects in nuclear γ-decay processes (see §X-7). Some of them have been evaluated by Coutinho (1972).

A rather different approach has been adopted by Clement and Heller (1971) who introduce a T-violating four particle interaction NNMγ, where M is a meson (see fig. X-3(a)). For an interaction of this kind, T-violation can occur even though the initial and final nucleons are on the mass shell. Given a specific form for such an interaction it is then straightforward to calculate the lowest order contributions to the effective T-violating electromagnetic interaction (fig. X-3(b)), the two-body T-violating potential (fig. X-3(c)) and the three-body T-violating potential (fig. X-3(d)). Assuming the simplest possible form for the NNMγ interaction, Clement and Heller calculate diagram X-3(b) and also obtain expressions for the corresponding E1 and M1 multipole operators. They also argue that because of the long range nature of the Coulomb potential resulting from the photon exchange in fig. X-3(d) the

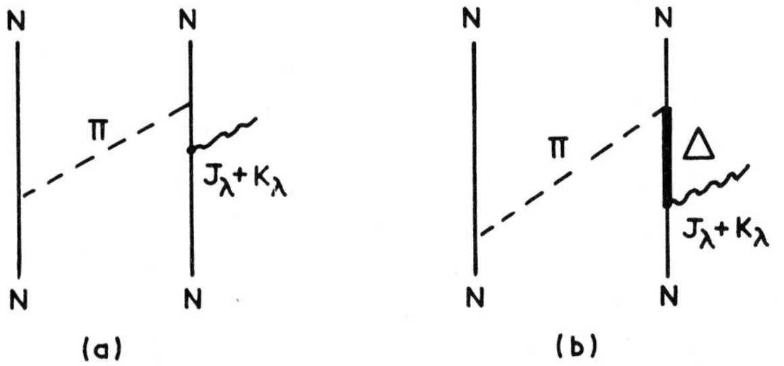

Fig. X-2. Diagrams leading to an effective two-body, T-violating electromagnetic interaction.

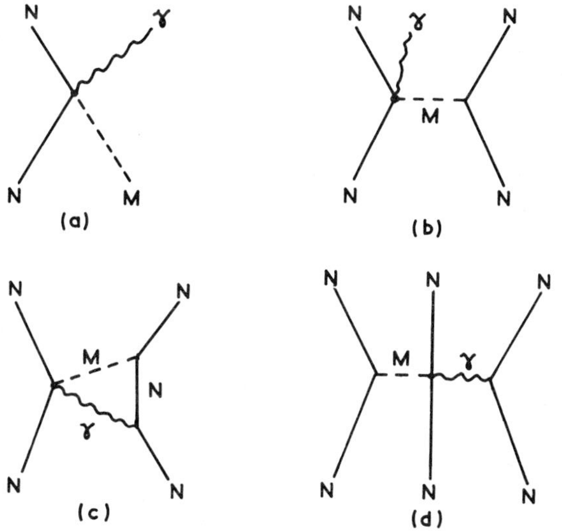

Fig. X-3. (a) T-violating NNMγ interaction, (b) two-body T-violating electromagnetic interaction, (c) two-body T-violating potential, (d) three-body T-violating potential.

three-body T-violating potential may be more important than the two-body (fig. X-3(c)).

Unfortunately, all of these calculations are preliminary and can only be taken as a rough guide as to the form and magnitude of the T-violating inter-nucleon potential and electromagnetic transition operators. Much more detailed work is needed and for our purposes, in the rest of this chapter, we shall proceed in a simple phenomenological fashion using, where necessary, the single particle T-violating potential given in eq. (X-1).

X-4. Low energy nuclear physics tests of T-violation

In the previous section we have discussed the form and possible magnitude of T-violating interactions involving nucleons and the electromagnetic field. For the remainder of this chapter we shall be concerned with the way in which such interactions can manifest themselves in low-energy nuclear physics processes. Possible tests of T-violation can be divided into four broad categories as follows:

(i) Tests of detailed balance.
(ii) Tests of the asymmetry – polarisation equality in nucleon–nucleon or nucleon–nucleus scattering.
(iii) Tests for T-violating asymmetries in γ-decay processes.
(iv) Tests for T-violating asymmetries in β-decay processes.

As stated earlier we shall not consider in this chapter category (iv) tests since these have been dealt with in §V-4. Thus our concern will be with nuclear reaction or scattering processes and γ-decay processes. In the following we deal with the theory of each type of test and give a summary of the associated experimental data.

X-5. Tests of detailed balance

If time reversal invariance holds then it is well-known that the S-matrix responsible for a nuclear reaction or scattering process is symmetrical (Coester (1953); see also Blatt and Weisskopf (1952 – Chapt. X) and Henley and Jacobsohn (1959)). This follows from the relationship

$$TST^{-1} = S^{\dagger} \tag{X-15}$$

where T is the time reversal operator (see App. B) and for the process

$$\mathrm{a} + \mathrm{A} \to \mathrm{b} + \mathrm{B} \tag{X-16}$$

can be stated explicitly as follows

$$\langle \boldsymbol{k}_\beta; s_\mathrm{b}, m_\mathrm{b}; s_\mathrm{B}, m_\mathrm{B} | S | \boldsymbol{k}_\alpha; s_\mathrm{a}, m_\mathrm{a}; s_\mathrm{A}, m_\mathrm{A}\rangle = (-1)^{s_\mathrm{a}+s_\mathrm{A}+s_\mathrm{b}+s_\mathrm{B}-m_\mathrm{a}-m_\mathrm{A}-m_\mathrm{b}-m_\mathrm{B}}$$

$$\times \langle -\boldsymbol{k}_\alpha; s_\mathrm{a}, -m_\mathrm{a}; s_\mathrm{A} - m_\mathrm{A} | S | -\boldsymbol{k}_\beta; s_\mathrm{b}, -m_\mathrm{b}; s_\mathrm{B}, -m_\mathrm{B}\rangle\,. \tag{X-17}$$

Here $\boldsymbol{k}_\alpha = \boldsymbol{k}_\mathrm{a} - \boldsymbol{k}_\mathrm{A}; \boldsymbol{k}_\beta = \boldsymbol{k}_\mathrm{b} - \boldsymbol{k}_\mathrm{B}$ are relative momenta and s and m refer to spin and 3-component of spin of the particle indicated. The relationship (X-17) is known as the *reciprocity* theorem.

It then follows from eq. (X-17) and rotational invariance that

$$\sum_{m_\mathrm{a}, m_\mathrm{A}, m_\mathrm{b}, m_\mathrm{B}} |\langle \boldsymbol{k}_\beta; s_\mathrm{b}, m_\mathrm{b}; s_\mathrm{B}, m_\mathrm{B} | S | \boldsymbol{k}_\alpha; s_\mathrm{a}, m_\mathrm{a}; s_\mathrm{A}, m_\mathrm{A}\rangle|^2$$

$$= \sum_{m_\mathrm{a}, m_\mathrm{A}, m_\mathrm{b}, m_\mathrm{B}} |\langle \boldsymbol{k}_\alpha; s_\mathrm{a}, m_\mathrm{a}; s_\mathrm{A}, m_\mathrm{A} | S | \boldsymbol{k}_\beta; s_\mathrm{b}, m_\mathrm{b}; s_\mathrm{B}, m_\mathrm{B}\rangle|^2 \tag{X-18}$$

or symbolically

$$\sum_{m_\alpha,m_\beta} |\langle\beta|S|\alpha\rangle|^2 = \sum_{m_\alpha,m_\beta} |\langle\alpha|S|\beta\rangle|^2 \,. \tag{X-19}$$

This relationship is a statement of the principle of detailed balance and relates the cross sections for the process a + A → b + B (symbolically $\alpha \to \beta$) and b + B → a + A (symbolically $\beta \to \alpha$) in the case that the final state spins are not measured and the target and bombarding particles are not polarised.

Since the reaction differential cross section $\mathrm{d}\sigma_{\alpha\beta}/\mathrm{d}\Omega$ for the process $\alpha \to \beta$ is given by

$$\frac{\mathrm{d}\sigma_{\alpha\beta}}{\mathrm{d}\Omega} = \frac{\pi}{k_\alpha^2}\,\frac{1}{(2s_\mathrm{a}+1)(2s_\mathrm{A}+1)} \sum_{m_\alpha,m_\beta} |\langle\beta|S|\alpha\rangle|^2 \tag{X-20}$$

and similarly for $\mathrm{d}\sigma_{\beta\alpha}/\mathrm{d}\Omega$ it follows from eq. (X-19) that

$$\frac{(2s_\mathrm{a}+1)(2s_\mathrm{A}+1)\,k_\alpha^2\,\mathrm{d}\sigma_{\alpha\beta}/\mathrm{d}\Omega}{(2s_\mathrm{b}+1)(2s_\mathrm{B}+1)\,k_\beta^2\,\mathrm{d}\sigma_{\beta\alpha}/\mathrm{d}\Omega} = 1 \,. \tag{X-21}$$

Any deviation of this ratio from unity indicates a breakdown of eq. (X-19) and therefore implies that time reversal invariance does not hold. Thus tests of this detailed balance relationship could be used to test for *T*-violation.

Unfortunately the test is not always effective. As Blatt and Weisskopf (1952 – Chapt. X) point out, if the process under consideration is well described by the first order Born approximation then

$$\langle\beta|S|\alpha\rangle \simeq \langle\beta|H'|\alpha\rangle \tag{X-22}$$

where H' is the (Hermitian) perturbing operator responsible for the transition. Under this circumstance the Hermitian character of H' ensures that

$$|\langle\beta|H'|\alpha\rangle|^2 = |\langle\alpha|H'|\beta\rangle|^2 \tag{X-23}$$

so satisfying the detailed balance relation independently of whether there is any *T*-violation. Similarly Henley and Jacobsohn (1959) demonstrate that if only two states are relevant to the S matrix (i.e. it is a 2 × 2 matrix or breaks up into 2 × 2 matrices), then unitarity alone leads to the principle of detailed balance. These authors conclude that the best test is likely to be a reaction in which there are many open competing channels and which is difficult to describe by a simple model. We therefore first address our attention to compound nucleus reactions.

X-5.1. *Detailed balance in compound nucleus reactions*

Various authors (Ericson (1966); Mahaux and Weidenmüller (1966); Moldauer (1968a)) extending the work of Henley and Jacobsohn (1959) have argued that most sensitivity to T-violation is obtained in reactions where the width of the compound nucleus excited states is much greater than their separation so that Ericson fluctuations (Ericson (1963)) result. In the presence of T-odd forces and taking the T-odd and T-even amplitudes to be uncorrelated then it is to be expected that if an effect is looked for in a deep minimum of the cross section, the T-odd amplitude should be relatively enhanced.

Ericson (1966) gives the following simple theoretical argument. For a nuclear reaction a + A → b + B (symbolically $\alpha \to \beta$) going through different compound nuclear states c, the S-matrix element can be written in the form

$$\langle \beta | S | \alpha \rangle \propto \sum_c \gamma_{\beta c} \frac{1}{E - W_c} \gamma_{c\alpha} \tag{X-24}$$

where the γ's are reduced widths, E is the energy and W_c the (complex) energy of a compound state. As mentioned above it is assumed that the reaction is taking place in an energy region where Γ (average compound level width) $\gg D$ (average level spacing). Denoting T-even and T-odd reduced widths by $\gamma^{(e)}$ and $\gamma^{(o)}$ respectively, writing $W_c = W_c^{(e)} + v$ where v is the T-violating interaction, we then have to first order in T-odd terms the following expressions for the T-even and odd parts of the S matrix element

$$\langle \beta | S^{(e)} | \alpha \rangle = \sum_c \gamma_{\beta c}^{(e)} \frac{1}{\Delta_c} \gamma_{c\alpha}^{(e)} \tag{X-25}$$

$$\langle \beta | S^{(o)} | \alpha \rangle = \sum_c \left[\gamma_{\beta c}^{(o)} \frac{1}{\Delta_c} \gamma_{c\alpha}^{(e)} + \gamma_{\beta c}^{(e)} \frac{1}{\Delta_c} \gamma_{c\alpha}^{(o)} + \gamma_{\beta c}^{(e)} \frac{1}{\Delta_c} v \frac{1}{\Delta_c} \gamma_{c\alpha}^{(e)} \right] \tag{X-26}$$

where $\Delta_c = E - W_c^{(e)}$.

Ericson (1966) assumes (and Moldauer (1968a) subsequently justifies) that the first two terms in eq. (X-26) can be neglected. Thus, on taking appropriate energy averages and making the usual statistical assumption of random sign for $\gamma_{\beta c}$ and $\gamma_{c\alpha}$, we obtain

$$\overline{|\langle\beta|S^{(\mathrm{o})}|\alpha\rangle|^2} \approx \overline{\gamma_{\beta c}^2\gamma_{c\alpha}^2} \sum_{cc'} \frac{\overline{|\langle c|v|c'\rangle|^2}}{|\Delta_c|^2\,|\Delta_{c'}|^2}$$

$$\approx \overline{|\langle\beta|S^{(\mathrm{e})}|\alpha\rangle|^2}\,\frac{2\pi}{\Gamma D}\,\overline{|\langle c|v|c'\rangle|^2}\,. \tag{X-27}$$

Denoting the ratio of the T-odd to the T-even amplitude by F gives

$$|F|^2 = \frac{\overline{|\langle\beta|S^{(\mathrm{o})}|\alpha\rangle|^2}}{\overline{|\langle\beta|S^{(\mathrm{e})}|\alpha\rangle|^2}} \approx \frac{2\pi}{\Gamma D}\,\overline{|\langle c|v|c'\rangle|^2} \approx \frac{W}{\Gamma}\,\mathcal{F}^2 \tag{X-28}$$

where W is the energy region over which the matrix elements of the T-even force V (say) are thoroughly mixed. $\mathcal{F}$ is essentially the ratio of the strengths of the T-odd and T-even forces (as distinct from amplitudes) defined by the equation

$$\overline{|\langle c|v|c'\rangle|^2} \approx \mathcal{F}^2\,\overline{|\langle c|V|c'\rangle|^2}$$

and is a measure of the relative strength of the T-violating interaction. According to Ericson (1966), in heavier elements $\sqrt{W/\Gamma}$ is commonly between one and two orders of magnitude so that quite a sizeable enhancement of T-violation might be obtained. However, Mahaux and Weidenmüller (1966) argue that W is associated with the width of doorway states which then gives $\sqrt{W/\Gamma}$ of the order unity.

During the last few years two experiments have been carried out seeking to test detailed balance in reactions for which the above circumstances hold. Von Witsch *et al.* (1966, 1967, 1968) have made a full study of the reaction $^{24}\mathrm{Mg} + \alpha \rightleftharpoons {}^{27}\mathrm{Al} + \mathrm{p}$ at bombarding energies in the region 10–15 MeV. At these energies the conditions just described are met. In addition at scattering angles of 0° and 180° angular momentum conservation requires that only one spin channel contributes. This is a distinct advantage since in the case of many spin channels, any T-violating effect which might show up clearly in one channel, would be obscured due to the incoherent mixture of the different uncorrelated spin channels. It should be noted, however, that there is always the possibility that T-violating effects might vanish at 0° and 180° because of some other symmetry.

In their series of experiments Von Witsch *et al.* placed counters symmetrically to the beam at laboratory angle $\theta_{\mathrm{lab}} \approx 172°$ where, it was estimated, the admixture of other spin amplitudes is less than 10%. The cross sections in both reactions were compared at two maxima and one minimum of the excitation function. In the former it was supposed *T*-violating effects are negligible whilst in the latter an enhancement of the type discussed earlier might be expected. Detailed analysis of the experimental results led the authors to conclude that an upper limit to the *T*-violating part of the reaction amplitude is 3×10^{-3} with 85% confidence, i.e.

$$|F| \lesssim 3 \times 10^{-3} \,. \tag{X-29}$$

Assuming an enhancement by a factor $(W/\Gamma)^{1/2}$ gives

$$|\mathcal{F}| \lesssim (W/\Gamma)^{-1/2}\, 3 \times 10^{-3} \,.$$

Mahaux and Weidenmüller (1966) estimate $(W/\Gamma)^{-1/2} \approx 1$ whilst Ericson (1966) estimates $(W/\Gamma)^{-1/2} \approx 1/6$ so that within this uncertainty

$$\mathcal{F} \lesssim 5 \times 10^{-4} \text{ to } 3 \times 10^{-3} \,. \tag{X-30}$$

Another reaction of this kind which has also been studied (Thornton *et al.* (1968, 1971) is $^{16}O + d \rightleftharpoons {}^{14}N + \alpha$ at deuteron energies in the region 4.3 – 4.8 MeV and corresponding α-particle energies. Measurements were made at angles corresponding to maxima in the angular distributions. Unfortunately a number of reaction channels contribute and so the analysis of the experimental results is complicated and rather uncertain. The authors conclude that $|F| \lesssim 2 \times 10^{-3}$ and make no estimate of $|\mathcal{F}|$.

These two experiments indicate that with improved experimental techniques it should be possible to put limits on $|F|$ of the order of magnitude predicted by a *T*-violating medium strong or electromagnetic interaction (see §X-1). However, the theoretical interpretation is clearly very complicated and at the present time there are many uncertainties in the attempts to relate F to $\mathcal{F}$. Improvement here will follow when our general understanding of complex nuclear reactions is increased.

X-5.2. *Detailed balance in direct reactions*

In the case of direct reactions, because of the relative simplicity of the reaction mechanism (see Hodgson (1971) for a general review) it is possible to relate any divergence from detailed balance more readily to *T*-violating terms in the basic Hamiltonian. It must be remembered, however, that in the extreme situation that the particular process under consideration is well described by the plane wave Born approximation, then Hermiticity alone ensures the principle of detailed balance (see §X-5). This, of course, is not the usual situation and generally the approach used is the Distorted Wave Born Approximation (DWBA). Various authors (Robson (1968); Henley and Huffman (1968); Moldauer (1968b); Huffman (1968, 1970b)) have studied in detail the effects of a T-violating potential on the DWBA theory of direct reactions and the essentials of their arguments can be summarised briefly as follows (cf. Henley (1969a)).

To be explicit we will consider a stripping reaction (A + d ⇌ B + p) in the zero range approximation. The S-matrix element $\langle \mathrm{B}, \mathrm{p}|S|\mathrm{A}, \mathrm{d}\rangle$ for the reaction can be written

$$\langle \mathrm{B}, \mathrm{p}|S|\mathrm{A}, \mathrm{d}\rangle \propto \int \psi^{(-)\times}_{\mathrm{B,p}}(\boldsymbol{k},\boldsymbol{r})\Phi(\boldsymbol{r})\,\psi^{(+)}_{\mathrm{A,d}}(\boldsymbol{K},\boldsymbol{r})\,\mathrm{d}^3 r \qquad \text{(X-31)}$$

where $\psi^{(-)}_{\mathrm{B,p}}(\boldsymbol{k},\boldsymbol{r})$ is an (ingoing) elastically scattered wave for scattering of the proton by the final nucleus B, $\psi^{(+)}_{\mathrm{A,d}}(\boldsymbol{K},\boldsymbol{r})$ is an (outgoing) elastically scattered wave for scattering of the deuteron by the initial nucleus A and $\Phi(\boldsymbol{r})$ is a single particle bound state wave function for the captured neutron.

The time reversal operation leads to the following intuitively obvious results

$$T\psi^{(-)}_{\mathrm{B,p}}(\boldsymbol{k},\boldsymbol{r}) = \psi^{(-)}_{\mathrm{B,p}}(\boldsymbol{k},\boldsymbol{r})^{\times} = \psi^{(+)}_{\mathrm{B,p}}(-\boldsymbol{k},\boldsymbol{r})$$

$$T\psi^{(+)}_{\mathrm{A,d}}(\boldsymbol{K},\boldsymbol{r}) = \psi^{(+)}_{\mathrm{A,d}}(\boldsymbol{K},\boldsymbol{r})^{\times} = \psi^{(-)}_{\mathrm{A,d}}(-\boldsymbol{K},\boldsymbol{r}) \qquad \text{(X-32)}$$

and requires $\Phi(\boldsymbol{r})$ to be real. This in turn leads to the relation

$$\langle \mathrm{A}, \mathrm{d}|S|\mathrm{B}, \mathrm{p}\rangle = \langle \mathrm{B}, \mathrm{p}|S|\mathrm{A}, \mathrm{d}\rangle \qquad \text{(X-33)}$$

and hence to the principle of detailed balance. However, even if there are T-violating interactions present the relationships given in eq. (X-32) still hold *outside* the nuclear interaction radius R where there are no forces or only the usual T-invariant Coulomb force (Robson (1968)). This result follows from the asymptotic boundary condition imposed on the scattered waves. Further, in the region $r > R$ the radial part of the single particle function $\Phi(\boldsymbol{r})$ has the form of a Hankel function $h\,(\mathrm{i}\beta r)$ where β is real and is related to the energy eigenvalue. Thus $\Phi(\boldsymbol{r})$ has a definite phase (ϵ say) independent of the value of r so that

$$\Phi(\boldsymbol{r})^{\times} = \mathrm{e}^{-2\mathrm{i}\epsilon}\Phi(\boldsymbol{r}) \quad \text{for} \quad r > R\,. \qquad \text{(X-34)}$$

Using eq. (X-31) and eqs. (X-32) and (X-34) which are valid in the external region $r > R$ we can therefore write

$$\langle \mathrm{B,p}|S|\mathrm{A,d}\rangle \propto \int_0^R \psi_{\mathrm{B,p}}^{(-)\times}(\boldsymbol{k},\boldsymbol{r})\,\Phi(\boldsymbol{r})\,\psi_{\mathrm{A,d}}^{(+)}(\boldsymbol{K},\boldsymbol{r})\,\mathrm{d}^3r$$

$$+\int_R^\infty \psi_{\mathrm{B,p}}^{(-)\times}(\boldsymbol{k},\boldsymbol{r})\,\Phi(\boldsymbol{r})\,\psi_{\mathrm{A,d}}^{(+)}(\boldsymbol{K},\boldsymbol{r})\,\mathrm{d}^3r$$

$$= \langle \mathrm{B,p}|S|\mathrm{A,d}\rangle_{\mathrm{int}} + \langle \mathrm{B,p}|S|\mathrm{A,d}\rangle_{\mathrm{ext}}$$

$$= \langle \mathrm{B,p}|S|\mathrm{A,d}\rangle_{\mathrm{int}} + \mathrm{e}^{-2i\epsilon}\langle \mathrm{A,d}|S|\mathrm{B,p}\rangle_{\mathrm{ext}}\,. \quad \text{(X-35)}$$

If now the contribution from the internal region is neglected (the "cut-off DWBA") then detailed balance follows from eq. (X-35) independent of any T-violation.

The question therefore arises as to the relative sizes of the contributions to the S-matrix element from the internal and external regions since it is the interference between these two contributions which can lead to a T-violating effect and the breakdown of detailed balance. Henley and Huffman (1968) and more recently Huffman (1970b) have investigated this problem in detail using a two-body T-violating potential of the form (see eq. (X-1))

$$U(r)_{\mathrm{t.v.}} = \tfrac{1}{2}\left[(\alpha_1(r) + \alpha_2(r)\tau_3)(\boldsymbol{r}\cdot\boldsymbol{p} + \boldsymbol{p}\cdot\boldsymbol{r}) + \mathrm{h.c.}\right]. \quad \text{(X-36)}$$

In the case of Huffman's (1970b) work the strength of the potential is related to that of the two-body T-violating potential derived by the author (Huffman (1970a)) assuming T-violation in the electromagnetic interaction (see §X-3). The results of these calculations show that the contribution of the internal region is such that the percentage violation of detailed balance is roughly the same as the percentage relative strengths of the T-violating and strong interaction potentials. More specifically for the particular model of T-violation used by Huffman (1970a) and for the stripping reaction $^{24}\mathrm{Mg(d,p)}^{25}\mathrm{Mg}$ Huffman finds an effect of the order 0.2%. His result depends sensitively on various assumptions made in the calculation and particularly on the strength of the isovector part of $U(\boldsymbol{r})_{\mathrm{t.v.}}$ and for this reason these calculations should be regarded as very preliminary. They do nevertheless show that a stripping reaction, for example, can be used as an effective detailed-balance tests of T-violation.

A precision study of detailed balance has been made by Bodansky *et al.* (1966) and Weitkamp *et al.* (1968) on the stripping reaction

$$^{24}\mathrm{Mg} + \mathrm{d} \rightleftharpoons {}^{25}\mathrm{Mg} + \mathrm{p}$$

using approximately 10 MeV deuterons and 15 MeV protons. The ratio of the differential cross sections was measured at angles corresponding to maxima in the angular distributions and on the basis of these measurements the authors concluded that $F < 0.3\%$. On the basis of the work of Huffman (1970b) it could perhaps be further, very tentatively, deduced that also $\mathcal{F} < 0.3\%$ but it should be noted here that the interpretation of this particular reaction is complicated by the many spin channels which contribute.

It is nevertheless encouraging that the present experimental limit on T-violation is close to the value predicted by one specific model.

X-5.3. *Detailed balance in photon reactions*

As mentioned in §X-3, Barshay (1966) suggested that T-violation might arise through the $\gamma N\Delta$ vertex and went on to examine the effects of such T-violation on detailed balance in the reactions $\gamma + \mathrm{d} \rightleftharpoons \mathrm{n} + \mathrm{p}$. Since this effect involves the first nucleon isobar Δ it is only expected to show itself at high energies and does not properly, therefore, fall within the terms of reference of this book. However, because a positive effect *might* have been observed, it has been thought worthwhile to mention the relevant experiments.

Barshay (1966) estimates that at about 290 MeV/c photon laboratory momentum the differential cross-section of the two reactions might differ by as much as 40% if T-violation is attributable to the mechanism considered. Early experiments (Buon *et al.* (1968); Sober *et al.* (1969); Anderson *et al.* (1969); Bartlett *et al.* (1969); Longo (1969)) showed deviations from detailed balance of 20% or more. However, more recent measurements by Schrock *et al.* (1971) and Bartlett *et al.* (1971) are in agreement with the predictions of T-invariance.

Von Wimmersperg *et al.* (1970) have studied detailed balance in the reactions $^{12}\mathrm{C} + \alpha \rightleftharpoons {}^{16}\mathrm{O} + \gamma$ at a γ-ray energy of 13.1 MeV. Here any T-violation could appear as a difference between the backward–forward asymmetry (which is strong due to E1–E2 interference) for the two reactions. The experiment showed that the asymmetries were the same within one standard deviation.

X-6. The asymmetry – polarisation equality in scattering

We consider here the elastic scattering of spin $\frac{1}{2}$ particles by a target consisting of unpolarised nucleons or nuclei. If time reversal invariance holds then for an unpolarised incident beam, the polarisation, p, of the scattered particles is equal to the "up–down" asymmetry, α, resulting from a scattering in which the incident beam is fully polarised. This has been demonstrated specifically for the nucleon–nucleon scattering problem by Dalitz (1952) and Wolfenstein and Ashkin (1952) and generalised to the case of an arbitrary nuclear reaction involving ingoing and outgoing spin $\frac{1}{2}$ particles by Blin-Stoyle (1952). The result has been further generalised to the case that parity is not conserved by Bell and Mandl (1958) and we use their approach to demonstrate the equality between p and α for the case of elastic scattering of spin $\frac{1}{2}$ particles.

Consider the scattering process

$$a + A \rightarrow a + A$$

in which the scattered particle a has spin $\frac{1}{2}$ and denote the cross section for scattering through an angle θ from the magnetic substate m_a to m'_a of the spin $\frac{1}{2}$ particle by unpolarised target nuclei A by $\sigma_\theta(m'_a|m_a)$. Here the axis of spin quantisation is taken to be $\boldsymbol{k}_\alpha \times \boldsymbol{k}'_\alpha$ (i.e. at right angles to the plane of scattering) where $\boldsymbol{k}_\alpha$ and $\boldsymbol{k}'_\alpha$ are the initial and final momentum transfers and $\cos\theta = \boldsymbol{k}_\alpha \cdot \boldsymbol{k}'_\alpha/k_\alpha^2$. In terms of the corresponding S-matrix element (see eqs. (X-15)–(X-17) and eq. (X-20)) we can write

$$\sigma_\theta(m'_a|m_a) = \frac{\pi}{k_\alpha^2}\frac{1}{(2s_A+1)}\sum_{m_A,m'_A} |\langle \boldsymbol{k}'_\alpha;\tfrac{1}{2}, m'_a; s_A, m'_A|S|\boldsymbol{k}_\alpha;\tfrac{1}{2}, m_a; s_A, m_A\rangle|^2 . \tag{X-37}$$

Using eq. (X-17) it then follows that if time reversal invariance holds

$$\sigma_\theta(m'_a|m_a) = \sigma_{-\theta}(-m_a|-m'_a) . \tag{X-38}$$

This relationship is indicated symbolically in fig. X-4.

In terms of $\sigma_\theta(m'_a|m_a)$, $\boldsymbol{p}$ is defined as follows

$$p = \frac{\sum\limits_{m_a}[\sigma_\theta(\tfrac{1}{2}|m_a) - \sigma_\theta(-\tfrac{1}{2}|m_a)]}{\sum\limits_{m_a}[\sigma_\theta(\tfrac{1}{2}|m_a) + \sigma_\theta(-\tfrac{1}{2}|m_a)]} . \tag{X-39}$$

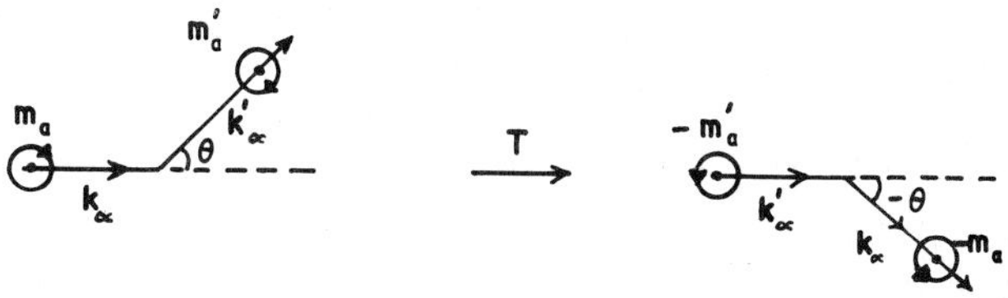

Fig. X-4. Scattering of spin $\frac{1}{2}$ particles and time reversal.

Similarly for a fully polarised incident beam

$$\alpha = \frac{\sum_{m'_a} [\sigma_\theta(m'_a|\tfrac{1}{2}) - \sigma_{-\theta}(m'_a|\tfrac{1}{2})]}{\sum_{m'_a} [\sigma_\theta(m'_a|\tfrac{1}{2}) + \sigma_{-\theta}(m'_a|\tfrac{1}{2})]} . \tag{X-40}$$

Assuming rotational invariance and rotating the system through 180° about the $\boldsymbol{k}_\alpha$ axis which effects the transformations $+\theta \to -\theta; m_a, m'_a \to -m_a, -m'_a$ gives

$$\sigma_\theta(m'_a|m_a) = \sigma_{-\theta}(-m'_a|-m_a) \tag{X-41}$$

or, using eq. (X-38)

$$\sigma_\theta(m'_a|m_a) = \sigma_\theta(m_a|m'_a) . \tag{X-42}$$

Thus, using eq. (X-42) in the first terms in the square bracket and eq. (X-38) in the second terms, the expression for p can be rewritten

$$p = \frac{\sum_{m'_a} [\sigma_\theta(m'_a|\tfrac{1}{2}) - \sigma_{-\theta}(-m'_a|\tfrac{1}{2})]}{\sum_{m'_a} [\sigma_\theta(m'_a|\tfrac{1}{2}) + \sigma_{-\theta}(-m'_a|\tfrac{1}{2})]} . \tag{X-43}$$

Changing the summation index of the second terms in the square brackets from m'_a to $-m'_a$ then makes eq. (X-40) and (X-43) identical and so establishes that

$$p = \alpha \tag{X-44}$$

as required. It should be noted, however, that if $s_A = 0$ then the above identity holds even with T-violation if parity is conserved since from the only dynamical variables s_a, $\boldsymbol{k}_\alpha$ and $\boldsymbol{k}'_\alpha$ it is impossible to construct a T-violating parity conserving quantity. This result does not hold, however, for the case $s_a > \frac{1}{2}$ (see Csonka and Moravcsik (1966) and references contained therein for a discussion of the limitations imposed by P- and T-invariance on scattering amplitudes in the case of arbitrary spins).

Any experimental violation of this identity then indicates the presence of T-violating effects. During the last few years two experiments have been carried out seeking to test this identity. The first by Handler *et al.* (1967) involved triple proton–proton scattering at 430 MeV where the first and last scatterings act as polariser and analyser in the measurement of α and p respectively. An analysis of the data implies that the T-violating amplitude is less than $\frac{1}{2}$% of the T-conserving amplitude, i.e. $F \lesssim 0.5\%$.

The second experiment (Gross *et al.* (1968)) involved the scattering of 32.9 MeV protons by ^{13}C nuclei and gave

$$p/\alpha = 0.992 \pm 0.025\ .$$

This can be crudely interpreted as implying $F \lesssim 2.5\%$. Unfortunately no detailed theoretical calculations of the deviation of p/α from unity for specific T-violating models have been carried out but unless an unexpected enhancement effect is operative it is unlikely that the foregoing experiments are of sufficiently high accuracy to rule out any of the current theories of T-violation.

X-7. *T*-violation in γ-decay

In this section we study the way in which T-violating interactions can manifest themselves to first order * in γ-decay angular correlation phenomena between nuclear states, some of which may be polarised. There are several different situations involving single, double or triple gamma transitions in which measurement of T-odd quantities could establish the presence of T-violating interactions. These different possibilities will be tabulated and discussed in §X-7.2. First, however, it is necessary to consider the implications

* T-violating interactions can, of course, contribute in higher (in particular, second) order to angular correlation terms which are invariant under time reversal. The forms of some such contributions have been studied by Henley and Jacobsohn (1959) and Kuebbing and Casper (1969). The effects are expected to be very small and will not be considered further here.

of T-invariance and T-violation for the matrix elements between nuclear states of electromagnetic multipole operators.

X-7.1. *Time reversal properties of electromagnetic multipole operators*

In §IX-3.1 an expression for the matrix element of the electromagnetic interaction $\mathcal{H}(A)$ between two nuclear substates $|IM\rangle$ and $|I'M'\rangle$ was given, namely,

$$\langle I'M'|\mathcal{H}(A)|IM\rangle = \pi \sum_{L\mu P} (2L+1)^{1/2} f(P) D^{(L)}_{\mu P}(\boldsymbol{k})[\langle I'M'|\mathcal{H}(A^{\mu}_{L}(m))|IM\rangle$$

$$+ (-P)\langle I'M'|\mathcal{H}(A^{\mu}_{L}(e))|IM\rangle] \qquad \text{(X-45)}$$

where $f(P)$ ($P \equiv +1$ or -1 corresponding to left or right circular polarisation) defines the polarisation of the gamma-ray emitted in the direction $\boldsymbol{k}$. $D^{(L)}_{\mu P}(\boldsymbol{k})$ is an element of the rotation matrix and $\mathcal{H}(A^{\mu}_{L}(m))$ and $\mathcal{H}(A^{\mu}_{L}(e))$ are the usual magnetic and electric multipole operators respectively. If time-reversal holds (see App. B) then these operators satisfy the following relations (e.g. Brink and Satchler (1962) – p. 91)

$$T\mathcal{H}(A^{\mu}_{L}(e))T^{-1} = (-1)^{\mu}\mathcal{H}(A^{-\mu}_{L}(e))$$

$$T\mathcal{H}(A^{\mu}_{L}(m))T^{-1} = (-1)^{\mu+1}\mathcal{H}(A^{-\mu}_{L}(m)) . \qquad \text{(X-46)}$$

Further, using eqs. (B8–B10) we can write

$$\langle I'M'|\mathcal{H}(A^{\mu}_{L}(\gamma))|IM\rangle = (-1)^{I'+I-M'-M}\langle I'-M'|T\mathcal{H}(A^{\mu}_{L}(\gamma))T^{-1}|I-M\rangle^{\times}$$

$$= (-1)^{I'+I-M'-M+\mu+\beta_{\gamma}}\langle I'-M'|\mathcal{H}(A^{-\mu}_{L}(\gamma)|I-M\rangle^{\times} \qquad \text{(X-47)}$$

where $\gamma = e$ or m and $\beta_e = 0, \beta_m = 1$. Finally, using the Wigner–Eckart theorem and the symmetry properties of Clebsch–Gordan coefficients it follows that

$$e_L = (-1)^L e^{\times}_{L}; \quad m_L = (-1)^{L+1} m^{\times}_{L} \qquad \text{(X-48)}$$

where e_L and m_L are the reduced matrix elements

$$e_L = \langle I'\|\mathcal{H}(A_L(e))\|I\rangle ; \quad m_L = \langle I'\|\mathcal{H}(A_L(m))\|I\rangle . \qquad \text{(X-49)}$$

Because of the parity selection rule (neglecting weak interaction effects) which requires non-vanishing e_L and m_L for a given transition to be those for which L is odd (even) and even (odd) respectively it follows that interfering multipoles in a gamma transition have a relative phase of either O or π. By suitable choice of phase for the nuclear states involved they can therefore be made relatively real. This result was first demonstrated by Lloyd (1951) (see also Rose and Brink (1967)).

Should however, time reversal invariance not hold the relations given in eq. (X-48) no longer apply and the phase difference between interfering multipoles can differ from O or π. Measurement of this phase difference through a T-violating angular correlation (see §X-7.2) can then give information about the magnitude of T-violating terms in the Hamiltonian. For most practical purposes the interference with which we are concerned is that between a magnetic multipole m_L (say) and an electric multipole e_{L+1}. The interference is normally expressed in terms of the quantity δ

$$\delta = \frac{e_{L+1}}{m_L} = |\delta| \exp\{\mathrm{i}(\eta_{L+1}^{(e)} - \eta_L^{(m)})\} = |\delta| \exp\{\mathrm{i}\eta\} \tag{X-50}$$

where η is the relative phase of the two multipoles and equals O or π if there is no T-violation but can differ from these values otherwise.

It is clearly important to relate the value of η to the actual T-violating terms in the nuclear Hamiltonian. Consider, therefore, a Hamiltonian of the form

$$H = H_0 + \mathcal{H}(A) + V_{\mathrm{t.v.}} + \mathcal{H}_{\mathrm{t.v.}}(A) \tag{X-51}$$

where H_0 is a time reversal invariant nuclear Hamiltonian with eigenstates ψ_i and corresponding eigenvalues E_i, $\mathcal{H}(A)$ is the usual electromagnetic interaction, $V_{\mathrm{t.v.}}$ is a parity conserving T-violating internucleon potential and $\mathcal{H}_{\mathrm{t.v.}}(A)$ is a T-violating electromagnetic interaction. For the moment we do not consider the origin or detailed form of the T-violating terms. The last three terms can all be treated as perturbations and to first order in $V_{\mathrm{t.v.}}$ the nuclear eigenstates can be written

$$\Psi_i = \psi_i + \sum_{j \neq i} \frac{\langle j | V_{\mathrm{t.v.}} | i \rangle}{E_i - E_j} \psi_j \tag{X-52}$$

where $\langle j | V_{\mathrm{t.v.}} | i \rangle$ signifies $\langle \psi_j | V_{\mathrm{t.v.}} | \psi_i \rangle$.

Allowing for T-violating effects, the matrix element of the electromagnetic interaction between two states Ψ_a and Ψ_b can be written

$$\langle\Psi_b|\mathcal{H}(\boldsymbol{A})+\mathcal{H}_{\text{t.v.}}(\boldsymbol{A})|\Psi_a\rangle=\langle b|\mathcal{H}(\boldsymbol{A})|a\rangle+\langle b|\mathcal{H}_{\text{t.v.}}(\boldsymbol{A})|a\rangle$$

$$+\sum_{j\neq a}\frac{\langle b|\mathcal{H}(\boldsymbol{A})|j\rangle\langle j|V_{\text{t.v.}}|a\rangle}{E_a-E_j}+\sum_{j\neq b}\frac{\langle b|V_{\text{t.v.}}|j\rangle\langle j|\mathcal{H}(\boldsymbol{A})|a\rangle}{E_b-E_j}. \tag{X-53}$$

Correspondingly, using a multipole expansion for $\mathcal{H}(\boldsymbol{A})$ and $\mathcal{H}_{\text{t.v.}}(\boldsymbol{A})$ (cf. eq. (X-45)), we can write

$$\langle\Psi_b|\mathcal{H}(A_L^\mu(\gamma))+\mathcal{H}_{\text{t.v.}}(A_L^\mu(\gamma))|\Psi_a\rangle=\langle b|\mathcal{H}(A_L^\mu(\gamma))|a\rangle\,(1+\mathrm{i}\,\epsilon_L(\gamma))$$

$$\approx\langle b|\mathcal{H}(A_L^\mu(\gamma))|a\rangle\,\mathrm{e}^{\mathrm{i}\,\epsilon_L(\gamma)} \tag{X-54}$$

where

$$\mathrm{i}\,\epsilon_L(\gamma)=\langle b|\mathcal{H}(A_L^\mu(\gamma))|a\rangle^{-1}\,[\langle b|\mathcal{H}_{\text{t.v.}}(A_L^\mu(\gamma))|a\rangle$$

$$+\sum_{j\neq a}\frac{\langle b|\mathcal{H}(A_L^\mu(\gamma))|j\rangle\langle j|V_{\text{t.v.}}|a\rangle}{E_a-E_j}+\sum_{j\neq b}\frac{\langle b|V_{\text{t.v.}}|j\rangle\langle j|\mathcal{H}(A_L^\mu(\gamma))|a\rangle}{E_b-E_j} \tag{X-55}$$

and $\gamma = e$ or m. Here, by following procedures analogous to those which led to eq. (X-48) it is straightforward to show that $\epsilon_L(\gamma)$ is real. It is also, clearly, first order in the T-violating terms. Thus, referring to eq. (X-50), we can write for this T-violating case

$$\delta=\frac{e_{L+1}}{m_L}=|\delta|\,\mathrm{e}^{\mathrm{i}(\eta_0+\epsilon)} \tag{X-56}$$

where $\eta_0 = 0$ or π and $\epsilon=\epsilon_{L+1}(e)-\epsilon_L(m)\approx\mathcal{F}$ where $\mathcal{F}$ is a measure of the relative strength of the T-violating part of the Hamiltonian. To proceed further requires the specific forms of $V_{\text{t.v.}}$ and $\mathcal{H}_{\text{t.v.}}(\boldsymbol{A})$ to be specified.

As a simple example, therefore, we consider here the case of a gauge invariant single particle T-violating Hamiltonian of the form

$$H=H_0+H_{\text{t.v.}} \tag{X-57}$$

where

$$H_0=\frac{p^2}{2m}-\frac{e}{2mc}(\boldsymbol{p}\cdot\boldsymbol{A}+\boldsymbol{A}\cdot\boldsymbol{p})+V(r)+f(r)\,\boldsymbol{l}\cdot\boldsymbol{s}-\frac{e}{c}f(r)\,\boldsymbol{r}\times\boldsymbol{A}\cdot\boldsymbol{s}$$

$$H_{\text{t.v.}}=[\alpha(r)\,\boldsymbol{r}\cdot\boldsymbol{p}+\boldsymbol{p}\cdot\boldsymbol{r}\,\alpha(r)]-\frac{2e}{c}\,\alpha(r)\,\boldsymbol{r}\cdot\boldsymbol{A}\,.$$

Here H_0 is a simple shell model type Hamiltonian including a spin-orbit force and corresponding electromagnetic interaction and $H_{\mathrm{t.v.}}$ consists of a single particle T-violating potential (see eq. (X-1)) together with a T-violating electromagnetic interaction deriving from the T-violating potential through gauge invariance. This Hamiltonian can be diagonalised to first order in $\alpha(r)$ through the transformation

$$H = \mathrm{e}^{\mathrm{i}S} H_0 \mathrm{e}^{-\mathrm{i}S}; \qquad \Psi_i = \mathrm{e}^{-\mathrm{i}S}\, \psi_i \tag{X-58}$$

where

$$S = \beta(r) = -\int_0^r \frac{2m}{\hbar}\, r'\alpha(r')\,\mathrm{d}r' \tag{X-59}$$

and the ψ_i are eigenstates of H_0. It then follows that the matrix element of an electromagnetic multipole operator $\mathcal{H}(A_L^\mu(m))$ or $\mathcal{H}(A_L^\mu(e))$ between states Ψ_b and Ψ_a is given by

$$\langle \Psi_b | \mathcal{H}(A_L^\mu(\gamma)) | \Psi_a \rangle \triangleq \langle b | \mathcal{H}(A_L^\mu(\gamma)) | a \rangle + \langle b | -\mathrm{i}\, [S, \mathcal{H}(A_L^\mu(\gamma))] | a \rangle\,. \tag{X-60}$$

However, since $S\ (= \beta(r))$ is a function of r only, the commutator in the second term vanishes. Thus, comparing with eq. (X-54) we have $\epsilon_L(\gamma) = 0$ and to this approximation there is no modification of the phase of δ due to T-violating effects *.

The preceding result generalises to a many body situation provided no two body forces are allowed and follows because of the simplicity of the Hamiltonian used. Should, however, the restriction to single particle potentials be removed and two body potentials be introduced into H, whether or not they violate time reversal invariance, then a finite value for $\epsilon_L(\gamma)$ is to be expected. The result therefore suggests that T-violating effects are most likely to show up in gamma transitions in a complex nucleus rather than in, for example, a nucleus like ^{17}O whose wavefunctions can be reasonably approximated by single particle states.

So far, no detailed calculations relating the value of $\epsilon(\gamma)$ to the T-violating potential have been carried out and there is obviously a need to remedy this. Nevertheless we can proceed further and study the way in which finite values for the $\epsilon_L(\gamma)$ can lead to manifestly T-violating effects in angular correlation processes. This is the concern of the next section.

* That there should be no effect due to the T-violating electromagnetic interaction in this single particle problem has already been demonstrated more generally in § X-3.

X-7.2. *T-violation in γ-decay angular correlations*

Before considering the detailed form of T-violating effects in angular correlations we first look at their general forms as predicted and limited by invariance arguments. Such a study was first made by Jacobsohn and Henley (1959) and has been extended and further discussed by Boehm (1968) and Henley (1969a).

Physically the situation can be described as one in which the gamma transition under study (i.e. involving interfering multipoles) is between two nuclear states, $a \rightarrow b$, about both of which orientation information is available either directly or indirectly. The "degree of orientation" of a nuclear state is denoted by Ω (= 0, 1, 2, ...) where Ω labels the representation of the rotation group according to which the spin statistical tensor transforms. Thus $\Omega = 0$ corresponds to a completely unoriented state, $\Omega = 1$ to a polarised state, $\Omega = 2$ to an aligned state and so on. For a given degree of orientation, Ω, the nuclear spin must satisfy $I \geq \Omega/2$. Referring to the orientation parameters $B_\nu(I)$ introduced in §IX-3.2, where I is the nuclear spin, a state with degree of orientation Ω is one for which $B_\Omega(I) \neq 0$. Put another way, Ω is also the power of the nuclear spin operator $\boldsymbol{I}$ which must be measured.

The orientation (polarisation) of the initial state a can be achieved by various means e.g. a strong magnetic field or a crystalline electric field at low temperatures, capture of polarised neutrons, a preceding β-ray or a preceding gamma ray. Similarly the orientation of the final state b can be detected (analysed) by appropriate measurements of the polarisation and momentum of a subsequent gamma-ray. In addition the linear polarisation ($\boldsymbol{\varepsilon}$) or circular polarisation ($\boldsymbol{P}$) of the gamma transition $a \rightarrow b$ can also be measured. Here it should be noted that since the direction of linear polarisation has no positive or negative sense the unit vector $\boldsymbol{\varepsilon}$ must occur an even number of times in any physically meaningful expression.

Boehm (1968) has gathered together in illuminating tabular form different T-violating vector combinations which can occur in angular correlation measurements in various different physical situations. Some elements of his tables are reproduced in table X-1 where each vector combination is explicitly T-violating (remember that spin $\boldsymbol{I}$, circular polarisation $\boldsymbol{\sigma}$ and momenta $\boldsymbol{k}$ and $\boldsymbol{p}$ are odd under time reversal). The table is not exhaustive, but indicates the simplest T-violating angular correlations. Ω_a and Ω_b denote the degrees of orientation necessary in order to achieve the desired angular correlation. It should be noted here that final state interactions can also lead to terms of the type included in table X-1 even without T-violation. These effects are, in general, small and will be discussed in more detail in §X-7.5.

The next step is to consider the detailed form of T-violating angular corre-

Table X-1

T-violating terms in γ-ray angular correlation measurements [1)]

Orientation Ω_a	Ω_b	Polarisation of gamma-ray $a \to b$	Gamma-ray $a \to b$ measured	Second gamma-ray (analyser) measured [2)]	Preceding gamma-ray (polariser) and subsequent gamma-ray (analyser) measured	Preceding beta-decay (polariser) and subsequent gamma-ray (analyser) measured
			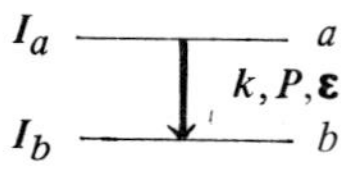	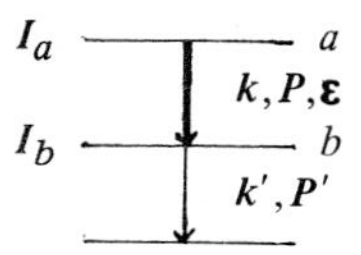	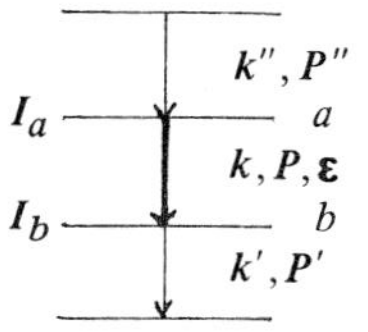	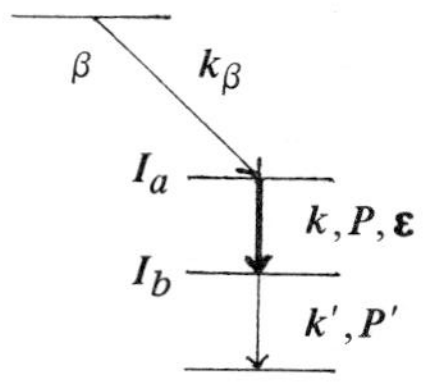
2	1	Polarisation	$(\boldsymbol{k}\cdot\boldsymbol{I}_a\times\boldsymbol{I}_b)(\boldsymbol{k}\cdot\boldsymbol{I}_a)$	$(\boldsymbol{k}\cdot\boldsymbol{I}_a\times\boldsymbol{P}')(\boldsymbol{k}\cdot\boldsymbol{I}_a)$	$(\boldsymbol{k}\cdot\boldsymbol{k}''\times\boldsymbol{P}')(\boldsymbol{k}\cdot\boldsymbol{k}'')$	—
1	2	not measured	$(\boldsymbol{k}\cdot\boldsymbol{I}_b\times\boldsymbol{I}_a)(\boldsymbol{k}\cdot\boldsymbol{I}_b)$	$(\boldsymbol{k}\cdot\boldsymbol{k}'\times\boldsymbol{I}_a)(\boldsymbol{k}\cdot\boldsymbol{k}')$	$(\boldsymbol{k}\cdot\boldsymbol{k}'\times\boldsymbol{P}'')(\boldsymbol{k}\cdot\boldsymbol{k}')$	$(\boldsymbol{k}\cdot\boldsymbol{k}'\times\boldsymbol{k}_\beta)(\boldsymbol{k}\cdot\boldsymbol{k}')$
1	1	Circular polarisation ***P***	$\boldsymbol{P}\cdot\boldsymbol{I}_a\times\boldsymbol{I}_b$ [3)]	$\boldsymbol{P}\cdot\boldsymbol{I}_a\times\boldsymbol{P}'$	$\boldsymbol{P}\cdot\boldsymbol{P}''\times\boldsymbol{P}'$	$\boldsymbol{P}\cdot\boldsymbol{k}_\beta\times\boldsymbol{P}'$
2	2	measured	$(\boldsymbol{P}\cdot\boldsymbol{I}_a\times\boldsymbol{I}_b)(\boldsymbol{I}_a\cdot\boldsymbol{I}_b)$	$(\boldsymbol{P}\cdot\boldsymbol{I}_a\times\boldsymbol{k}')(\boldsymbol{I}_a\cdot\boldsymbol{k}')$	$(\boldsymbol{P}\cdot\boldsymbol{k}''\times\boldsymbol{k}')(\boldsymbol{k}''\cdot\boldsymbol{k}')$	–
3	0		$(\boldsymbol{k}\cdot\boldsymbol{I}_a\times\boldsymbol{\varepsilon})(\boldsymbol{I}_a\cdot\boldsymbol{\varepsilon})(\boldsymbol{I}_a\cdot\boldsymbol{k})$	–	$(\boldsymbol{k}\cdot\boldsymbol{k}''\times\boldsymbol{\varepsilon})(\boldsymbol{k}''\cdot\boldsymbol{\varepsilon})(\boldsymbol{P}''\cdot\boldsymbol{k})$ [4)]	$(\boldsymbol{k}\cdot\boldsymbol{k}_\beta\times\boldsymbol{\varepsilon})(\boldsymbol{k}_\beta\cdot\boldsymbol{\varepsilon})(\boldsymbol{k}_\beta\cdot\boldsymbol{k})$ [5)]
2	1	Linear	$(\boldsymbol{k}\cdot\boldsymbol{I}_a\times\boldsymbol{\varepsilon})(\boldsymbol{I}_a\cdot\boldsymbol{\varepsilon})(\boldsymbol{I}_b\cdot\boldsymbol{k})$	$(\boldsymbol{k}\cdot\boldsymbol{I}_a\times\boldsymbol{\varepsilon})(\boldsymbol{I}_a\cdot\boldsymbol{\varepsilon})(\boldsymbol{P}'\cdot\boldsymbol{k})$	$(\boldsymbol{k}\cdot\boldsymbol{k}''\times\boldsymbol{\varepsilon})(\boldsymbol{k}''\cdot\boldsymbol{\varepsilon})(\boldsymbol{P}'\cdot\boldsymbol{k})$	–
1	2	polarisation	$(\boldsymbol{k}\cdot\boldsymbol{I}_a\times\boldsymbol{\varepsilon})(\boldsymbol{I}_b\cdot\boldsymbol{\varepsilon})(\boldsymbol{I}_b\cdot\boldsymbol{k})$	$(\boldsymbol{k}\cdot\boldsymbol{k}'\times\boldsymbol{\varepsilon})(\boldsymbol{I}_a\cdot\boldsymbol{\varepsilon})(\boldsymbol{k}'\cdot\boldsymbol{k})$	$(\boldsymbol{k}\cdot\boldsymbol{k}'\times\boldsymbol{\varepsilon})(\boldsymbol{P}''\cdot\boldsymbol{\varepsilon})(\boldsymbol{k}'\cdot\boldsymbol{k})$	$(\boldsymbol{k}\cdot\boldsymbol{k}'\times\boldsymbol{\varepsilon})(\boldsymbol{k}_\beta\cdot\boldsymbol{\varepsilon})(\boldsymbol{k}'\cdot\boldsymbol{k})$
0	3	**ε** measured	$(\boldsymbol{k}\cdot\boldsymbol{I}_b\times\boldsymbol{\varepsilon})(\boldsymbol{I}_b\cdot\boldsymbol{\varepsilon})(\boldsymbol{I}_b\cdot\boldsymbol{k})$	$(\boldsymbol{k}\cdot\boldsymbol{k}'\times\boldsymbol{\varepsilon})(\boldsymbol{k}'\cdot\boldsymbol{\varepsilon})(\boldsymbol{P}'\cdot\boldsymbol{k})$	–	–

1) Notation: ***I*** = nuclear spin, ***k*** = gamma-ray momentum, ***P*** = gamma-ray circular polarisation, **ε** = gamma-ray linear-polarisation, $\boldsymbol{k}_\beta$ = beta-ray momentum.

2) The experimental situation would involve orienting the state ***a*** by some means e.g. by electric or magnetic fields at low temperature, neutron capture etc.

3) This term can also be written $(\boldsymbol{k}\cdot\boldsymbol{I}_a\times\boldsymbol{I}_b)(\boldsymbol{k}\cdot\boldsymbol{P})$.

4) This does not involve a measurement of the gamma-ray from level *b*.

5) Since $\Omega_a = 3$ it is necessary for the preceding β-decay to be first-forbidden and for $I_a \geq \frac{3}{2}$.

lations, i.e. to relate the strength of the different terms listed in table X-1 to the value of ϵ (see eq. X-56) where ϵ measures the difference of the relative phase of interfering multipoles from 0 or π.

X-7.3. *Detailed form of T-violating angular correlations*

The calculation of the different possible T-violating angular correlations is straightforward using standard methods, but extremely tedious. They have been carried out by Stichel (1958), Jacobsohn and Henley (1959), de Sabbata (1961), Fuschini *et al.* (1964a), Lobov (1965), Krupchitsky and Lobov (1969), Coutinho and Ridley (1971).

Physically the nature of these calculations can be described simply as follows (Jacobsohn and Henley (1959)). Considering the basic transition $a \to b$ under study, denote the amplitude for a transition between the magnetic substates M_a and M_b in which a photon of momentum $\boldsymbol{k}$ and circular polarisation P_0 $(= \pm 1)$ is emitted by $\langle I_b M_b \boldsymbol{k} P_0 | R | I_a M_a \rangle$. Referring to eq. (X-45) and taking $f(P) = \delta_{PP_0}$ it is clear that this amplitude is given explicitly by

$$\langle I_b M_b \boldsymbol{k} P_0 | R | I_a M_a \rangle = \pi \sum_{L\mu} (2L+1)^{1/2} D^{(L)}_{\mu P_0}(\boldsymbol{k}) [\langle I_b M_b | \mathcal{H}(A^{\mu}_L(m)) | I_a M_a \rangle$$

$$+ (-P_0) \langle I_b M_b | \mathcal{H}(A^{\mu}_L(e)) | I_a M_a \rangle] \,. \tag{X-61}$$

Further, a corresponding "probability" for such photon emission when the states $\boldsymbol{a}$ and $\boldsymbol{b}$ have orientations (Ω_a, ω_a) and (Ω_b, ω_b) respectively * can be defined, namely

$$W(I_b \Omega_b \omega_b; P' \boldsymbol{k} P; I_a \Omega_a \omega_a) \equiv \sum_{M_b' M_b M_a' M_a} (-1)^{I_b + M_b} C(I_b I_b \Omega_b; \omega_b - M_b M_b)$$

$$\times \langle I_b \omega_b - M_b \boldsymbol{k} P' | R | I_a \omega_a - M_a \rangle^{\times} \langle I_b M_b \boldsymbol{k} P' | R | I_a M_a \rangle (-1)^{I_a + M_a}$$

$$\times C(I_a I_a \Omega_a; \omega_a - M_a M_a) \,. \tag{X-62}$$

Substituting from eq. (X-61) into eq. (X-62) then gives after some manipulation

* The labels (Ω, ω) for a state of angular momentum J occur in the expansion of the density matrix $\langle IM' | \rho | IM \rangle$ as follows

$$\langle IM' | \rho | IM \rangle = \sum_{\Omega,\omega} (-1)^{I+M'} (2\Omega+1)^{1/2} \begin{pmatrix} I & I & \Omega \\ -M & M & -\omega \end{pmatrix} \rho_I(\Omega,\omega)$$

where $\begin{pmatrix} I & I & \Omega \\ -M & M & -\omega \end{pmatrix}$ is a $3-j$ coefficient and $\rho_I(\Omega, \omega)$ a statistical tensor.

$$W(I_b\Omega_b\omega_b;P'kP;I_a\Omega_a\omega_a)$$

$$= N_{ba}^2 \sum_{\lambda,\nu,\mu} (2\lambda+1)D_{\nu\mu}^{(\lambda)}(k)[(2\Omega_b+1)(2\Omega_a+1)]^{\frac{1}{2}} \begin{pmatrix} \Omega_b & \Omega_a & \lambda \\ -\omega_b & \omega_a & \nu \end{pmatrix}$$

$$\times \sum_{LL'} [(2L'+1)(2L+1)]^{\frac{1}{2}} \begin{pmatrix} L' & L & \lambda \\ P' & -P & -\mu \end{pmatrix} (-)^{L'+\lambda+\Omega_a+\omega_a+1} \begin{Bmatrix} I_b & I_a & L' \\ I_b & I_a & L \\ \Omega_b & \Omega_a & \lambda \end{Bmatrix}$$

$$\times (e_{L'}+(-P')m_{L'})(e_L+(-P)m_L)^{\times} \tag{X-63}$$

where { } is a 9–j symbol, the e_L and m_L are the usual reduced matrix elements and N_{ba} is a normalisation constant independent of the orientations of the states a and b. The expression for W obviously depends on the relative phases of the interfering multipoles. Further, if parity is conserved, for a given L either e_L or m_L is identically zero.

The different possible angular correlations referred to in table X-1 can be constructed assuming that intermediate states are unperturbed, by multiplying W's for each transition together and summing over all intermediate (unobserved) orientations. For example, the angular correlation function for two successive gamma rays $\boldsymbol{k}$ and $\boldsymbol{k}'$ in the sequence $a \xrightarrow{k} b \xrightarrow{k'} c$ and for which the polarisation P of the first is not measured and that of the second is P', is given by

$$W_{\text{corr}}(I_c\Omega_c\omega_c;P'k'I_bk;I_a\Omega_a\omega_a) = \sum_{\Omega_a,\omega_a,\Omega_b,\omega_b,P=\pm1} \rho_{I_a}(\Omega_a,\omega_a)$$

$$\times W(I_c\Omega_c\omega_c;P'k'P';I_b\Omega_b\omega_b)\, W(I_b\Omega_b\omega_b;PkP;I_a\Omega_a\omega_a) \tag{X-64}$$

where $\rho_{I_a}(\Omega_a,\omega_a)$ is the statistical tensor (see footnote on page 225) describing the orientation of the state a. If further the orientation of the final state c is not measured, then $\Omega_c = \omega_c = 0$.

As stated earlier and as must be clear from the foregoing paragraphs extremely detailed and tedious work is necessary in order to arrive at final forms for the different angular correlations. Here, therefore, we just quote as an example an expression obtained by Krupchitsky and Lobov (1969). This describes in detail the situation just referred to when a polarised nucleus (i.e. $\Omega = 1$) emits two cascade photons, but where the polarisation of neither photon is measured. It is assumed that in the first transition ($a \rightarrow b$) only the multipoles ML and $E(L+1)$ contribute and in the second ($b \rightarrow c$) only ML'

and $E(L'+1)$. The following angular correlation expression is obtained.

$$W = 1 + \{(1+|\delta|^2)(1+|\delta'|^2)\}^{-1} \Bigg[\sum_{\nu=2,4,\ldots} A_\nu(ab) A_\nu(bc) P_\nu(\boldsymbol{k}\cdot\boldsymbol{k}')$$

$$- 2\sqrt{3}\, B_1(I_a)(\hat{\boldsymbol{k}}\cdot\hat{\boldsymbol{k}}'\times\hat{\boldsymbol{n}}_a)|\delta| \sin\eta \sum_{\nu=2,4,\ldots} \left(\frac{2\nu+1}{\nu(\nu+1)}\right)^{1/2}$$

$$\times F^{\nu}_{1\nu}(LL+1I_bI_a) A_\nu(bc) P'_\nu(\boldsymbol{k}\cdot\boldsymbol{k}')\Bigg] \qquad \text{(X-65)}$$

where $P'_\nu(x) = (\mathrm{d}/\mathrm{d}x)\, P_\nu(x)$,

$$\delta = \frac{e_{L+1}}{m_L} = |\delta|\,\mathrm{e}^{\mathrm{i}\eta} = |\delta|\,\mathrm{e}^{\mathrm{i}(\eta_0+\epsilon)} \qquad \text{(see eq. X-56)}$$

$$\delta' = \frac{e_{L'+1}}{m_{L'}}$$

and $\hat{\boldsymbol{k}}, \hat{\boldsymbol{k}}'$ and $\hat{\boldsymbol{n}}_a$ are unit vectors in the direction of emission of the two gamma-rays and the spin polarisation of the state a respectively.

$$A_\nu(ab) = F_\nu(LLI_aI_b) + 2\delta\cos\eta\, F_\nu(LL+1I_aI_b) + |\delta|^2 F_\nu(L+1\; L+1\, I_aI_b)$$

$$A_\nu(bc) = F_\nu(L'L'I_cI_b) + 2\delta'\cos\eta'\, F_\nu(L'L'+1I_cI_b) + |\delta'|^2 F_\nu(L'+1\; L'+1\, I_cI_b)$$

$$F^{\gamma}_{\alpha\beta}(L_1L_2I_bI_a) = (-)^{L_1-1}\{(2I_b+1)(2I_a+1)(2L_1+1)(2L_2+1)\}^{1/2}$$

$$\times\, C(L_1L_2\gamma; 1-1)\, X(I_aI_a\alpha; L_1L_2\gamma; I_bI_b\beta)\,.$$

$B_1(I_a)$ is the nuclear-polarisation orientation parameter for the state a defined by (see eq. (IX-27))

$$B_1(I_a) = \frac{\sqrt{3}\sum_{M_a} M_a\, p(M_a)}{[I_a(I_a+1)]^{1/2}} \qquad \text{(X-66)}$$

where $p(M_a)$ is the probability of finding the nucleus in the substate M_a. $B_1(I_a)$ is related to the magnitude of the polarisation P_a of the state by

$$P_a = \frac{\sum_{M_a} M_a\, p(M_a)}{I_a} = \left(\frac{I_a+1}{3I_a}\right)^{\frac{1}{2}} B_1(I_a)\,. \tag{X-67}$$

In the above formulae the F's are angular correlation parameters referred to in §IX-3.2 and tabulated, for example, by Frauenfelder and Steffen (1966) and Alder *et al.* (1967), and X is a Fano coefficient. It should be stressed that the formula given in eq. (X-65) holds only if $\Omega_a = 1$ and that additional terms (which may, however, be negligible in practice) arise if higher order orientation occurs. Further the expression is only correct if $\boldsymbol{n}_a$ is parallel or anti-parallel to the vector $\boldsymbol{k} \times \boldsymbol{k}'$. The generalisation to arbitrary values of Ω_a and arbitrary relative directions of the vectors $\boldsymbol{k}, \boldsymbol{k}'$ and $\boldsymbol{n}_a$ has been carried out by Coutinho and Ridley (1971). One final point to be underlined is that the T-violating effect is shown explicitly to be proportional to $\sin\eta$, i.e. to the phase difference occurring in the first gamma-transition and it must always be remembered that this is the transition under investigation. The second transition is measured solely to obtain information about the orientation of the state b. The actual structure of the T-violating term is clearly of the form given in column 4 of table X-1 for the case "$\Omega_a = 1, \Omega_b = 2$, polarisation of gamma-ray $a \to b$ not measured".

It is also possible to reproduce directly the equivalent result given in column 6 when the polarisation of the state a is achieved by means of a preceding β-decay of momentum $\boldsymbol{k}_\beta$. In this case $\hat{\boldsymbol{n}}_a = \hat{\boldsymbol{k}}_\beta$ and in §IX-3.4 we have shown that for an allowed β-decay

$$B_1(I_a) = \frac{k_\beta}{E_\beta}\left(\frac{I_a+1}{3I_a}\right)^{\frac{1}{2}} \left[\pm\lambda_{I_\beta I_a} + 2\delta_{I_\beta I_a}|y|\left(\frac{I_a}{I_a+1}\right)^{\frac{1}{2}}\right][1+|y|^2]^{-1} \tag{X-68}$$

where I_β is the spin of the β-emitting state, $y = (G_V\langle t_\pm\rangle)/(G_A\langle \boldsymbol{\sigma} t_\pm\rangle)$, $\langle t_\pm\rangle$ and $\langle \boldsymbol{\sigma} t_\pm\rangle$ are the usual allowed β-decay matrix elements and

$$\begin{aligned} \lambda_{I_\beta I_a} &= 1 &&\text{for } I_\beta = I_a - 1 \\ &= 1/(I_a+1) &&\text{for } I_\beta = I_a \\ &= -I_a/(I_a+1) &&\text{for } I_\beta = I_a + 1\,. \end{aligned}$$

In eq. (X-68) the upper sign refers to β^- decay and the lower to β^+ decay. Similarly the state a could be polarised by means of the capture of thermal neutrons having polarisation P_n in the direction $\hat{\boldsymbol{n}}_a$. In this case (see §IX-3.5)

$$B_1(I_a) = \frac{\frac{3}{4} + I_a(I_a+1) - I_i(I_i+1)}{[3I_a(I_a+1)]^{\frac{1}{2}}} P_n$$

where I_i is the spin of the initial capturing nuclear state.

The foregoing examples illustrate the complexity of the T-violating angular correlations * occurring even in the simplest experimental configurations. Clearly it is important to choose cascades in which there is large mixing in the transition under test so that $|\delta| \approx 1$, and such that the other coefficients of $\sin\eta$ are as large as possible. But of paramount importance is the requirement that $\sin\eta$ (= $\sin(\eta_0 - \epsilon) = \pm \sin\epsilon$) should be large. Inspection of eqs. (X-55) and (X-56) shows that this is likely to occur if, consistent with $|\delta| \approx 1$, there is inhibition of one or other of the multipoles concerned, i.e. $\langle 2|\mathcal{H}(A_L^{\mu}(\gamma))|1\rangle$ is relatively small. In the following sections some recent experimental studies using suitable cascades are discussed.

X-7.4. *Experimental studies of T-violating angular correlations*

A number of experiments have been carried out during the last few years seaching for T-violating angular correlations and these are listed in table X-2. These experiments are of four types. Firstly, there are those (Fuschini *et al.* (1964b); Garrell *et al.* (1969); Perkins and Ritter (1968)) in which the initial state a of the transition being studied is polarised by means of a preceding β-decay and in which the orientation of the final state b is measured by means of a subsequent gamma-ray. Secondly, there are those (Kajfosz *et al.* (1966, 1968); Eichler (1968, 1969); Bulgakov *et al.* (1972)) in which the initial state a is polarized by means of a captured polarised thermal neutron and in which, again, the orientation of the final state b is measured by means of a subsequent γ-ray. Most of these experiments have so far enabled upper limits to be imposed on $\sin\epsilon$ of the order a few times 10^{-2}. More effective in this respect has been the work of Kistner (1967) and Atac *et al.* (1968) who, using the Mössbauer effect, studied the angular correlation of linearly polarised photons in a gamma transition between specific initial and final nuclear substates (i.e. between oriented nuclear states). Most recently Holmes *et al.* (1971) have achieved similar accuracies using a dilution refrigerator to polarise the initial state a of the transition being studied.

In the case of Kistner's experiment gamma rays emitted by a single line ^{99}Ru source were passed (in the direction $\boldsymbol{k}$) through a polarising filter

* An interesting experimental variant on the correlation described by eq. (X-68) is to be noted which makes use of resonance fluorescence on a polarised scatterer to define the vectors $\hat{k}, \hat{k}'$ and $\hat{n}_a$ (Burgov and Lobov (1967)).

Table X-2
Experimental tests of *T*-violation in γ-decay

Nucleus	Transition	δ	E_γ(MeV)	Polariser	Analyser	Measured correlation	$\sin\epsilon \times 10^3$	Reference
^{36}Cl	$2^+ \xrightarrow{E2+M1} 3^+$	+0.21	7.79	Captured polarised thermal neutrons [1]	Subsequent γ-decay (E2 + M1)	$(\boldsymbol{k}\cdot\boldsymbol{k}'\times\boldsymbol{s}_n)(\boldsymbol{k}\cdot\boldsymbol{k}')$	4 ± 12 −0.9 ± 2.9	Eichler (1968,1969) Bulgakov *et al.* (1972)
^{49}Ti	$\frac{1}{2}^- \xrightarrow{E2+M1} \frac{3}{2}^-$	+2.2	0.34	Captured polarised thermal neutrons [1]	Subsequent γ-decay (E2)	$(\boldsymbol{k}\cdot\boldsymbol{k}'\times\boldsymbol{s}_n)(\boldsymbol{k}\cdot\boldsymbol{k}')$	17 ± 25	Kajfosz *et al.* (1966, 1968)
^{56}Mn	$2^+ \xrightarrow{E2+M1} 2^+$	+0.18	1.81	Beta-decay	Subsequent γ-decay (E2)	$(\boldsymbol{k}\cdot\boldsymbol{k}'\times\boldsymbol{k}_\beta)(\boldsymbol{k}\cdot\boldsymbol{k}')$	−45 ± 27	Garrell *et al.* (1969)
^{56}Mn	$2^+ \xrightarrow{E2+M1} 2^+$	−0.28	2.12	Beta-decay	Subsequent γ-decay (E2)	$(\boldsymbol{k}\cdot\boldsymbol{k}'\times\boldsymbol{k}_\beta)(\boldsymbol{k}\cdot\boldsymbol{k}')$	−26 ± 14	Garrell *et al.* (1969)
^{99}Ru	$\frac{3}{2}^+ \xrightarrow{E2+M1} \frac{5}{2}^+$	−1.64	0.090	Ru absorber	Resonant (Möss-bauer) absorption	$(\boldsymbol{k}\cdot\boldsymbol{I}\times\boldsymbol{\varepsilon})(\boldsymbol{I}\cdot\boldsymbol{\varepsilon})(\boldsymbol{I}\cdot\boldsymbol{k})$	1.0 ± 1.7 [0.0 ± 1.7] [2]	Kistner (1967)
^{106}Rh	$2^+ \xrightarrow{E2+M1} 2^+$	−0.2	1.045	Beta-decay	Subsequent γ-decay (E2)	$(\boldsymbol{k}\cdot\boldsymbol{k}'\times\boldsymbol{k}_\beta)(\boldsymbol{k}\cdot\boldsymbol{k}')$	30 ± 40	Fuschini *et al.* (1964b)
^{106}Pd	$2^+ \xrightarrow{E2+M1} 2^+$	−0.2	1.050	Beta-decay	Subsequent γ-decay (E2)	$(\boldsymbol{k}\cdot\boldsymbol{k}'\times\boldsymbol{k}_\beta)(\boldsymbol{k}\cdot\boldsymbol{k}')$	4.0 ± 18	Perkins and Ritter (1968)
^{192}Pt	$3^+ \xrightarrow{E2+M1} 2^+$	−2.2 and [3] +8.2	0.604 and [3] 0.308	Dilution refrigerator	Subsequent γ-decay (E2)	$(\boldsymbol{I}\cdot\boldsymbol{k}\times\boldsymbol{k}')$	[5 ± 6] [3]	Holmes *et al.* (1971)
^{193}Ir	$\frac{1}{2}^+ \xrightarrow{E2+M1} \frac{3}{2}^+$	+0.556	0.073	Ir absorber	Resonant emission	$(\boldsymbol{k}\cdot\boldsymbol{I}\times\boldsymbol{\varepsilon})(\boldsymbol{I}\cdot\boldsymbol{\varepsilon})(\boldsymbol{I}\cdot\boldsymbol{k})$	1.1 ± 3.8 [0.2 ± 3.8] [2]	Atac *et al.* (1968)

Notes: 1) Neutron spin is denoted by $\boldsymbol{s}_n$ in column 7
2) Corrected for phase shift due to internal conversion (see § X-7.5)
3) Measurements do not distinguish between two cascades each involving a $3^+ \to 2^+$ mixed transition. The value for $\sin\epsilon$ is the sum of contributions from both cascades.

(a Ru-Fe absorber magnetised in the transverse direction) so that the transmitted radiation was linearly polarised ($\boldsymbol{\varepsilon}$). The radiation then passed through a second absorber in a magnetic field and a comparison was made of the absorption intensities of the two hyperfine transitions $M_a = \frac{5}{2} \rightleftharpoons M_b = \frac{3}{2}$ and $M_a = -\frac{5}{2} \rightleftharpoons M_b = -\frac{3}{2}$. In addition the change in intensity for each transition was measured when the latter magnetic field was reversed. Effectively such procedures test for the presence of an angular correlation of the form $(\boldsymbol{k} \cdot \boldsymbol{I} \times \boldsymbol{\varepsilon})(\boldsymbol{I} \cdot \boldsymbol{\varepsilon})(\boldsymbol{I} \cdot \boldsymbol{k})$. A similar procedure to this was used by Atac *et al.* (1968) and both experiments impose upper limits on $\sin \epsilon$ of the order a few times 10^{-3}, an order of magnitude better than experiments using β-decay or polarised neutrons as the nuclear polarising agents.

The experiment by Holmes *et al.* (1971) is simple in concept and the experimental result quoted is a preliminary result of a continuing experiment so that higher accuracy may be expected in due course. As pointed out in table X-2 it is complicated through the inability of the experimental set up to distinguish between two similar (energetically) cascades. The value quoted, therefore, for $\sin \epsilon$ is the sum of contributions from both cascades.

With accuracies of the order 10^{-3} it is important to evaluate carefully the magnitude of terms of the same form which stem from initial or final state interactions and which could therefore influence the interpretation of the experimental results. This is the concern of the next section.

X-7.5. *Final and initial state interactions and T-violating angular correlations*

Henley and Jacobsohn * (1966) were the first authors to study in detail the magnitude of those terms which mimic T-violating effects. They point out that in calculating angular correlations as in the preceding sections, the electromagnetic matrix elements are evaluated in Born approximation and are therefore Hermitian. However, if higher order terms are taken into account then the transition operator is itself no longer Hermitian and apparent T-violating terms can energe. The reason for this can be understood simply as follows.

Let S denote the transition operator for a transition $a \to b$. Under the successive operations of time-reversal (T) and Hermitian conjugation (H), we have

$$\langle b|S|a\rangle \to \langle a^t|S^t|b^t\rangle \to \langle b^t|S^{t\dagger}|a^t\rangle^{\times} \qquad \text{(X-70)}$$

where the superfix t indicates time reversed. Briefly, under the operation TH, $S \to S^{t\dagger}$. Further we can write S as a sum of a TH-even part (S_+) and a

* See also Karpman *et al.* (1968) for a rather general discussion of this issue.

TH-odd part (S_-), thus

$$S = S_+ + S_- \tag{X-71}$$

where $S_+ = \frac{1}{2}(S + S^{t\dagger})$, $S_- = \frac{1}{2}(S - S^{t\dagger})$.

Now in an angular correlation process in which no measurement is made with initial and final states interchanged (as distinct from tests of detailed balance) it is the S_- component of S which leads to the T-violating terms. In Born approximation (Hermitian operators) an S_- component can only arise from a T-violating potential and $S_- = V_{\text{t.v.}}$. However, even if there are no T-violating terms, higher order (non-Hermitian) contributions to S can lead to a non-vanishing component S_- and hence to simulated T-violation. Such contributions are shown diagramatically in fig. X-5 (diagrams (i) – (iv)) in relation to the basic photon emission process of diagram (i). Their total magnitude has been estimated by Jacobsohn and Henley (1966) who find in the case of an interfering E2 + M1 transition that the contribution to the phase ϵ is $\epsilon_{\text{spoiling}} < 10^{-6}$ and is therefore negligible. The reason why the contribution is so small (compared with the fine structure constant α) is because the imaginary part (i.e. the part contributing to S_-) of the propagator $1/(E - H_0 + i\epsilon)$ for an intermediate state involves the δ-function $\delta(E - H_0)$. This in turn requires at least one intermediate state to be on the energy shell and so introduces a further retardation factor of the order $(kR)^L$ where k is the wave number of the virtual photon, R is a typical dimension of the system and L is multipolarity.

More important in the case of low energy transitions ($\approx$ 100 keV) is the contribution from diagrams of the type (v) shown in fig. X-5 where the dashed loop represents an excited atomic electron together with a hole in a normally filled electronic level. Hannon and Trammell (1968) have considered in detail the effect of currents induced in the K and L electronic shells (for which $r < \lambda_{\text{electronic}}$) due to this effect in the case of interfering M1 and E2 transitions. As in the case of diagrams (ii)–(iv) these currents come mainly from the energy conserving (internal conversion) pole terms and are 90° out of phase with the "driving" nuclear currents. Thus a "spoiling" phase shift, $\epsilon_{\text{spoiling}}$, unrelated to T-violation, will again be introduced. Because atomic dimensions are much larger than nuclear dimensions, the retardation effect referred to earlier is much less inhibiting and Hannon and Trammell (1968) (see also Hannon (1971)) obtain the following values for $\epsilon_{\text{spoiling}}$ in the case of the M1 + E2 γ-decays in ^{99}Ru and ^{193}Ir

$$\epsilon_{\text{spoiling}}(\text{Ru}) = -6.5 \times 10^{-3}; \qquad \epsilon_{\text{spoiling}}(\text{Ir}) = +0.9 \times 10^{-3}.$$

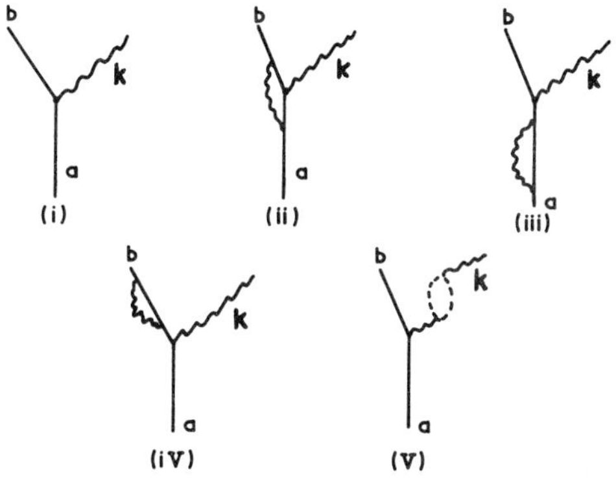

Fig. X-5. Diagrams illustrating higher order corrections to the radiative transition $a \to b + \gamma$.

The smaller effect in Ir arises because K-shell internal conversion is not allowed by energy conservation and only L-shell conversion contributes.

Clearly these effects are of the same order of magnitude as the experimental results given for these decays in table X-2 and as the T-violation effect being searched for. They must therefore be taken into account in deducing limits on T-violation from experiment and this has been done in the table.

In summary simulated T-violating terms can be introduced into γ-decay angular correlations through higher order effects. In the case of higher energy γ-transitions they seem to be negligible ($\epsilon_{\text{spoiling}} \lesssim 10^{-6}$) compared with the foreseeable accuracy of experimental data. On the other hand, in the Mössbauer transition range (≈ 100 keV) they can be important ($\epsilon_{\text{spoiling}} \sim 10^{-3}$) and have to be taken into account fully in determining the magnitude of T-violating effects from experiment.

X-8. **Simultaneous P- and T-violating effects**

As mentioned in the introduction to this chapter effects which are both P- and T-violating are expected to be very small in magnitude in low energy physics processes. Thus, if there is T-violation in the weak interaction they should be no greater than 10^{-6} to 10^{-9} (except in the case of a weak interaction process) whilst if the P-violation stems from the weak interaction and T-violation from a medium strong or electromagnetic interaction, effects are expected of the order 10^{-9} or 10^{-10}. Sensitivities of this order of magnitude are beyond the reach of present day nuclear physics experiments.

It should nevertheless be noted that some theoretical studies have been made of the form that PT-violating effects should take in certain cases. For example, Szymanski (1968) (see also Boehm (1968)) has set out the form of possible PT-violating gamma-ray angular correlations in a tabular form equivalent to table X-1 (which refers only to T-violating angular correlations). He has also estimated the strength of a single particle TP-violating potential of the form $G_{PT}\,\boldsymbol{\sigma}\cdot\boldsymbol{r}$ deriving from the particular T-violating theory of Zachariasen and Zweig (1965). He then considers the case of a linear polarisation correlation experiment in which the measured correlation has the (PT-violating) form $(\boldsymbol{k}\cdot\boldsymbol{k}''\times\boldsymbol{\varepsilon})(\boldsymbol{k}''\cdot\boldsymbol{\varepsilon})$ where $\boldsymbol{k}$ and $\boldsymbol{\varepsilon}$ refer to the transition under test and $\boldsymbol{k}''$ to a preceding transition (see fig. X-6). For the purposes of his calculation the transition under test is taken to be M1 + E2 with an irregular E1 contribution. The PT-violating term is proportional to $G_{PT}R\sin\eta_{(\mathrm{E1,M1})}$ where G_{PT} is the strength of the PT-violating potential, R is the relative amplitude of the irregular E1 to the regular M1 transition and $\eta_{(\mathrm{E1,M1})}$ is the phase different between the E1 and M1 multipole matrix elements. If T invariance holds, $G_{PT}=0$, $\eta_{(\mathrm{E1,M1})}=\pm\pi/2$ * and no effect is observed. If there is PT-violation, $G_{PT}\neq 0$ but assuming the effect is small $\eta_{(\mathrm{E1,M1})}$ will still be close to $\pm\pi/2$. Szymanski (1968) studies the 482 keV $\frac{5}{2}^+\to\frac{7}{2}^+$ decay in ^{181}Ta (see table IX-2) together with the preceding gamma decay and estimates the strength of the PT-violating asymmetry to be $\approx\pm 1\times 10^{-6}$ as predicted by order of magnitude arguments. Such an effect is not measurable at the present time. Other PT-violating angular correlations are in principle measurable but, because of the smallness of the anticipated effect are not treated further here.

The most promising study of PT-violation arises in a field strictly outside the concern of low energy nuclear physics namely the measurement of the electric dipole moment of the neutron ** or a neutral atom ** of spin $\frac{1}{2}$. That the existence of such an electric dipole moment requires PT-violation can be demonstrated simply from invariance arguments (see App. A) as follows.

The electric dipole operator $\boldsymbol{d}(\sim \mathrm{e}\,\boldsymbol{r})$ for symmetry reasons must be proportional to the relevant spin operator $\boldsymbol{s}$. But, under the P- and T-transforma-

* This result follows by consideration of the transformation properties of the corresponding multipole operators under time reversal as in §X-7.1. Thus eq. (X-48) implies that the E1 and M1 reduced matrix elements e_1 and m_1 are relative imaginary.

** Experimentally it is only possible to detect a small electric dipole moment when the system being dealt with has no net charge. Further, complications from higher electric multipole moments must be avoided and so the spin of the system is restricted to $\frac{1}{2}$.

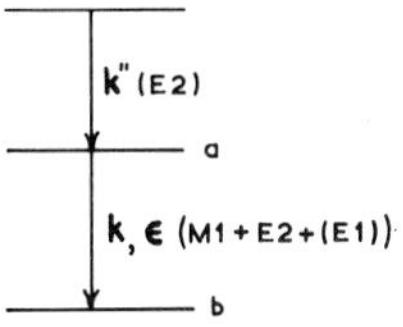

Fig. X-6. PT-violating γ–γ angular correlation.

tions $\boldsymbol{d}$ and $\boldsymbol{s}$ transform in opposite fashion as follows:

$$PdP^{-1} = -d, \quad PsP^{-1} = +s$$

$$TdT^{-1} = +d, \quad TsT^{-1} = -s. \qquad \text{(X-72)}$$

Thus both P and T violation must occur if $\boldsymbol{d}$ is to be non-vanishing. This being so it is straightforward to attempt an order of magnitude estimate for the electric dipole moment of the neutron along the following lines.

$$d_{\text{neutron}} \sim e\langle r\rangle \times GM^2 \times 10^{-3} \sim 10^{-22} \text{ e-cm} \qquad \text{(X-73)}$$

where $\langle r\rangle$ is the "size" of the neutron charge distribution ($\approx 10^{-14}$ cm), $GM^2(\approx 10^{-5})$ is the dimensionless weak interaction coupling constant (leading to P-violation) and 10^{-3} is the measured relative strength of T-violation in K_L^0-decay. It could further be argued that if T-violation stems from the electromagnetic interaction then since we are dealing with the measurement of an electromagnetic quantity the factor 10^{-3} should not be included so that

$$d_{\text{neutron}} \sim e\langle r\rangle \times GM^2 \sim 10^{-19} \text{ e-cm}. \qquad \text{(X-74)}$$

Finally, however, if T-violation stems from a superweak interaction then

$$d_{\text{neutron}} \sim e\langle r\rangle \times GM^2 \times 10^{-8} \sim 10^{-27} \text{ e-cm} \qquad \text{(X-75)}$$

which is far too small to be detectable at the present time.

Measurements of d_{neutron} by beam-resonance experiments have been carried out by, in particular, Ramsey and his co-workers and these are summarised in table X-3.

It is clear that these values are smaller than some of the order of magnitude

Table X-3
Measured values of the neutron electric dipole moment.

d_{neutron} (e-cm)	Reference
$d = (-0.1 \pm 2.4) \times 10^{-20}$	Smith *et al.* (1957)
$\|d\| \lesssim 3 \times 10^{-23}$	Miller *et al.* (1967); Dress *et al.* (1968)
$d = (+2.4 \pm 3.9) \times 10^{-22}$	Shull and Nathans (1967)
$d = (2 \pm 2) \times 10^{-23}$	Miller *et al.* (1968)
$\|d\| \lesssim 1 \times 10^{-21}$	Cohen *et al.* (1969)
$\|d\| < 5 \times 10^{-23}$	Baird *et al.* (1969)

estimates just given. However, more detailed theoretical calculations show that simple order of magnitude estimates can be misleading. For example, Barton and White * (1969) using sideways dispersion relations show that for a milliweak type theory an upper limit on d_{neutron} lies between $10^{-23} - 10^{-24}$ e-cm whilst for a *CP*-violating electromagnetic theory an optimistic upper limit of 6×10^{-22} e-cm might hold although a "safe" expectation would be in the region 10^{-23} e-cm (Broadhurst (1970) finds $d_{\text{neutron}} < 4 \times 10^{-23}$ e-cm). The situation then is that the present experimental limits on d_{neutron} do not conflict with either milli-weak or electromagnetic theories of *CP*-violation. Broadhurst (1972) has also pointed out that, using the Clement–Heller model of a *T*-violating NNMγ interaction (Clement and Heller (1971) – see §X-3) then the present experimental limit on d_{neutron} is an order of magnitude *more* sensitive to *T*-violation than the best nuclear measurement (see table X-2).

In the context of these remarks it is also relevant to note that an upper limit has been set on the electric dipole moment of the proton by Harrison *et al.* (1969) (see also Sandars (1967)) by means of a high precision molecular-beam-resonance experiment. They obtain

$$|d_{\text{proton}}| = (7 \pm 9) \times 10^{-21} \text{ e-cm} . \qquad \text{(X-76)}$$

A further experiment by Weisskopf *et al.* (1968) using similar techniques sets an upper limit on the electric dipole moment of the cesium atom and hence of the electron (see Sandars (1965, 1966)). The authors obtain

* This paper contains references to earlier calculations of d_{neutron}.

$$|d_{\text{electron}}| < 3 \times 10^{-24}\ \text{e-cm}\ . \tag{X-77}$$

There are no detailed calculations with which to make comparisons but neither of the limits seem to conflict with current theories of T-violation.

X-9. Conclusions

Low energy nuclear physics experiments aimed at testing for T-violation are clearly tantalisingly near in accuracy to values predicted by certain theories of T-violation. Thus to achive an order of magnitude better accuracy in detailed balance experiments or in γ-decay angular correlations measurements could well confirm or rule out the presence of a T-violating electromagnetic or medium-strong interaction. As far as angular correlation studies are concerned such accuracy could be achieved by experiments underway using helium dilution refrigerators to polarise an initial nuclear state and measuring the angular correlation $\boldsymbol{I}_a \cdot (\boldsymbol{k} \times \boldsymbol{k}')$. If a positive effect is obtained then detailed calculations will be necessary to ensure that it does not arise from initial or final state interactions (see §X-7.5). If no effect is obtained then before ruling out a particular theory reliable calculations will be needed to establish the precise size of the expected effect according to that theory.

CHAPTER XI

ELECTROMAGNETIC AND CHARGE DEPENDENT INTERACTIONS

XI-1. Introduction

The general form of the electromagnetic interaction Hamiltonian density can be written as follows

$$\mathcal{H}_{\text{e.m.}} = -e\,\mathcal{J}_\lambda^{(\text{e.m.})}(x)A_\lambda(x) \tag{XI-1}$$

where $\mathcal{J}_\lambda^{(\text{e.m.})}$ is the total electromagnetic current and A_λ is a four-vector potential describing the electromagnetic field. $\mathcal{J}_\lambda^{(\text{e.m.})}$ can be decomposed into hadronic and leptonic parts

$$\mathcal{J}_\lambda^{(\text{e.m.})} = J_\lambda + \ell_\lambda^{(\text{e.m.})} \tag{XI-2}$$

and our concern in the nucleus will be with the hadronic part (J_λ) of the interaction.

As described in §I-5, the hadronic current is normally regarded as a member of an octet of polar vector currents so that

$$J_\lambda = \mathcal{F}_\lambda^{(3)} + \frac{1}{\sqrt{3}}\mathcal{F}_\lambda^{(8)} \tag{XI-3}$$

where $\mathcal{F}_\lambda^{(8)}$ is an isoscalar and $\mathcal{F}_\lambda^{(3)}$ is the 3-component of an isovector. The above relation follows naturally as a microscopic generalisation of the Gell-Mann–Nishijima relation

$$Q = T_3 + \tfrac{1}{2}Y \tag{XI-4}$$

where Q, Y and T_3 are the charge, hypercharge and 3-component of isospin respectively. This isospin structure of the hadrònic electromagnetic current reveals itself explicitly in the forms already used in §I-4 for the matrix element of J_λ between nucleon states, thus

$$\langle N'|J_\lambda|N\rangle = \mathrm{i}(\bar{u}_{N'}|(F_1^{(s)} + F_1^{(v)}\tau_3)\gamma_\lambda - (F_2^{(s)} + F_2^{(v)}\tau_3)\,\sigma_{\lambda\nu}k_\nu|u_N) \qquad \text{(XI-5)}$$

where the $F_i^{(s)}$ and $F_i^{(v)}$ are the usual electromagnetic isoscalar and isovector form factors respectively. Of course, and this is a point we shall return to, the matrix element of an operator between nucleons (isospin $\frac{1}{2}$ states) can lead at most to isoscalar and isovector components.

In general, the treatment of electromagnetic effects in nuclei assumes that the electromagnetic interactions has only isoscalar and isovector parts as in the foregoing discussion. However, during recent years, the question has been raised (e.g. Dombey and Kabir (1966); Grishin *et al.* (1966); see also Adler (1967); Shaw (1967); Veltman (1967); Bergmann (1968); Gittelman and Schmidt (1968); Divakaran *et al.* (1968)) as to whether the electromagnetic interaction might contain an isotensor part. There is, in fact, littke evidence one way or the other on this point and in the following sections we shall consider the extent to which nuclear studies can help.

The incentive for such a consideration stems from the fact that some theories of *CP*-violation via the electromagnetic interaction (e.g. Veltman (1967)) require the presence of higher isospin components in the electromagnetic interaction. Further, analysis of the single-pion photoproduction process ($\gamma + \mathrm{N} \rightarrow \pi + \mathrm{N}$) provides some evidence for the existence of an isotensor component (Sanda and Shaw (1970, 1971a, b) although this has been questioned by Schwela (1971).

XI-2. The nuclear electromagnetic Hamiltonian

Taking the matrix element of $\int \mathcal{H}_{\mathrm{e.m.}}(x)\,\mathrm{d}^4x$ between single nucleon states, ignoring the possibility of an isotensor component for the present, and going to a non-relativistic approximation (cf. §II-2.2 for the corresponding situation in the case of the β-decay interaction), we arrive at the usual effective electromagnetic interaction Hamiltonian for a single nucleon, i, namely

$$H_{\mathrm{e.m.}}^{(i)} = -\frac{e_i}{2M_i c}\,[\boldsymbol{A}(\boldsymbol{r}_i)\cdot\boldsymbol{p}_i + \boldsymbol{p}_i\cdot\boldsymbol{A}(\boldsymbol{r}_i)] - \frac{e\hbar}{2Mc}\,\boldsymbol{\mu}^{(i)}\cdot\mathrm{curl}\,\boldsymbol{A}(\boldsymbol{r}_i) + e_i\phi(\boldsymbol{r}_i) \qquad \text{(XI-6)}$$

where the four vector potential A_λ has been written $(\boldsymbol{A}, \mathrm{i}\phi)$; e_i and $\boldsymbol{\mu}_i$ are the charge and magnetic moment respectively of the nucleon under consideration and higher order terms in A_λ have been neglected. Quantities such as e_i and μ_i

(say O_i) can also be written in an isospin notation as follows:

$$O_i = O_p \tfrac{1}{2}(1 + \tau_3^{(i)}) + O_n \tfrac{1}{2}(1 - \tau_3^{(i)}) \qquad \text{(XI-7)}$$

where O_p and O_n refer to the values of these quantities for a proton and neutron respectively and $\tau_3^{(i)}$ is the 3-component of the isospin operator $\boldsymbol{\tau}^{(i)}$ for the ith nucleon. For example, the intrinsic nucleon magnetic moment operator $\boldsymbol{\mu}^{(i)}$ can be written

$$\boldsymbol{\mu}^{(i)} = \boldsymbol{\mu}_s^{(i)} + \boldsymbol{\mu}_v^{(i)} \qquad \text{(XI-8)}$$

where

$$\boldsymbol{\mu}_s^{(i)} = \tfrac{1}{2}(\mu_p + \mu_n)\boldsymbol{\sigma}^{(i)}$$

$$\boldsymbol{\mu}_v^{(i)} = \tfrac{1}{2}(\mu_p - \mu_n)\boldsymbol{\sigma}^{(i)}\tau_3^{(i)} .$$

Here μ_p and μ_n are the intrinsic magnetic moments of the proton and neutron respectively and the suffices s and v refer to the isoscalar and isovector parts of $\boldsymbol{\mu}^{(i)}$ respectively. Such a decomposition obviously leads to isoscalar and isovector contributions to $H_{\text{e.m.}}^{(i)}$.

It is further straightforward to extend the Hamiltonian to deal with the case of a nucleus consisting of A nucleons, thus,

$$H_{\text{e.m.}} = \sum_{i=1}^{A} H_{\text{e.m.}}^{(i)} \qquad \text{(XI-9)}$$

In the absence of external electric fields the scalar potential $\phi(\boldsymbol{r}_i)$ can be transformed away and the terms in e_i replaced by the usual Coulomb potential. $H_{\text{e.m.}}$ can then be written

$$H_{\text{e.m.}} = \sum_i \left\{ \frac{e_i}{2M_i c} [A(r_i) \cdot p_i + p_i \cdot A(r_i)] - \frac{e\hbar}{2M_p c} \boldsymbol{\mu}_i \cdot \text{curl}\, A(r_i) \right\}$$

$$+ \sum_{i<j} \frac{e_i e_j}{r_{ij}} \qquad \text{(XI-10)}$$

where $\boldsymbol{A}$ refers both to the transverse part of the quantised electromagnetic field and any external magnetic field.

In an isospin formulation the Coulomb term can be expressed as follows

$$V_C = \sum_{i<j} \frac{e_i e_j}{r_{ij}} = \sum_{i<j} \frac{e^2(1+\tau_3^{(i)})(1+\tau_3^{(j)})}{4r_{ij}} \tag{XI-11}$$

or, writing it as a sum of zero (0), first (1) and second rank tensors (2) in isospin space

$$V_C = V_C^{(0)} + V_C^{(1)} + V_C^{(2)}$$

where

$$V_C^{(0)} = \sum_{i<j} \frac{1}{4}\frac{e^2}{r_{ij}}\left(1 + \frac{1}{3}\boldsymbol{\tau}^{(i)} \cdot \boldsymbol{\tau}^{(j)}\right)$$

$$V_C^{(1)} = \sum_{i<j} \frac{1}{4}\frac{e^2}{r_{ij}}(\tau_3^{(i)} + \tau_3^{(j)}) \tag{XI-12}$$

$$V_C^{(2)} = \sum_{i<j} \frac{1}{4}\frac{e^2}{r_{ij}} T^{(i,j)}$$

and $T^{(i,j)} = \tau_3^{(i)}\tau_3^{(j)} - \frac{1}{3}\boldsymbol{\tau}^{(i)} \cdot \boldsymbol{\tau}^{(j)}$.

In addition, for some purposes, the electromagnetic spin-orbit coupling resulting from the Larmor precession of the nucleons in the nuclear electric field due to their intrinsic magnetic moments and from the Thomas precession experienced by protons because of their charge needs to be included. In an isospin notation the corresponding potential V_{so} can be written (e.g. Bethe and Salpeter (1957))

$$V_{\mathrm{so}} = V_{\mathrm{so}}^{(0)} + V_{\mathrm{so}}^{(1)} \tag{XI-13}$$

where

$$V_{\mathrm{so}}^{(0)} = \sum_i (g_{\mathrm{p}} + g_{\mathrm{n}} - 1)\frac{e}{4M^2c^2}\frac{1}{r_i}\frac{\mathrm{d}U_C(r_i)}{\mathrm{d}r_i}\boldsymbol{l}_i \cdot \boldsymbol{s}_i$$

$$V_{\mathrm{so}}^{(1)} = \sum_i \tau_3^{(i)}(-g_{\mathrm{p}} + g_{\mathrm{n}} + 1)\frac{e}{4M^2c^2}\frac{1}{r_i}\frac{\mathrm{d}U_C(r_i)}{\mathrm{d}r_i}\boldsymbol{l}_i \cdot \boldsymbol{s}_i .$$

$U_C(r_i)$ is the electrostatic potential experienced by the ith nucleon, g_{p} and g_{n} are the proton and neutron g-factors and $\boldsymbol{l}_i$ and $\boldsymbol{s}_i$ are orbital and spin angular momentum operators.

Sometimes it is useful to write the part of $H_{\text{e.m.}}$ depending on $\boldsymbol{A}$ in terms of an effective nucleon current. Thus, in an isospin notation we can write

$$H_{\text{e.m.}} = \frac{e}{c} \int \boldsymbol{J}_{\text{N}}(\boldsymbol{r}) \cdot \boldsymbol{A}(\boldsymbol{r})\, \mathrm{d}^3 r + V_{\text{C}} \tag{XI-14}$$

where

$$\boldsymbol{J}_{\text{N}}(\boldsymbol{r}) = \frac{1}{2} \sum_i (1 + \tau_3^{(i)}) \left[\frac{\boldsymbol{p}}{2M} \delta(\boldsymbol{r} - \boldsymbol{r}_i) + \delta(\boldsymbol{r} - \boldsymbol{r}_i) \frac{\boldsymbol{p}}{2M} \right]$$

$$+ \frac{1}{2} c \left(\frac{\hbar}{2Mc} \right) \sum_i [\mu_{\text{p}} + \mu_{\text{n}}) + (\mu_{\text{p}} - \mu_{\text{n}}) \tau_3^{(i)}]\, \nabla \times \boldsymbol{\sigma}^{(i)} \delta(\boldsymbol{r} - \boldsymbol{r}_i) \tag{XI-15}$$

and V_{C} is given by eq. (XI-11). The first part of $\boldsymbol{J}_{\text{N}}$ is the "convection" current and the second the "spin" current. Clearly $\boldsymbol{J}_{\text{N}}$ can also be further decomposed into an isoscalar part, $\boldsymbol{J}_{\text{N}}^{(\text{s})}$ (independent of the $\tau_3^{(i)}$) and an isovector part, $\boldsymbol{J}_{\text{N}}^{(\text{v})}$ (linearly dependent on the $\tau_3^{(i)}$)

$$\text{i.e.} \quad \boldsymbol{J}_{\text{N}} = \boldsymbol{J}_{\text{N}}^{(\text{s})} + \boldsymbol{J}_{\text{N}}^{(\text{v})} \tag{XI-16}$$

where

$$\boldsymbol{J}_{\text{N}}^{(\text{s})} = \frac{1}{2} \sum_i \left[\frac{\boldsymbol{p}}{2M} \delta(\boldsymbol{r} - \boldsymbol{r}_i) + \delta(\boldsymbol{r} - \boldsymbol{r}_i) \frac{\boldsymbol{p}}{2M} \right]$$

$$+ \frac{1}{2} c \left(\frac{\hbar}{2Mc} \right) \sum_i (\mu_{\text{p}} + \mu_{\text{n}}) \nabla \times \boldsymbol{\sigma}^{(i)} \delta(\boldsymbol{r} - \boldsymbol{r}_i) \tag{XI-17}$$

and

$$\boldsymbol{J}_{\text{N}}^{(\text{v})} = \frac{1}{2} \sum_i \tau_3^{(i)} \left[\frac{\boldsymbol{p}}{2M} \delta(\boldsymbol{r} - \boldsymbol{r}_i) + \delta(\boldsymbol{r} - \boldsymbol{r}_i) \frac{\boldsymbol{p}}{2M} \right]$$

$$+ \frac{1}{2} c \left(\frac{\hbar}{2Mc} \right) \sum_i \tau_3^{(i)} (\mu_{\text{p}} - \mu_{\text{n}}) \nabla \times \boldsymbol{\sigma}^{(i)} \delta(\boldsymbol{r} - \boldsymbol{r}_i) . \tag{XI-18}$$

The above many body expressions (eqs. (XI-6 – XI-18)), although satisfactory for most purposes, do assume that the nucleons in the nucleus are non-interacting other than through the Coulomb field. This is, of course, an approximation and neglects the effects of what have been called exchange (or interaction) currents. Physically, these currents derive from the presence of nuclear forces and qualitatively can be attributed to meson exchanges taking place between nucleons (cf. the situation in β-decay, §II-2.3). They enter into the nuclear electromagnetic Hamiltonian as 2- and higher body terms so that more generally we should add further terms to $H_{(\mathrm{e.m.})}$. How some such terms appear can be seen by considering briefly the required properties of the current $J_\mathrm{N}(r)$ defined in eq. (XI-15). $J_\mathrm{N}(r)$ should satisfy the usual current-conservation equation

$$\nabla \cdot J_\mathrm{N} = -\frac{\partial \rho_\mathrm{N}}{\partial t} = -\frac{\mathrm{i}}{\hbar}[H, \rho_\mathrm{N}] \tag{XI-19}$$

where H is the nuclear many body Hamiltonian and ρ_N is the charge density operator given by

$$\rho_\mathrm{N}(r) = \tfrac{1}{2}\sum_i (1 + \tau_3^{(i)})\,\delta(r - r_i)\,. \tag{XI-20}$$

So long as H contains no velocity dependent potentials or exchange forces then J_N as defined in eq. (XI-15) satisfies eq. (XI-19). However, if this restriction is relaxed then this is no longer the case. In particular if there is a charge exchange potential V_τ in H having the form

$$V_\tau = \sum_{i<j} f(r_{ij})\,\boldsymbol{\tau}^{(i)} \cdot \boldsymbol{\tau}^{(j)} \tag{XI-21}$$

then an additional (exchange) current, $J_\mathrm{N}^\mathrm{exch}$, has to be introduced satisfying

$$\nabla \cdot J_\mathrm{N}^\mathrm{exch} = -\frac{\mathrm{i}}{\hbar}[V_\tau, \rho_\mathrm{N}]\,. \tag{XI-22}$$

This current is obviously two-body in nature and is related directly to the charge exchange potential. It is unfortunately not determined completely by eq. (XI-22) and is uncertain to the extent of the curl of an arbitrary vector V; i.e. for any $J_\mathrm{N}^\mathrm{exch}$ satisfying eq. (XI-22) the current $J_\mathrm{N}^\mathrm{exch} + \nabla \times V$ also satisfies it.

The most important exchange contributions to $H_\mathrm{e.m.}$ are the two-body terms since these dominate compared with 3 and higher body interactions and these will be a function of the separation r_{ij} of the two nucleons (i and j) concerned. Since it is an interaction effect the range of $H_{(\mathrm{e.m.})}^\mathrm{exch}$ is expected to be similar to that of nuclear forces.

These additional two-body terms in $H_{(\mathrm{e.m.})}$ generate corresponding two-body terms in the electric and magnetic multipole operators used in calculating static moments of nuclei and *EL* and *ML* multipole transition probabilities *. Here the well known Siegert (1937) theorem should be noted which states that to a good approximation exchange terms do not arise in the case of electric multipole operators. This theorem which was first formulated in connection with the E1 operator was generalised to all electric multipoles by Sachs, Austern and Foldy [Sachs and Austern (1951); Austern and Sachs (1951); Foldy (1953); see also Sachs (1953)].

In the case of magnetic multipoles exchange effects can be important and, in particular, much study has been made of the magnetic dipole (M1) exchange operator. From a phenomenological point of view the general form of this operator can be written down at once (e.g. Osborne and Foldy (1950)), assuming that it has the same transformation properties as the corresponding single particle operator. Since the magnetic dipole operator $\boldsymbol{\mu}$ has both isoscalar and isovector parts, $\boldsymbol{\mu}^{\text{exch}}$ can be similarly decomposed and the following general forms are obtained in the simplest static approximation (i.e. no momentum dependent or non-local terms):

$$\boldsymbol{\mu}_{\text{S}}^{\text{exch}} = \frac{e}{4M}\sum_{i<j}\left\{\left[k_{\text{I}}(\boldsymbol{\sigma}^{(i)}+\boldsymbol{\sigma}^{(j)}) + k_{\text{II}}\frac{\boldsymbol{r}_{ij}\boldsymbol{r}_{ij}\cdot(\boldsymbol{\sigma}^{(i)}+\boldsymbol{\sigma}^{(j)})}{r_{ij}^2}\right]\tau_3^{(i)}\tau_3^{(j)}\right.$$
$$+\left[\ell_{\text{I}}(\boldsymbol{\sigma}^{(i)}+\boldsymbol{\sigma}^{(j)}) + \ell_{\text{II}}\frac{\boldsymbol{r}_{ij}\boldsymbol{r}_{ij}\cdot(\boldsymbol{\sigma}^{(i)}+\boldsymbol{\sigma}^{(j)})}{r_{ij}^2}\right]$$
$$\left.+\left[m_{\text{I}}(\boldsymbol{\sigma}^{(i)}+\boldsymbol{\sigma}^{(j)}) + m_{\text{II}}\frac{\boldsymbol{r}_{ij}\boldsymbol{r}_{ij}\cdot(\boldsymbol{\sigma}^{(i)}+\boldsymbol{\sigma}^{(j)})}{r_{ij}^2}\right]\boldsymbol{\tau}^{(i)}\cdot\boldsymbol{\tau}^{(j)}\right\} \qquad \text{(XI-23)}$$

$$\boldsymbol{\mu}_{\text{V}}^{\text{exch}} = \frac{e}{4M}\sum_{i<j}\left\{\left[g_{\text{I}}(\boldsymbol{\sigma}^{(i)}\times\boldsymbol{\sigma}^{(j)}) + g_{\text{II}}\frac{\boldsymbol{r}_{ij}\boldsymbol{r}_{ij}(\boldsymbol{\sigma}^{(i)}\times\boldsymbol{\sigma}^{(j)})}{r_{ij}^2}\right](\boldsymbol{\tau}^{(i)}\times\boldsymbol{\tau}^{(j)})_3\right.$$
$$+\left[h_{\text{I}}(\boldsymbol{\sigma}^{(i)}-\boldsymbol{\sigma}^{(j)}) + h_{\text{II}}\frac{\boldsymbol{r}_{ij}\boldsymbol{r}_{ij}\cdot(\boldsymbol{\sigma}^{(i)}-\boldsymbol{\sigma}^{(j)})}{r_{ij}^2}\right](\tau_3^{(i)}-\tau_3^{(j)})$$
$$\left.+\left[j_{\text{I}}(\boldsymbol{\sigma}^{(i)}+\boldsymbol{\sigma}^{(j)}) + j_{\text{II}}\frac{\boldsymbol{r}_{ij}\boldsymbol{r}_{ij}\cdot(\boldsymbol{\sigma}^{(i)}+\boldsymbol{\sigma}^{(j)})}{r_{ij}^2}\right](\tau_3^{(i)}+\tau_3^{(j)})\right\} \qquad \text{(XI-24)}$$

* See footnote on page 245.

where s and v refer to the isoscalar and isovector parts of the exchange interaction and the functions k ... j are arbitrary functions of r_{ij}. $\boldsymbol{\mu}_{\text{v}}^{\text{exch}}$ is of course similar in form to the axial vector β-decay exchange interaction discussed in §II-2.3 but the functions g, h and j are unrelated to those relevant to β-decay.

The detailed forms of the different spatial functions in $\boldsymbol{\mu}_{\text{s,v}}^{\text{exch}}$ can only be obtained on the basis of some model capable of describing meson–nucleon interactions. Such calculations were first performed in a perturbation approach by Villars (1947) and subsequently by many other authors with ever-increasing degrees of sophistication. Most recently (e.g. Chemtob (1969); Rho (1970); Chemtob and Rho (1971)) the calculations have been carried out simulating multi-pion exchanges by heavy meson exchanges along the lines of nuclear force calculations using one-boson-exchange models (e.g. Signell (1969)). In particular the one-pion exchange contribution can be evaluated fairly accurately using the low energy theorems associated with PCAC and current algebra in describing the π-N interaction (e.g. Adler and Dashen (1968); see also §IV-5.1 on exchange effects in β-decay). Thus specific forms are available for the radial functions which can be used to evaluate values for exchange contributions to nuclear magnetic dipole moments (see §XI-4).

Finally, in connection with the nuclear electromagnetic Hamiltonian, we have to consider the modifications which would result if the fundamental electromagnetic interaction contained an isotensor part (see §XI-1). All that can be said on this point is that inevitably further terms would be introduced into $H_{(\text{e.m.})}$ of, for example, the form

$$H_{\text{e.m.}}^{(\text{t})} = \text{e} \sum_{i<j} \boldsymbol{J}^{(ij)} \cdot \boldsymbol{A} \tag{XI-25}$$

where $\boldsymbol{J}^{(i,j)}$ is a 2-particle electromagnetic current having an isotensor character (i.e. proportional to $T^{(i,j)}$ $(= \tau_3^{(i)}\tau_3^{(j)} - \frac{1}{3}\boldsymbol{\tau}^{(i)} \cdot \boldsymbol{\tau}^{(j)})$. (As mentioned earlier, an isotensor cannot be constructed from a single particle current.) No calculations have been carried out on possible forms for $\boldsymbol{J}^{(i,j)}$ or of other possible isotensor contributions to $H_{\text{e.m.}}$ (e.g. coupling through a $\Delta\text{N}\gamma$ vertex) and all we can do at this stage is to note the possible presence of such terms and that they can lead to three- and four-body charge dependent potentials including third and fourth rank isotensor components.

* Some exchange operators can be derived directly through the operation of making the internucleon potential gauge invariant (e.g. Osborne and Foldy (1950)). Dalitz (1954) has pointed out the ambiguities in this procedure and shown how to define unambiguously the term arising specifically from the exchange parts of the internucleon potential using eq. (XI-22).

XI-3. Charge dependence of the internucleon potential

It is well known that to a good approximation the nuclear potential is charge independent. That is to say the nuclear potentials in the pp, np and nn systems are the same when the two nucleons are in the same state. This is evidenced, for example, by nucleon–nucleon scattering data, the properties of mirror nuclei and isospin multiplets and, overall, by the successful use of isospin in classifying nuclear states. Of course charge dependent effects do appear in these different phenomena stemming particularly from the Coulomb interaction and it is only after account has been taken of these that the approximate charge independence of the internucleon potential itself is really apparent. However, it *is* an approximate independence and not an exact one (see the review by Henley (1969b) for a full account of the situation). For example the experimental values for the singlet S-state scattering length, a, and effective range, r_0, listed in table XI-1 (taken from Henley (1969b) (see also Noyes and Lipinski (1971)) differ between the systems np, pp and nn *. The pp and np results derive from low energy scattering data whilst the nn results derive from an analysis of final state interaction effects in the processes $\pi^- + d \to \gamma + n + n$ and $d + {}^3H \to {}^3He + 2n$. The differences in the scattering lengths are particularly revealing and can be simply related to differences in the strength and range of the internucleon potential. For example, for square, exponential and Yukawa potential wells of depth V_0 and range b changes in V_0 and b of magnitude ΔV_0 and Δb respectively give a corresponding change in a of Δa (Jackson and Blatt (1950); Moravcsik (1964))

$$\frac{\Delta a}{a} \approx c\,\frac{\Delta V_0}{V_0} + d\,\frac{\Delta b}{b} \tag{XI-26}$$

where c and d are constants depending on the form of potential well. A similar relation holds for $\Delta r_0/r_0$.

It is usual to describe the extent of the charge dependence of the internucleon potential in terms of (i) a breakdown of charge symmetry (CS) and (ii) a breakdown of charge independence (CI). The former measures the difference between the nn and pp potentials and the latter the difference between the np potential on the one hand and the nn and pp potentials on the other. Using eq. (XI-26) and the results of table XI-1 Henley (1969b)

* Note that the values for a_{pp} and r_{0pp} have been corrected for Coulomb and vacuum polarisation effects.

Table XI-1
Singlet S-state scattering lengths, a, and effective ranges, r_0, for the nucleon–nucleon system.

Systems	a (fm)	r_0 (fm)
pp	−(17.0 ± 0.2)	2.83 ± 0.03
np	−(23.715 ± 0.013)	2.76 ± 0.07
nn	−(17.6 ± 1.5)	3.2 ± 1.6

obtains

$$\frac{\Delta \bar{V}_{\mathrm{CS}}}{V} = \frac{|\bar{V}_{\mathrm{nn}}| - |\bar{V}_{\mathrm{pp}}|}{\frac{1}{2}(|\bar{V}_{\mathrm{nn}}| + |\bar{V}_{\mathrm{pp}}|)} \approx 0.25 \pm 0.80\% \tag{XI-27}$$

$$\frac{\Delta \bar{V}_{\mathrm{CI}}}{V} = \frac{|\bar{V}_{\mathrm{np}}| - |\bar{V}_{\mathrm{nn}}|}{\frac{1}{2}(|\bar{V}_{\mathrm{np}}| + |\bar{V}_{\mathrm{nn}}|)} \approx 2.10 \pm 0.51\% \tag{XI-28}$$

where the bar indicates a spatial average over the internucleon potentials indicated. The above results indicate that charge symmetry holds to within about 0.8% but that charge independence is definitely violated with the np potential stronger than both the nn and the pp.

The above results, however, do not allow for charge dependent effects arising from the difference in mass of the neutron and proton, the magnetic forces between nucleons and the finite size of nucleons. These have been taken into account by Downs and Nogami (1967) – n-p mass difference; Schwinger (1950), Salpeter (1953), Breit (1962), Lovitch (1964, 1965) – magnetic forces; Riazuddin (1958), Schneider and Thaler (1965) – finite size effects. They modify the pp, np and nn scattering lengths respectively by −0.26 fm, +0.2 fm and +0.22 fm so that, taking them into account, eqs. (XI-27) and (XI-28) now read (Henley (1969b))

$$\frac{\Delta \bar{V}_{\mathrm{CS}}}{V} \approx 0 \pm 0.8\% \tag{XI-29}$$

$$\frac{\Delta \bar{V}_{\mathrm{CI}}}{V} \approx 2.13 \pm 0.52\% \,. \tag{XI-30}$$

These modifications do not change the overall picture and, in particular, a clear breakdown in charge independence in the internucleon potential is indicated. In addition, a small breakdown in charge symmetry is allowed. We now consider the possible origin of these divergences of the short range internucleon potential from charge independence and charge symmetry.

XI-3.1. *Theories of charge dependence of the internucleon potential*

One of the most successful and illuminating ways of calculating the internucleon potential has been in terms of one-boson-exchange potential (OBEP) models (e.g. Signell (1969)). Here the long-range one-pion exchange contribution can be derived fairly unambiguously and multipion exchanges are simulated by the exchange of heavier bosons (ρ, ω, η, ...). Typical diagrams contributing to the potential are shown in fig. XI-1.

In making these calculations it is normally assumed that for $T \neq 0$ mesons (π, ρ, ...) the masses of the different members of the meson isospin multiplets (e.g. π^+, π^0, π^-) are the same. It is further assumed that the different meson–nucleon coupling constants (e.g. $G_{\pi^+ \mathrm{np}}$, $G_{\pi^0 \mathrm{nn}}$ etc.) are identical. In other words charge dependent effects are ignored and the resulting potential is thus, of course, charge independent. This is, however, an approximation. For example the charged and neutral pions have different masses, thus

$$m_{\pi^\pm} = 139.579 \pm 0.014 \text{ MeV}$$

$$m_{\pi^0} = 134.975 \pm 0.014 \text{ MeV}$$

so that

$$\frac{m_{\pi^\pm} - m_{\pi^0}}{\frac{1}{2}(m_{\pi^\pm} + m_{\pi^0})} \approx 3.4\% \,. \tag{XI-31}$$

In the case of the ρ-mesons there is an uncertainty of the order 20 MeV in their masses, although theoretical calculations (e.g. Schwinger (1967) and

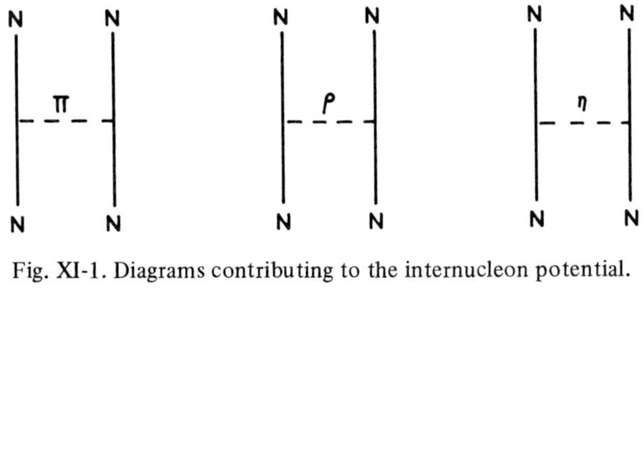

Fig. XI-1. Diagrams contributing to the internucleon potential.

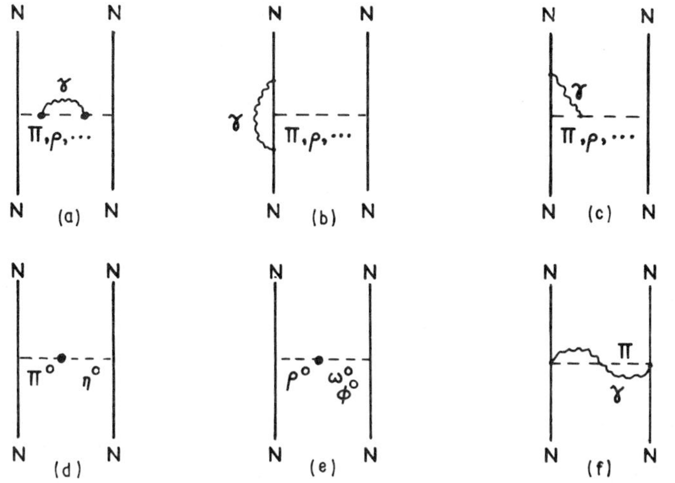

Fig. XI-2. Charge dependent contributions to the internucleon potential.

Lee (1968)) suggest that $m_{\rho^0} > m_{\rho^\pm}$. Both of these mass differences can be attributed to the electromagnetic interaction and their effect on calculations of the internucleon potential are illustrated in diagram (a) of fig. XI-2.

Because of the *CPT* theorem the masses of charged members of the isospin multiplets are the same, hence charge symmetry is maintained in the resulting potential. Assuming pseudoscalar pion–nucleon coupling and taking this mass difference into account, the central component of the one-pion exchange potential between nucleons has the following basic form (e.g. Blin-Stoyle and Kearsley (1960))

$$V_{\mathrm{NN}} = \frac{1}{3}\left(\frac{G_{\pi\mathrm{NN}}}{2M}\right)^2 \boldsymbol{\sigma}(1)\cdot\boldsymbol{\sigma}(2)\left\{\boldsymbol{\tau}(1)\cdot\boldsymbol{\tau}(2)\mu^2\frac{e^{-\mu r}}{r}\right.$$

$$\left. + \frac{\delta\mu}{\mu}\tau_3^{(1)}\tau_3^{(2)}\mu^2\left[\mu e^{-\mu r} - \frac{2e^{-\mu r}}{r}\right]\right\} \qquad \text{(XI-32)}$$

where $\mu = m_{\pi^\pm}$ and $\delta\mu = m_{\pi^\pm} - m_{\pi^0}$. For the pp and nn systems $\tau_3^{(1)}\tau_3^{(2)} = +1$ whilst for the np system $\tau_3^{(1)}\tau_3^{(2)} = -1$. Charge symmetry is thus manifest in the expression for V_{NN} and, in addition, study of the radial dependence of the charge dependent term shows that V_{np} is stronger than V_{nn} and V_{pp}.

More detailed study of the effect of this mass splitting on V_{NN} and, in turn, on the nucleon–nucleon scattering length and effective range has been carried out by Riazuddin (1958); Heller *et al.* (1964); Lin (1964); Schneider and Thaler (1965); Henley and Morrison (1965) and Yalçin (1971). Henley and Morrison (1965), for example, find that, due to this effect in the one-pion exchange potential Δa (= $a_{\mathrm{np}} - a_{\mathrm{nn}}$) = $-$ 3.5 fm. Taking into account two-pion, ρ, ω and η exchanges the magnitude of Δa is somewhat increased so as to be able to account for about $\frac{2}{3}$ (or even more – see Yalçin (1971)) of the experimental value $\Delta a \approx -6.1$ fm (see table XI-1).

The next correction to be taken into account is that due to electromagnetic radiative corrections to meson–nucleon vertices. Typical diagrams corresponding to this correction are (b) and (c) in fig. XI-2. Estimates of this effect, which destroys both charge independence and charge symmetry, have been made for the pion–nucleon vertex by Riazuddin (1958) and Stevens (1965) using Feynman diagram techniques, by Goldberg (1965) using dispersion relations and by Morrison (1967, 1968) using current algebra. It turns out that there is considerable uncertainty here and that the magnitude and size of the radiative corrections are sensitive to the method used to calculate them. However, it does seem to emerge as a common feature that the breakdown of charge independence in the internucleon potential due to this effect is larger than that of charge symmetry and is in the right direction to help with the value of Δa.

A further correction to the internucleon potential and one which leads to a breakdown of charge symmetry is due to isospin mixing of the different mesons. This effect is illustrated in diagram (d) of fig. XI-2 which shows a mixing of the π^0 ($T = 1$) and η^0 ($T = 0$) mesons. Similar contributions can arise from the mixing of the ρ^0 ($T = 1$) meson with the ω^0 ($T = 0$) and ϕ^0 ($T = 0$) (diagram (e)). This mixing is due to the electromagnetic interaction and its magnitude has been estimated by various authors (Dalitz and Von Hippel (1964); Barrett and Barton (1964); Stevens (1965); Downs and Nogami (1967); Yalçin and Birsen (1970); Okamoto and Pask (1971b)). The mixing is small but because of the smaller mass differences, is larger among the vector mesons than among the pseudoscalar mesons. Its effect on the internucleon potential is readily calculated. For example, in the case of fig. XI-2(d) an additional charge dependent term is introduced having the form

$$V = (\tau_3^{(1)} + \tau_3^{(2)}) \frac{G_{\pi\mathrm{NN}} G_{\eta\mathrm{NN}}}{4\pi} (c_\eta V_\pi + c_\pi V_\eta)$$

where c_η and c_π are the amplitudes (≈ 0.01) with which the η^0 and the π^0 are admixed into the π^0 and η^0 states respectively and $V_{\pi,\eta}$ are the normal pseudoscalar π and η exchange potentials. The isospin dependence of V given in eq. (XI-33) is characteristic of a charge asymmetric potential (note that $(\tau_3^{(1)} + \tau_3^{(2)})$ takes the values +2, 0, −2 respectively for the pp, np and nn systems). Using the above expression and equivalent expressions stemming from vector meson exchange it is found (e.g. Downs and Nogami (1967)) that the effect on $a_{pp} - a_{nn}$ is at most ≈ 1 fm and is of the right sign to agree with the experimental situation. In a more recent calculation Henley and Keliher (1972) obtain $a_{pp} - a_{nn} \approx 0.8$ fm.

A final correction to be taken into account is that due to the exchange of meson + photon as illustrated in fig. XI-2 (diagram f). This effect has been estimated by Leung and Nogami (1968) within the framework of a difficult and complicated calculation. It leads to a violation of charge independence (which seems to be opposite in sign to that stemming from the pion mass splitting) but there is little violation of charge symmetry.

In summary, there is a fair measure of uncertainty about the theoretical strength of charge dependent components of the internucleon potential arising from the electromagnetic interaction. However, the calculations so far carried out do seem to suggest that the effects operate in the right direction to account for the charge dependence of a and r_0 as indicated by low-energy scattering data.

In general the calculated charge dependent potentials are too complicated and uncertain to use in nuclear structure calculations. At this point, therefore, we introduce a phenomenological central charge dependent potential (V_{CD}) of the form

$$V_{CD} = [L(r_{12}) + \boldsymbol{\sigma}^{(1)} \cdot \boldsymbol{\sigma}^{(2)} M(r_{12})](\tau_3^{(1)} + \tau_3^{(2)}) + [N(r_{12}) + \boldsymbol{\sigma}^{(1)} \cdot \boldsymbol{\sigma}^{(2)} K(r_{12})]\, T^{(1,2)} \tag{XI-33}$$

where $T^{(1,2)}$ is the isotensor operator

$$T^{(1,2)} = \tau_3^{(1)}\tau_3^{(2)} - \tfrac{1}{3}\boldsymbol{\tau}^{(1)} \cdot \boldsymbol{\tau}^{(2)} . \tag{XI-34}$$

In the above expression L, M, N and K are arbitrary scalar functions of the internucleon separation r_{12}. They are expected to have a range characteristic of ordinary nuclear forces and we shall frequently make use of the following phenomenological form for them (Blin-Stoyle and Le Tourneux (1961))

$$L(r_{12}) = pV_0\, \mathrm{e}^{-r_{12}^2/\mu^2}, \quad M(r_{12}) = rV_0\, \mathrm{e}^{-r_{12}^2/\mu^2}$$

$$N(r_{12}) = qV_0\, \mathrm{e}^{-r_{12}^2/\mu^2}, \quad K(r_{12}) = sV_0\, \mathrm{e}^{-r_{12}^2/\mu^2} \tag{XI-35}$$

where V_0 is the strength of a typical charge independent interaction which we take to be -50 MeV. The parameters p, q, r and s measure the strength of the charge dependence and the range parameter μ, which is not critical, is taken to be $\mu = 1.73$ fm.

As far as the values of $p, \ldots, s$, are concerned the evidence from low energy scattering data (§XI-3) is that q and s (which are responsible for the breakdown of charge independence) are significantly larger than p and r (which are responsible for the breakdown of charge symmetry). Explicitly, using eqs. (XI-33)–(XI-35), we have in the singlet S-state ($\boldsymbol{\sigma}^{(1)} \cdot \boldsymbol{\sigma}^{(2)} = -3$)

$$\Delta V_{\mathrm{CS}} = V_{\mathrm{nn}} - V_{\mathrm{pp}} = -2V_0(p - 3r)\,\mathrm{e}^{-r_{12}^2/\mu^2} \tag{XI-36}$$

$$\Delta V_{\mathrm{CI}} = V_{\mathrm{np}} - V_{\mathrm{nn}} = -2V_0[(q - 3s) - (p - 3r)]\,\mathrm{e}^{-r_{12}^2/\mu^2}\,. \tag{XI-37}$$

It then follows that (see eqs. (XI-29) and (XI-30))

$$\frac{\Delta \overline{V}_{\mathrm{CS}}}{V} \approx -2(p - 3r) \approx 0 \pm 0.8\% \tag{XI-38}$$

$$\frac{\Delta \overline{V}_{\mathrm{CI}}}{V} \approx -2[(q - 3s) - (p - 3r)] \approx 2.13 \pm 0.52\%\,. \tag{XI-39}$$

Thus, roughly speaking,

$$3r - p \approx 0 \pm 0.4\% \tag{XI-40}$$

$$3s - q \approx 1\%\,. \tag{XI-41}$$

These relations can be taken as no more than a guide to the order of magnitude of the different parameters. Yalçin and Yap (Yalçin (1971); Yalçin and Yap (1970); see also Blin-Stoyle and Yalçin (1965)) have obtained values of p, q, r and s from a theoretical approach by requiring the volume integrals of the different parts of the phenomenological potential and of a simple OBEP (which allows for mass splitting and isospin mixing) to be equal in magnitude. In particular they find $3r - p \approx 0.1\%$ and $3s - q \approx 1\%$ provided that $m_{\rho^0} - m_{\rho^\pm} \approx 1.5$ to 5 MeV which is in agreement with the discussion given earlier in this section (see also Yalçin (1972)).

Later on in this chapter we shall consider the extent to which studies in complex nuclei can throw light on the nature of charge dependent potentials and hence on their theoretical bases.

XI-4. Magnetic dipole exchange effects *

In §XI-2 it has been pointed out that in addition to the usual single body contribution to nuclear magnetic moments there can be further contributions from exchange effects. It is important to establish experimentally the form and magnitude of these effects in order to test out the fundamental ideas used in deducing them theoretically. As always nuclear wave function uncertainties are generally larger than the effect being studied so that only in the case of very simple nuclei (^{2}H, ^{3}H, ^{3}He) it is possible to draw significant conclusions.

In the case of the deuteron we have to do with a $T = 0, I = 1$ state consisting of an admixture of $L = 0$ and $L = 2$ parts of the form

$$\psi = (1 - p(\mathrm{D}))^{1/2}\psi_{\mathrm{S}} + \alpha(\mathrm{D})\psi_{\mathrm{D}} \tag{XI-42}$$

where $\alpha(\mathrm{D})$ is the amplitude and $p(\mathrm{D})$ the probability of the D-state admixture. Since $T = 0$, only the isoscalar part (μ_{s}) of the usual single body magnetic moment operator contributes and, it is straightforward to obtain the following expression for the deuteron magnetic moment

$$\mu_{\mathrm{s}} = \mu_{\mathrm{p}} + \mu_{\mathrm{n}} - \tfrac{3}{2}p(\mathrm{D})(\mu_{\mathrm{p}} + \mu_{\mathrm{n}} - \tfrac{1}{2})\,. \tag{XI-43}$$

Although there is some uncertainty about $p(\mathrm{D})$ its most likely value is (Hartman (1967)) $p(\mathrm{D}) = 6–7\%$ so that taking $p(\mathrm{D}) = (6.5 \pm 0.5)$ gives $\mu_{\mathrm{s}} = (0.843 \pm 0.003)$ n.m. This is to be compared with $\mu_{\mathrm{exp}} = 0.857$ n.m. which implies a small discrepancy of about 1.6%

$$\delta\mu = \mu_{\mathrm{exp}} - \mu_{\mathrm{s}} = (0.014 \pm 0.003)\ \text{n.m.}\,. \tag{XI-44}$$

Recent estimates of the isoscalar meson exchange magnetic moment operator by Chemtob and Rho (1971) indicate that effects of this order of magnitude probably occur, but there are too many uncertainties in the calculations for firm conclusions to be drawn. Further, relativistic effects (Breit and Bloch (1947)) may be important as may the admixture of baryon resonances within the ^{2}H wave function. On this latter point Kisslinger (1969) and Arenhövel *et al.* (1970, 1971) have estimated the size of this effect and find contribu-

* Sometimes in the literature use of the word "exchange" is restricted to those phenomena which in diagramatic terms correspond to a photon interacting with an exchanged meson line. Other effects (e.g. a photon interacting with a nucleon line) are referred to as "interaction" phenomena. We use the word "exchange" to cover all such effects.

tions to $\delta\mu$ of the order 0.01 n.m. However, it must be realised that including baryon resonances in this way is to some extent taking into account effects already included within the exchange magnetic moment calculations (see e.g. fig. IV-1(c) in connection with β-decay exchange contributions).

In the case of $A = 3$ nuclei the situation is less ambiguous and exchange effects turn out to be much greater in magnitude. Using the expression for the isoscalar and isovector magnetic moment operators given in eq. (XI-8) together with the orbital contribution it is again straightforward to obtain theoretical expressions for the corresponding 3-body magnetic moments μ_s and μ_v. The following results are obtained

$$\mu_s = \frac{\mu_p + \mu_n}{2}\,[p(S) + p(S') - p(D)] + \frac{1}{2}p(D) \qquad \text{(XI-45)}$$

$$\mu_v = \frac{\mu_p - \mu_n}{2}\,[p(S) - \frac{1}{3}p(S') + \frac{1}{3}p(D)] - \frac{1}{6}p(D) \qquad \text{(XI-46)}$$

where, as in eq. (IV-37), p is the probability with which the state indicated is admixed into the 3-body wave functions. S refers to the dominant ^{2}S state, S′ to the mixed symmetry ^{2}S state and D to the ^{4}D state (see e.g. Blatt (1952) or Delves and Phillips (1969)). In the above expression contributions from other states having negligible amplitudes in the 3-body wave functions have been ignored.

The isoscalar and isovector magnetic moments are related to those of the mirror nuclei ^{3}H ($T_3 = -\frac{1}{2}$) and ^{3}He ($T_3 = +\frac{1}{2}$) as follows

$$\mu_s = \frac{\mu(^3\mathrm{H}) + \mu(^3\mathrm{He})}{2}$$

$$\mu_v = \frac{\mu(^3\mathrm{H}) - \mu(^3\mathrm{He})}{2} \qquad \text{(XI-47)}$$

and experimentally $\mu_s^{\exp} = 0.426$ n.m. and $\mu_v^{\exp} = 2.553$ n.m.

In table XI-2 theoretical values for μ_s and μ_v are given for various assumptions about p(D) and p(S′) together with the corrections to μ_s ($\delta\mu_s$) and μ_v ($\delta\mu_v$) needed to reconcile theory with experiment. As discussed in §IV-5 in connection with β-decay exchange effects in the $A = 3$ system there is some theoretical uncertainty (e.g. Gibson (1965); Delves *et al.* (1969); Nunberg *et al.* (1972)) about the values of p(D) and p(S′) but most likely $p(\mathrm{D}) \approx 6\%$ and $p(\mathrm{S}') \approx 1$–2%. This means that corrections of the order 0.009 n.m. to μ_s and

Table XI-2
Theoretical values for μ_S and μ_V for $A = 3$ nuclei.

p(D) (%)	p(S′) (%)	μ_S (n.m.)	μ_V (n.m.)	$\delta\mu_S \left(\frac{\delta\mu_S}{\mu_S^{exp}}\%\right)$ (n.m.)	$\delta\mu_V \left(\frac{\delta\mu_V}{\mu_V^{exp}}\%\right)$ (n.m.)
4	0	0.423	2.284	0.003 (0.70%)	0.269 (10.5%)
6	0	0.417	2.249	0.009 (2.11%)	0.304 (11.9%)
6	2	0.417	2.185	0.009 (2.11%)	0.368 (14.4%)
9	2	0.406	2.134	0.020 (4.69%)	0.419 (16.4%)

0.35 n.m. to μ_V are needed in order to obtain agreement with experiment. The question is, can exchange contributions account for the above discrepancies?

As mentioned earlier Chemtob and Rho (1971) have made a very detailed calculation of exchange corrections to magnetic moments and, in particular, have evaluated the size of these corrections for the $A = 3$ nuclei. Neglecting the effects of S and D states and allowing for a reasonable spread for the hard core radius built into the 3-body wave function through a short-range correlation function and in the overall size parameter of the 3-body system they obtain

$$\delta\mu_S = 0.0093\,^{+0.0077}_{-0.0053}\ \text{n.m.}$$
$$\delta\mu_V = 0.193\,^{+0.041}_{-0.041}\ \text{n.m.}. \qquad \text{(XI-48)}$$

These results are somewhat sensitive to the values and signs taken for the meson nucleon coupling constants in the case of ρ- and ω-exchange and coupled with other uncertainties in the calculation mean that too great a faith should not be placed in them. Nevertheless the result for $\delta\mu_V$ is significantly in disagreement with the values indicated by experiment in table XI-2. A first possible cause for the disagreement, assuming that $p(\text{D}) \approx 6\%$, $p(\text{S}) \approx 1\text{–}2\%$ is correct, is the neglect of exchange operator matrix elements coupling S and D states (Riska and Brown (1972)). Contributions of this kind have been shown to be important in the case of β-decay exchange effects (Blomqvist (1970); Riska and Brown (1970); (see §IV-5.1)). A second possible cause is the admixture of N^* resonances into $A = 3$ wave function although here, as with the deuteron, there is ambiguity about the extent to which effects of this kind are included within the exchange operator. Arenhövel and Danos (1968) estimate that a 1% admixture of the N* (1470) into the 3-body wave function could account for the bulk of $\delta\mu_V$ (see also Green and Shucan (1972)). A recent comprehensive calculation by Harper *et al.* (1972) allowing for both S–D coupling and the N* contribution leads to virtually complete agreement with experiment. Thirdly, there might be contribution from 3-body currents as suggested by the work of Padgett *et al.* (1965a,b). Fourthly there may be contributions from other diagrams not normally considered (e.g. Gerstenberger and Nogami (1972)). Finally, the assumption made in all calculations of exchange effects in the $A = 3$ system about the radial dependence of the wave functions and, in particular, the form of short range correlations may not be good. Undoubtedly there is a problem here whose resolution is most likely going to lead to a better understanding of exchange effects and the fundamental interactions which cause them.

As stated earlier, wave function uncertainties preclude detailed information being obtained about exchange effects in the case of the magnetic moments of heavy nuclei. Nevertheless the effects are not by any means negligible and they have been estimated theoretically by a number of authors. One approach is to calculate the "quenching" of nucleon intrinsic magnetic moments due to the presence of other nucleons in the nucleus (e.g. Drell and Walecka (1960); Nyman (1967, 1970)). This is an exclusion principle effect which prevents a given nucleon recoiling into occupied nucleon states when emitting or reabsorbing, say, a virtual π-meson. What appears to be a one-body phenomenon then becomes a two-body process. Indeed ignoring the effect of the exclusion principle on intermediate states (Feynman (1949)) the effect then appears as the result of a two-body magnetic moment operator (that derived from the charge exchange part of the one-pion exchange potential by making it gauge invariant (Nyman (1970)). The actual magnitude of the effect is a few tenths of a nuclear magneton but its precise value and sign is critically dependent on the form of nucleon–nucleon correlations introduced into the nuclear wave function.

Other authors (e.g. Ross (1952); Russek and Spruch (1952)) have used a phenomenological approach in order to analyse the general trend of magnetic moment exchange effects in different nuclei. And most recently Chemtob (1969) has calculated the magnitude of exchange currents in the one-meson exchange approximation and evaluated the corresponding magnetic moment contributions in ^{15}N, ^{17}O, ^{39}K and ^{41}Ca. Again the contributions are of the order a few tenths of a nuclear magneton and are certainly smaller than the uncertainty in the magnetic moment due to lack of knowledge about the detailed form of the nuclear wave function.

One other phenomenon in the area of low energy nuclear physics which gives information about exchange effects is the radiative capture process $n + p \rightarrow d + \gamma$. For many years it has been known that a discrepancy of the order 10% exists between the measured cross section and that calculated theoretically. Thus for slow neutrons Cox *et al.* (1965) obtain σ_{exp} = (334.2 ± 0.5) mb as against σ_{theor} = (302.5 ± 4.0) mb obtained (Austern and Rost (1959); Noyes (1965)) using the usual one-body magnetic moment operators. Attempts to resolve this discrepancy by appealing to exchange effects have failed in the past (e.g. Noyes (1965)). However, these attempts have always calculated the effects taking the matrix element of the exchange operator only between the $^1S_0$ neutron–proton continuum state and the $^3S_1$ deuteron state; i.e. the $^3D_1$ deuteron state is neglected. Recently Riska and Brown (1972) have shown that the contribution from the $^1S_0 \rightarrow {}^3D_1$ transition is in fact nearly as large as that from the $^1S_0 \rightarrow {}^3S_1$ transition and that taking the one-pion exchange contribution to the exchange operator (Chemtob and Rho (1971) and including the Δ intermediate state) a total contribution to the cross section of the order 10% is obtained. They conclude, therefore, that one-pion exchange effects *are* sufficient to explain the discrepancy between theory and experiment *.

XI-5. Isospin multiplet energies

As mentioned in §XI-3 the existence of isobaric spin multiplets in which the components have essentially the same properties as far as strong interactions are concerned is strong evidence for the near charge independence and charge symmetry of nuclear forces. The energies of these components, however, are not exactly the same and it is well known that the bulk of the differences in energy (displacement energy) between them can be attributed to the Coulomb interaction between protons (V_C) and to the mass difference between the neutron and proton (ΔM) (see Jänecke (1969b) for a detailed review and also Brink (1965)). Additional contributions to displacement energies can arise from any charge dependence of the nuclear forces themselves (V_{CD}) and from terms in the electromagnetic interaction not normally considered. In the latter category, for example, would be any contribution from a possible isotensor component of the electromagnetic current. A

* Breit and Rustgi (1971) have suggested another possible contribution to the discrepancy deriving from the assumptions made about the energy independent nature of the internucleon potential.

programme of study can therefore be carried out which compares experimental isospin multiplet displacement energies with those calculated theoretically in terms of ΔM and V_C and other small but well known electromagnetic effects and seeks to determine information about the charge dependence of the internucleon potential etc. from any discrepancies which come to light.

Studies of this kind have been carried out at two levels. Firstly, there are those which look at the systematics of groups of multiplets, search for trends in displacement energies and then draw rather general conclusions. Secondly there have been very detailed studies of single multiplets (particularly in the $A = 3$ nuclei) using all available information on wave function, shape, radius etc. with a view to making any discrepancy as precise and reliable as possible. In the following sections we shall deal with both of these aspects (for general reviews see Garvey (1969) and Jänecke (1969b)).

XI-5.1. *The isospin multiplet mass formula (theory)*

Many years ago it was pointed out (Wigner (1957); Weinberg and Treiman (1959)) that the energies (masses) of the $2T + 1$ members of an isospin multiplet are related to a good approximation by *

$$M(A, T, T_3) = a(A, T) - b(A, T) T_3 + c(A, T) T_3^2 \tag{XI-49}$$

where T_3 $(= \frac{1}{2}(Z-N))$ ranges from T to $-T$. The approximations involved in obtaining this result are (a) that the Coulomb and other charge dependent potentials are treated in first order perturbation theory only and (b) that no isotensor of rank greater than 2 arises within the potential. As far as (b) is concerned, because nucleons have isospin $\frac{1}{2}$, two body forces can lead at most to a second rank isotensor contribution (see eqs. (XI-12) and (XI-33)). However, if the fundamental electromagnetic current has an isotensor component then the resulting 3 and 4-body potentials could include third and fourth rank isotensors which would show up as T_3^3 and T_3^4 terms in eq. (XI-49).

The expression given in eq. (XI-49) can be derived straightforwardly by first order perturbation theory as follows (e.g. Jänecke (1969b)). Denote the isospin multiplet eigenstates of the charge independent nuclear Hamiltonian by $|\alpha T T_3\rangle$ where α signifies all other quantum numbers needed to specify the states. Ignoring all but the Coulomb interaction, V_C $(= V_C^{(0)} + V_C^{(1)} + V_C^{(2)}$; see eq. (XI-12)), we can write for the Coulomb energy $E_C(A, T, T_3)$ of the state $|\alpha T T_3\rangle$

* Note that because of the sign convention used in this book terms odd in T_3 in eq. (XI-49) and subsequent equations have opposite sign to those generally used.

$$E_C(A, T, T_3) = \langle \alpha T T_3 | V_C | \alpha T T_3 \rangle$$

$$= \sum_{t=0}^{2} \langle \alpha T T_3 | V_C^{(t)} | \alpha T T_3 \rangle$$

$$= \sum_{t=0}^{2} C(T t T; T_3 0) \langle \alpha T \| V_C^{(t)} \| \alpha T \rangle$$

$$= E_C^{(0)}(A, T) + T_3 E_C^{(1)}(A, T) + (3T_3^2 - T(T+1)) E_C^{(2)}(A, T) \qquad \text{(XI-50)}$$

where scalar, vector and tensor Coulomb energies are given by

$$E_C^{(0)}(A, T) = \langle \alpha T \| V_C^{(0)} \| \alpha T \rangle \qquad \text{(XI-51)}$$

$$E_C^{(1)}(A, T) = -\frac{1}{\sqrt{T(T+1)}} \langle \alpha T \| V_C^{(1)} \| \alpha T \rangle \qquad \text{(XI-52)}$$

$$E_C^{(2)}(A, T) = \frac{1}{\sqrt{T(T+1)(2T-1)(2T+3)}} \langle \alpha T \| V_C^{(2)} \| \alpha T \rangle \,. \qquad \text{(XI-53)}$$

In addition we have to take account of the difference in mass between the neutron and proton noting that in an isospin notation we can write for the nucleon mass

$$M = \tfrac{1}{2}(M_p + M_n) + (M_p - M_n)\, t_3 \,. \qquad \text{(XI-54)}$$

Including this mass difference effect and the charge independent part of the nuclear binding energy E_0 (which, of course, depends on α and T) we obtain finally an expression of the form given in eq. (XI-49) with

$$a(A, T) = \tfrac{1}{2}(M_p + M_n)A + E_0 + E_C^{(0)}(A, T) - T(T+1) E_C^{(2)}(A, T) \qquad \text{(XI-55)}$$

$$b(A, T) = (M_n - M_p) - E_C^{(1)}(A, T) \qquad \text{(XI-56)}$$

$$c(A, T) = 3 E_C^{(2)}(A, T) \,. \qquad \text{(XI-57)}$$

Note that inclusion of other one- or two-body charge dependent forces would not alter the general form of $M(A, T, T_3)$ and would just lead to additional terms in the $E_C^{(k)}$.

The terms $E_C^{(t)}(A, T)$ can be evaluated on the basis of some nuclear model wave function and brief mention will be made of such calculations when reviewing the extent to which experimental data agree with eq. (XI-49). However, before looking into this question it is important to consider the effect on eq. (XI-49) of including higher order perturbations.

In general it is expected that because of the relative weakness of Coulomb forces compared with nuclear forces first order perturbation theory should suffice. There are obvious exceptions however. For example, when two states of different isospin lie close to one another leading to a small energy denominator within the perturbation theory large isospin mixing will take place (e.g. the well known mixing of the $T=0$ and $T=1$ 2^+ states in ^8Be). Similarly, if the binding energy of the proton rich member of an isospin multiplet should be negative (because of Coulomb effects) then its wave function will be very different from that of other members of the multiplet and there will be large mixtures of unbound states resulting in an anomalous (Thomas–Ehrman) shift in energy (Thomas (1951, 1952); Ehrman (1951); see also Weidenmüller (1965); Humblet and Lebon (1967)).

Jänecke (1969a, b) has studied the effect of including second order perturbation terms on the isospin mass formula (eq. (XI-49)) and finds, as is to be expected, that under these circumstances the equation is now a quartic in T_3, namely

$$M(A, T, T_3) = a(A, T) - b(A, T)\, T_3 + c(A, T)\, T_3^2 - d(A, T)\, T_3^3 + e(A, T)\, T_3^4 \,. \tag{XI-58}$$

Calculation within the $A = 9$, $J = \frac{3}{2}^-$, $T = \frac{3}{2}$, quartet of states assuming that the main second order terms arise from perturbations of the $J = \frac{3}{2}^-$, $T = \frac{3}{2}$, $T_3 = \pm\frac{1}{2}$ states by nearby $J = \frac{3}{2}^-$, $T = \frac{1}{2}$, $T_3 = \pm\frac{1}{2}$ states show that a and b are changed by only a percent or so, c is reduced by $\approx 15\%$ and small values for d and e are predicted namely $d = 5.8 \pm 4.2$ keV; $e = -0.17 \pm 0.18$ keV. In other words in this case the mass formula is still essentially quadratic in nature, even though second order perturbations are taken into account.

Henley and Lacy (1969) have estimated the size of the T_3^3 term for nuclei in the range $A = 7$ to $A = 37$ using a simple non-perturbative model for the nucleus consisting of a core plus three valence nucleons. Each valence nucleon is assumed to interact with an average central potential (including a Woods–Saxon nuclear potential, a $\boldsymbol{t} \cdot \boldsymbol{T}$ symmetry potential and a Coulomb potential) produced by the core plus the other two valence nucleons. Again, the main effects seem to be absorbed into changes in b and c and only a small value for $d(\sim (1/A)\,\Delta c$ where Δc is the change in c) is predicted. Numerically $d \lesssim 1$ keV.

Bertsch and Kahana (1970) have also estimated the value of d taking certain second order perturbation terms into account for a nuclear model consisting of a core plus three valence nucleons. They find $d \approx 1$ keV due to Coulomb effects.

As far as the Thomas–Ehrman shift is concerned in unbound or weakly bound states its effect can be large (e.g. ≈ 700 keV in $A = 13$ nuclei). It can be understood qualitatively but where it does occur is likely to mask any small charge dependent or other interesting effects of the type in which we are primarily interested. We shall not consider further, therefore, nuclei in which it is significant.

Apart from Thomas–Ehrman shifts, then, the theoretical situation as far as it is understood at present is that with Coulomb forces the quadratic mass formula should fairly accurately describe the energy levels of isospin multiplets and that any deviations from it (e.g. the presence of a T_3^3 term) could indicate the presence of additional charge dependent interactions. We now go on to consider the relevant experimental data.

XI-5.2. *The isospin multiplet mass formula (experiment)*

In the case of mirror nuclei ($T = \frac{1}{2}$ doublets) and $T = 1$ triplets the isospin multiplet mass formula cannot be tested since at least four measurements are needed for this purpose, i.e. a multiplet is needed with $T \geq \frac{3}{2}$. During the last few years some progress has been made in achieving this for $T = \frac{3}{2}$ and $T = 2$ multiplets (see Cerny (1968) for a review) and energies are now available for seven complete isospin quadruplets in $A = 7, 9, 13, 17, 21, 25$ and 37 nuclei. In table XI-3 (taken from Cerny (1968) for $A = 7$, from Mosher *et al.* (1971) for $A = 9$, from Mendelson *et al.* (1970) for $A = 15, 17, 21$ and 37 and from Trentelman and Proctor (1971) for $A = 25$ – see also Trentelman *et al.* (1970)) values of b, c and d which give the best fit to the experimental data are given together with the predicted and experimental values of the mass excess of the $T_3 = \frac{3}{2}$ (i.e. proton rich) member of the multiplet. These predictions are made using the energies of the $T_3 = \frac{1}{2}, -\frac{1}{2}, -\frac{3}{2}$ members assuming a strictly quadratic mass formula. The extent to which the agreement between these predictions and the experimental values is good is then a measure of the success of the simple quadratic formula. As far as the values of b and c are concerned there have been many calculations on the basis of one nuclear model or another (see Jänecke (1969b) for a review and also Auerbach *et al.* (1971) for calculations for $T = \frac{3}{2}$ multiplets in $4n + 1$ nuclei with $n = 2$ to $n = 10$) and they can reasonably be accounted for bearing in mind the usual nuclear wave function uncertainties.

Of particular interest for our purposes are the values of d. They are all

Table XI-3
Experimental data on $T = \frac{3}{2}$ isospin multiplets.

A	b (MeV)	c (MeV)	d (keV)	$T_3 = +\frac{3}{2}$ Mass excess (Expt) (MeV)	$T_3 = +\frac{3}{2}$ Mass excess (Predicted) (MeV)
7	-0.588 ± 0.050	0.255 ± 0.045	-11 ± 30	27.940 ± 0.100	27.870 ± 0.150
9	-1.3160 ± 0.0020	0.2645 ± 0.0020	8.0 ± 3.7	28.907 ± 0.004	28.956 ± 0.022
13	-2.1810 ± 0.0034	0.2560 ± 0.0027	-0.8 ± 3.4	23.107 ± 0.015	23.102 ± 0.014
17	-2.8780 ± 0.0054	0.2380 ± 0.0081	4.8 ± 9.2	16.479 ± 0.050	16.508 ± 0.023
21	-3.6580 ± 0.0068	0.2410 ± 0.0064	8.5 ± 7.8	10.889 ± 0.040	10.940 ± 0.024
25	-4.3780 ± 0.0070	0.2220 ± 0.0040	-7.9 ± 3.9	3.832 ± 0.012	3.785 ± 0.026
37	-6.1810 ± 0.0146	0.1740 ± 0.0146	$5.3. \pm 17.3$	-13.230 ± 0.050	-13.198 ± 0.091

essentially consistent with zero apart from the A = 9 and A = 25 quadruplets. For the A = 9 case, d = 8.0 ± 3.7 keV is in agreement with the value (d = 5.8 ± 4.2 keV) obtained theoretically by Jänecke (1969a, b) assuming second order perturbation contributions but not with the results of Henley and Lacy (1969) or Bertsch and Kahana (1970). However, subsequent work by Hardy *et al.* (1971) suggests that Jänecke's result over-estimates the size of d. There is therefore perhaps a significant discrepancy here and it is noteworthy that Bertsch and Kahana (1970) obtain a larger value for d (≈ 3.6 keV) if a short range charge dependent force of the strength discussed in §XI-3 (eq. (XI-30)) is included in the calculation. The discrepancy might also indicate the presence of a three body charge dependent force deriving from an isotensor component in the electromagnetic current (Adler (1967); Shaw (1967); Divakaran *et al.* (1968)). This is unlikely, however, since for a 3-body phenomenon roughly equivalent contributions would be expected in other nuclei.

No firm conclusion can be drawn in any case until further detailed calculations have been made on the effects of second order perturbations and other strictly nuclear phenomena (e.g. level shifts due to nearby thresholds for isospin allowed decays). Nevertheless this is obviously a profitable area for further study.

XI-5.3. *Studies of specific isospin multiplets*

A number of detailed theoretical studies of $T = \frac{1}{2}$ and T = 1 multiplets have been made for particularly simple nuclei which throw some light on the charge dependence of nuclear forces. Of particular importance here is the work that has been done on the mirror nuclei ^{3}H and ^{3}He. In the case of these nuclei fine details of the wave functions are gradually becoming available (see Delves and Phillips (1969) for a review) so that it is possible to obtain a quite reliable estimate of the Coulomb energy difference between the two nuclei.

Experimentally the difference in binding energy (B) of ^{3}He and ^{3}H is given by (Mattauch *et al.* (1965))

$$\Delta_{\rm exp} = \mathrm{B}(^3\mathrm{He}) - \mathrm{B}(^3\mathrm{H}) = 0.76384 \pm 0.00026 \text{ MeV} . \qquad \text{(XI-59)}$$

The most detailed theoretical calculations of electromagnetic contributions to Δ are by Okamoto and Lucas (Okamoto (1964, 1965, 1966); Okamoto and Lucas (1967a, b, 1968)) extending earlier work of Pappademos (1963, 1964). They (Okamoto and Lucas (1967b)) use a variational wave function of Gaussian form for the dominating spatially symmetric S-state part of the three body wave function together with a soft-core correlation function with the Gaussian parameters so adjusted as to give the observed rms radii for ^{3}He and ^{3}H. They also take account of the finite proton size, S′ (i.e. mixed symmetry S state) and D-state admixtures, vacuum polarisation effects, electric polarisation of the nucleon charge distributions and finally conclude that

$$\Delta_{\text{e.m.}} = 0.63 \pm 0.03 \text{ MeV} . \tag{XI-60}$$

More recently, Okamoto and Pask (1971b) have modified this result to read

$$\Delta_{\text{e.m.}} \approx 0.62 \sim 0.66 \text{ MeV} . \tag{XI-61}$$

This result is in general agreement with those of other authors (Pappademos (1964); Delves (1964); Krueger and Goldberg (1964); Tang and Herndon (1965); Levinger and Srivastava (1965); Srivastava (1965); Rosati and Barbi (1966); Lovitch and Rosati (1967); Khana (1967); Delves and Blatt (1967); Delves *et al.* (1969); Friar (1970); Delves and Hennell (1971)) using local potentials if these calculations are modified where necessary to account for the finite proton size. On the other hand it does not agree with a calculation by Gupta and Mitra (1967) who use an "exact" three-body wave function obtained by means of separable potentials. Taking account of the S′ state, the tensor force and the proton charge distribution they obtain $\Delta_{\text{e.m.}} = 0.83$ MeV which should be reduced by ≈ 10% to allow for hard core effects. Commenting on this work Okamoto and Lucas (1968) point out that the non-local potentials used by Gupta and Mitra do not tie in well with the 2-body scattering data and they estimate that on correcting for this a value of $\Delta_{\text{e.m.}}$ similar to that given in eq. (XI-61) would be obtained.

There thus seems to be a significiant difference between Δ_{exp} and $\Delta_{\text{e.m.}}$ of the order 0.1 MeV. This difference can be resolved by appealing to a charge dependence of the nuclear force itself (Blin-Stoyle and Yalçin (1965)). Since ^{3}He and ^{3}H are mirror nuclei the difference has to be attributed to a breakdown of charge symmetry and can be analysed in terms of the charge asymmetric part of the charge dependent phenomenological potential introduced in eqs. (XI-33) and (XI-35) namely

$$V_{\text{CD}} = [p + r\boldsymbol{\sigma}^{(1)} \cdot \boldsymbol{\sigma}^{(2)}](\tau_3^{(1)} + \tau_3^{(2)})\, V_0 f(r_{12}) \tag{XI-62}$$

where p and r measure the deviation from charge symmetry and $f(r_{12})$ is a convenient radial function.

In terms of the above potential and neglecting other states than the spatially symmetric S-state gives

$$\Delta_{\text{c.d.}} = 4V_0(p-3r)\langle\Psi_s|f(r_{12})|\Psi_s\rangle \,. \tag{XI-63}$$

where Ψ_s is the S-state wave function. On the basis of this equation Blin-Stoyle and Yalçin (1965) estimate that if $\Delta_{\text{exp}} - \Delta_{\text{e.m.}} \approx 0.1$ to 0.2 MeV, this requires

$$3r - p \approx (0.1 \text{ to } 0.3)\% \tag{XI-64}$$

which is certainly consistent with the limits imposed by the low energy 2-body scattering data given in eq. (XI-40) and with the results of OBEP calculations (see §XI-3.1). Explicit calculations of $\Delta_{\text{c.d.}}$ have also been carried out by various authors (e.g. Stevens (1965); Okamoto and Lucas (1967a); Yalçin and Birsen (1970); Okamoto and Pask (1971a)) using specific pion and photon exchange models of the charge dependent potential and broad agreement between theory and experiment is generally obtained. This work perhaps sets the firmest limit on the magnitude of charge asymmetry in the internucleon potential.

In the case of isospin multiplets in heavier nuclei, wave function uncertainties preclude any but qualitative conclusions being drawn. Early measurements and calculations (e.g. Wilkinson (1956b); Fairbairn (1961, 1963); Sengupta (1962); Altman and MacDonald (1962); Lovitch (1964)) centred on the 1p-shell nuclei (in particular the 0^+, $T = 1$ multiplet in $A = 14$) and gave some indication of the need for a charge dependence of nuclear forces of the size we now know it to be. However, the calculations were based on simple shell model wave functions and the results could be very sensitive to admixtures of higher configurations (Blin-Stoyle and Nair (1963)) so that it was not possible to draw firm conclusions other than that the breakdown of charge symmetry was significantly smaller than the breakdown of charge independence. Similar reservations apply to the more recent work of Murthy (1969) on nuclei in the region $A = 6$ to $A = 32$.

More sophisticated shell-model calculations on 1p-shell nuclei have been carried out by Wilkinson and his collaborators (Wilkinson (1964); Wilkinson and Hay (1966); Wilkinson and Mafethe (1966)) using wave functions computed in simple Woods–Saxon potentials adjusted to agree with the r.m.s. charge radius and the mean 1p-shell proton binding energy. The calculations use fractional parentage techniques, antisymmetrised proton wave functions and allow for both Coulomb and magnetic charge dependent effects. A comparison is made between the theoretical and experimental energy differences between the $(A, Z+1)$ and (A, Z) nuclei for $A = 9 - 16$. The agreement is good and the authors set an upper limit of 0.7% on the departure from charge symmetry of the internucleon potential. There is also some suggestion that V_{np} is some 2% greater than V_{pp} or V_{nn} in agreement with the nucleon–nucleon scattering data (§XI-3, eq. (XI-30)). Again the importance of using proper wave functions in this sort of work emerged as it does in related work by other authors (e.g. Nemirovskii (1966) on mirror nuclei from $A = 16$ to $A = 40$; Perey and Schiffer (1966); Elton (1967) and Nolen *et al.* (1967) on Ca and Sc isotopes).

Recently, in this field, considerable interest has developed in the binding energy differences of some $T = \frac{1}{2}$ mirror nuclei and implicitly on the information which can thereby be derived about the charge asymmetry of nuclear forces (Nolen and Schiffer (1969a, b); Schiffer *et al.* (1969); Auerbach *et al.* (1969a, b, c); Wong (1970); Damgaard *et al.* (1970); Van Giai *et al.* (1971); Negele (1971)). Nolen and Schiffer (1969a, b) and Schiffer *et al.* (1969) calculate the Coulomb energy difference between pairs of mirror nuclei by using a method which is largely model independent. The approach uses a phenomenological independent particle model in which the parameters of the Woods–Saxon potential are so chosen as to yield an r.m.s. charge radius consistent with electron scattering and muonic X-ray results and sensible single particle energies.

The electromagnetic energy difference $\Delta_{\text{e.m.}}$ can be written

$$\Delta_{\text{e.m.}} = \Delta_{\text{e.m.}}(\text{direct}) + \Delta_{\text{e.m.}}(\text{exchange}) + \Delta_{\text{e.m.}}(\text{spin-orbit})\,. \qquad \text{(XI-65)}$$

Here

$$\Delta_{\text{e.m.}}(\text{direct}) = e \int_0^\infty V_{\mathrm{C}}(r)\,\rho_{\text{exc}}(r)\,4\pi r^2\,\mathrm{d}r \qquad \text{(XI-66)}$$

where $V_{\mathrm{C}}(r)$ is the Coulomb potential due to the measured charge distribution and $\rho_{\text{exc}}(r)$ is the density distribution of the neutron excess.

$$\Delta_{\text{e.m.}}(\text{exchange}) = \sum_{\substack{\text{all protons,}\\ \text{neutron excess}}} \iint \psi_{\mathrm{p}}^{\times}(r_1)\,\psi_{\text{exc}}^{\times}(r_2)\,\frac{e^2}{r_{12}}\,\psi_{\mathrm{p}}(r_2)\,\psi_{\text{exc}}(r_1)\,\mathrm{d}^3r_1\,\mathrm{d}^3r_2 \qquad \text{(XI-67)}$$

where ψ_{p} and ψ_{exc} are single particle wave functions for protons and excess neutrons respectively.

$\Delta_{\text{e.m.}}$(spin-orbit)

$$=\frac{1}{N-Z}\sum_i \int \rho_i(r_i)\frac{1}{r_i}\frac{\mathrm{d}V_C}{\mathrm{d}r_i}4\pi r_i^2\,\mathrm{d}r_i(\tfrac{1}{2}-g_\mathrm{p}+g_\mathrm{n})\frac{e\hbar^2}{2m^2c^2}\boldsymbol{\sigma}^{(i)}\cdot \boldsymbol{l}^{(i)} \quad \text{(XI-68)}$$

where the sum i is taken over the excess neutrons. This latter term arises from the differences in g-factor (g_p and g_n) of the proton and neutron and Thomas precession effects (see §XI-2). These energy differences are then compared with the experimental binding energy difference (Δ_{exp}) of the mirror nuclei concerned (see table XI-4). There are clearly significant discrepancies between theory and experiment and the other papers listed above have introduced refinements into the calculation which take into account such factors as compound (internal) and continuum (external) mixing, nuclear two body correlations, isospin mixing in the mirror states, core polarisation by valence nucleons, finite proton size, dynamic effect of the neutron–proton mass difference. The general conclusion of all this work is that although these different effects taken together are non-negligible there is still an outstanding discrepancy between the theoretical and experimental values of the binding energy differences provided agreement with measured nuclear sizes is insisted upon. However, no account has so far been taken of a possible charge asymmetry in the nuclear force itself and it is noteworthy that the discrepancy has the same sign as that in $A = 3$ nuclei which is attributed to just this effect. Negele (1971) has calculated the strength of a spin independent Gaussian charge asymmetric potential needed to resolve the discrepancy in Δ for the ^{41}Sc–^{41}Ca mirror nuclei and then, using a symmetric S-state Gaussian wave function for ^{3}H and ^{3}He, calculated the contribution to Δ for these latter nuclei. He obtains a value 0.16 MeV which is in complete agreement with the

Table XI-4
Theoretical and experimental binding energy differences of mirror nuclei.

Mirror pair	$\Delta_{\text{e.m.}}$ (MeV)	Δ_{exp} (MeV)
^{13}N – ^{13}C	2.73	3.00
^{17}F – ^{17}O	3.29	3.54
^{29}P – ^{29}Si	5.53	5.73
^{33}Cl – ^{33}S	6.00	6.35
^{41}Sc – ^{41}Ca	6.67	7.28

conclusion given earlier that $\Delta_{exp} - \Delta_{e.m.} \approx 0.1$ to 0.2 MeV for the $A = 3$ nuclei.

Finally, mention should be made of an extensive calculation (Kahana (1972)) of the lowest $T = 1$ states in ^{18}O, ^{18}F and ^{18}N using a realistic force and including the Coulomb interaction. Particle–particle, hole–particle and hole–hole correlations are included and good agreement with experiment is obtained provided that an additional n–p attractive force is introduced. This force breaks charge independence but not charge symmetry and is consistent in strength and sign (i.e. $|\bar{V}_{np}| > |\bar{V}_{nn}|$) with that required to account for the nucleon–nucleon scattering data (see eq. (XI-30)).

XI-5.4. *Isospin multiplet energies. Conclusions*

In spite of the nuclear complexities inherent in calculating the energies of different members of isospin multiplets a situation is emerging which calls for the presence of a nucleon–nucleon potential violating both charge independence and charge symmetry.

Further, the signs and magnitudes of the different parts of this potential as far as they can be isolated are in agreement with those required to account for nucleon–nucleon scattering data. Unfortunately there is little evidence one way or the other from isospin multiplet energy data about the presence of an isotensor component in the electromagnetic current but further theoretical and experimental study of the T_3^3 term in the isospin multiplet mass formula might pay-off in this respect.

XI-6. **Isospin in nuclear γ-transitions**

Electromagnetic gamma transitions in a nucleus are induced by the term

$$\mathcal{H}(A) = \frac{e}{c}\int \boldsymbol{J}_{\mathrm{N}}(\boldsymbol{r}) \cdot \boldsymbol{A}(\boldsymbol{r})\, \mathrm{d}^3 r \tag{XI-69}$$

in the nuclear Hamiltonian given in eq. (XI-14). Since $\boldsymbol{J}_{\mathrm{N}}$ can be decomposed into an isoscalar part, $\boldsymbol{J}_{\mathrm{N}}^{(s)}$ and an isovector part, $\boldsymbol{J}_{\mathrm{N}}^{(v)}$, (see eq. (XI-16)) it follows that $\mathcal{H}(\boldsymbol{A})$ can be similarly decomposed, thus

$$\mathcal{H}(A) = \mathcal{H}^{(s)}(A) + \mathcal{H}^{(v)}(A) \tag{XI-70}$$

where

$$\mathcal{H}^{(s)}(A) = \frac{e}{c}\int J_N^{(s)}(r)\cdot A(r)\,d^3r$$

and

$$\mathcal{H}^{(v)}(A) = \frac{e}{c}\int J_N^{(v)}(r)\cdot A(r)\,d^3r\,. \tag{XI-71}$$

As far as isospin is concerned then, the matrix element for a gamma transition between two states of isospin T and T' has the form

$$\begin{aligned}\langle T'T_3'|\mathcal{H}(A)|TT_3\rangle &= \langle T'T_3'|\mathcal{H}^{(s)}(A)|TT_3\rangle + \langle T'T_3'|\mathcal{H}^{(v)}(A)|TT_3\rangle \\ &= C(T\,0\,T';T_3\,0)\,\langle T'\|\mathcal{H}^{(s)}(A)\|T\rangle \\ &\quad + C(T\,1\,T';T_3\,0)\,\langle T'\|\mathcal{H}^{(v)}(A)\|T\rangle\end{aligned} \tag{XI-72}$$

where $\langle T'\|\mathcal{H}^{(s)}(A)\|T\rangle$ and $\langle T'\|\mathcal{H}^{(v)}(A)\|T\rangle$ are reduced matrix elements. The first Clebsch–Gordan coefficient is independent of T_3 and vanishes unless $T' = T$ and the second vanishes unless $T' = T$ or $T \pm 1$. Further, for both parts of $\mathcal{H}(A)$ we must have $T_3' = T_3$ corresponding to charge conservation. This means that for $\mathcal{H}^{(s)}(A)$ the isospin selection rule is $\Delta T = 0$, $\Delta T_3 = 0$ and for $\mathcal{H}^{(v)}(A)$ the rule is $\Delta T = 0, \pm 1$ $(0 \not\to 0)$, $\Delta T_3 = 0$. Overall, then, we have the well known isospin selection rule for electromagnetic transitions (Radicati (1952); Gell-Mann and Telegdi (1953))

$$\begin{aligned}\Delta T &= 0, \pm 1 \\ \Delta T_3 &= 0\,.\end{aligned} \tag{XI-73}$$

If J_N contained an isotensor part as in eq. (XI-25) then gamma transitions would be allowed for which $\Delta T = \pm 2$. Violation of the selection rules given in eq. (XI-73) could therefore lead to information about the possibility of there being an isotensor component in the electromagnetic interaction. We shall return to this point in §XI-6.1.

Further general results emerge on more detailed consideration of the properties of the Clebsch–Gordan coefficients in eq. (XI-72). For example, in self conjugate nuclei (i.e. nuclei with T_3 $(= T_3') = 0$) the coefficient $C(T\,1\,T';00)$ vanishes for $T = T'$ so that there is no isovector contribution to a $\Delta T = 0$ transition. Further, since

$$C(T\,1\,T';T_3\,0) = (-)^{T+1-T'}C(T\,1\,T';-T_3\,0) \tag{XI-74}$$

it follows that for corresponding $\Delta T = \pm 1$ transitions (for which there is no isoscalar contribution) in conjugate nuclei (T_3 and $-T_3$) the transition matrix elements are identical and so, therefore, are all physical phenomena associated with the transitions (Morpurgo (1959)).

Another important limitation can be imposed in the case of E1 transitions (Radicati (1952); Trainor (1952)) by considering the detailed form of the E1 multipole operator as follows. This operator arises in the usual way in making a multipole expansion of $\mathcal{H}(\boldsymbol{A})$, namely (see e.g. Rose (1955))

$$\mathcal{H}(\boldsymbol{A}) = \pi \sum_{L\mu P} (2L+1)^{1/2} f(P) D^{(L)}_{\mu P}(\boldsymbol{k}) [\mathcal{H}(A^{\mu}_{L}(m)) + (-P)\mathcal{H}(A^{\mu}_{L}(e))] \qquad \text{(XI-75)}$$

where $f(P)$ ($P = +1$ or -1 corresponding to left or right circular polarisation) defines the polarisation of the gamma-ray emitted in the direction $\boldsymbol{k}$. $\mathcal{H}(A^{\mu}_{L}(m))$ and $\mathcal{H}(A^{\mu}_{L}(e))$ are the magnetic and electric multipole operators respectively (e.g. Moszkowski (1965)) and $D^{(L)}_{\mu P}(\boldsymbol{k})$ is an element of the rotation matrix. In an E1 transition the relevant operator is $\mathcal{H}(A^{\mu}_{1}(e))$ whose form in the long wavelength approximation (e.g. Moszkowski (1965)) can be written

$$\mathcal{H}(A^{\mu}_{1}(e)) = (-)^{\mu} \frac{1}{\sqrt{6\pi}} \frac{e}{2Mc} \hat{n}_{-\mu} \cdot \sum_{i} (1 + \tau^{(i)}_{3}) \boldsymbol{p}^{(i)} \qquad \text{(XI-76)}$$

where $\hat{n}_{-\mu}$ is a unit vector in the direction $-\mu$. Now the isoscalar part of $\mathcal{H}(A^{\mu}_{1}(e))$ (i.e. the part independent of $\tau^{(i)}_{3}$) can be written

$$\mathcal{H}^{(s)}(A^{\mu}_{1}(e)) = \boldsymbol{b}^{\mu} \cdot \boldsymbol{P} \qquad \text{(XI-77)}$$

where $\boldsymbol{b}^{\mu}$ is a constant vector and $\boldsymbol{P}$ ($= \Sigma_i \boldsymbol{p}^{(i)}$) is the centre of mass momentum. The matrix element of such an external operator taken between two orthogonal nuclear states clearly vanishes so that there is no isoscalar contribution to an E1 transition. This means that such a transition with $\Delta T = 0$ is completely forbidden in a self-conjugate nucleus since, as demonstrated earlier, there can be no isovector contribution either. Also the identity of corresponding E1 transitions in conjugate nuclei now holds for $\Delta T = \pm 1$ transitions.

Weaker and less precise selection rules (quasi-rules) can also be derived for ML transitions (e.g. Warburton and Weneser (1969)) which follow basically from the large difference between the strengths of the isoscalar and isovector parts of the "spin" current (see eq. (XI-15) and note that

$(\mu_p + \mu_n)/(\mu_p - \mu_n) \approx \frac{1}{6}$; see also Morpurgo (1958)). They will not be discussed here, however, since their rather qualitative nature means that they cannot be used to clarify fundamental properties of the electromagnetic interaction.

Summarising, then, a number of strong isospin selection rules can be specified for electromagnetic transition as follows (we follow here the forms of Warburton and Weneser (1969)).

Rule 1: *Electromagnetic transitions are forbidden unless* $\Delta T = 0, \pm 1$.

Rule 2: *Corresponding* $\Delta T = \pm 1$ *transitions in conjugate nuclei are identical in all properties.*

Rule 3: *Corresponding* E1 *transitions in conjugate nuclei – whether* $\Delta T =$ 0 *or* ± 1 *– have equal strengths.*

Rule 4: $\Delta T = 0$ E1 *transitions in self-conjugate nuclei are forbidden.*

These rules are all based on the standard form of the electromagnetic interaction and Rule 1 for its complete validity requires charge independence. Rules 2–4 on the other hand will still hold if charge independence is violated but charge symmetry holds. This is because these latter rules derive from consideration of the properties of matrix elements under the charge symmetry operation $T_3 \to -T_3$. Correspondingly, any observed violation of the rules can in principle give information about the form of the electromagnetic interaction and about the goodness of charge independence and charge symmetry.

Finally, in this general section, a result derived by Ram (1970) for gamma decays from members of isospin multiplets should be noted. He points out that assuming only isoscalar and isovector components in $\mathcal{H}(A)$ then a relation similar to the isospin multiplet mass formula (eq. (XI-49)) should hold for the total decay rates of members of isospin multiplets namely,

$$\Gamma_\gamma(A, T, T_3) = a(A, T) - b(A, T)\, T_3 + c(A, T)\, T_3^2 \,. \qquad \text{(XI-78)}$$

This quadratic dependence on T_3 is intuitively obvious given that the corresponding matrix elements are at most linear in T_3 (this follows directly from eq. (XI-72)). As with the isospin mass formula deviations from eq. (XI-78) could indicate the presence of an isotensor component in $\mathcal{H}(A)$. Unfortunately at the present time insufficient experimental data is available to make such a test.

We now go on to consider what information experimental tests of the different selection rules are able to give about the electromagnetic interaction.

XI-6.1. *Violation of the* $\Delta T = 0, \pm 1$ *selection rule*

The $\Delta T = 0, \pm 1$ selection rule can be tested in principle by looking for $\Delta T \geq 2$ gamma transitions between nuclear states. The presence of such a transition could then be interpreted as evidence for the presence of an isotensor term in the electromagnetic current (Dombey and Kabir (1966); Divakaran *et al.* (1968)). In the last few years $T = 2$ states in $T_3 = 0$ nuclei have been identified in various nuclei (e.g. Cerny *et al.* (1964); Adelberger and McDonald (1967); Riess *et al.* (1967); Kuan *et al.* (1967); Bloch *et al.* (1967); McGrath *et al.* (1968); Snover *et al.* (1969)) so that the observation of $\Delta T = 2$ transitions is within the realms of possibility. Such a transition involving, as it does, a short range two-particle current (see §XI-2), is expected to be suppressed compared with the usual $\Delta T = 0, \pm 1$ single particle current transition and, in general $\Delta T = 1$ transitions are expected to dominate. Nevertheless one case has been reported in the literature (Snover *et al.* (1969)) in which a limit has been set on a $\Delta T = 2$ γ-decay width ($\Gamma_\gamma(\Delta T = 2)$). This is a decay from the 0^+, $T = 2$ state at 15.2 MeV to the 2^+, $T = 0$ state at 1.78 MeV in ^{28}Si. An upper limit $\Gamma_\gamma(\Delta T = 2) \lesssim 0.03\, \Gamma_\gamma$ is obtained where Γ_γ is the total width of the 15.2 MeV state. This result implies that the $T = 2$ γ-decay amplitude for the 15.2 MeV state is less than about 20% of the $T = 1$ amplitude. Hanna (1969) also reports a measurement by Snover *et al.* (1968) for a $0^+, T = 2 \rightarrow 2^+, T = 0$ transition in ^{24}Mg which imposes a similar upper limit on the $T = 2$ amplitude. Of course if a positive effect should be found then a study would need to be made of the isospin mixing effects due to Coulomb and other charge dependent effects which would enable the transition to take place even if an isotensor electromagnetic interaction were not present.

XI-6.2. *Corresponding* γ*-transitions in conjugate nuclei*

The identity of corresponding $\Delta T = \pm 1$ gamma transitions in conjugate nuclei (Rule 2) follows on the assumption that there is no isotensor component in the electromagnetic interaction and that charge symmetry holds. An observed violation of this equality can therefore lead to interesting information about the electromagnetic interaction. Unfortunately, because of the high excitation of $T = |T_3| + 1$ states in $T_3 \neq 0$ nuclei they can usually decay by particle emission and gamma decay is very unfavoured. Nevertheless measurements on such decays are possible either using (particle, γ) capture reactions or (ee) or ($\gamma\gamma$) inelastic scattering processes. The widths of one such pair of corresponding gamma transitions ($T = \frac{3}{2} \rightarrow T = \frac{1}{2}$) have been measured in ^{13}C ($T_3 = -\frac{1}{2}$) and ^{13}N ($T_3 = +\frac{1}{2}$). The results obtained are as follows:

^{13}C (Peterson (1967); Cocke *et al.* (1968))

$$\Gamma_\gamma(+\tfrac{1}{2}) \equiv \Gamma_\gamma(^{13}\mathrm{C};\, T=\tfrac{3}{2}, I=\tfrac{3}{2}^-,\ 15.11\ \mathrm{MeV} \to T=\tfrac{1}{2}, I=\tfrac{1}{2}^-,\ 0.00\ \mathrm{MeV})$$
$$= 25 \pm 7\ \mathrm{eV}$$

^{13}N (Cocke *et al.* (1968); Dietrich *et al.* (1968))

$$\Gamma_\gamma(-\tfrac{1}{2}) \equiv \Gamma_\gamma(^{13}\mathrm{N};\, T=\tfrac{3}{2}, I=\tfrac{3}{2}^-,\ 15.07\ \mathrm{MeV} \to T=\tfrac{1}{2}, I=\tfrac{1}{2}^-,\ 0.00\ \mathrm{MeV})$$
$$= 27 \pm 5\ \mathrm{eV}\ .$$

Since the energy releases in both cases are essentially equal the results are clearly in agreement with selection rule 2 given in §XI-6.

It is interesting, however, to investigate what limits these experimental results place on the amount of isotensor interaction which chould be present. This has been done by Blin-Stoyle (1969d). The matrix element for a $T=\frac{3}{2} \to T=\frac{1}{2}$ transition can be written

$$M_\gamma(T_3) = \langle \tfrac{1}{2}, T_3 | \sum_{t=0,1,2} H_\gamma^{(t)} | \tfrac{3}{2}, T_3 \rangle \qquad \text{(XI-79)}$$

where the electromagnetic interaction H_γ responsible for the transition has been decomposed into its isoscalar ($t = 0$), isovector ($t = 1$) and isotensor ($t = 2$) parts. Application of the Wigner–Eckart theorem then gives

$$M_\gamma(\tfrac{1}{2}) = -\frac{1}{\sqrt{3}} M_1 - \frac{1}{\sqrt{5}} M_2$$

$$M_\gamma(-\tfrac{1}{2}) = -\frac{1}{\sqrt{3}} M_1 + \frac{1}{\sqrt{5}} M_2 \qquad \text{(XI-80)}$$

where M_1 and M_2 are reduced matrix elements for the isovector and isotensor parts of H_γ respectively. Assuming $M_2 \ll M_1$ it then follows that the γ-decay widths for the two decays [$\Gamma_\gamma(+\frac{1}{2})$ and $\Gamma_\gamma(-\frac{1}{2})$] are related by

$$R = \frac{\Gamma_\gamma(-\frac{1}{2})}{\Gamma_\gamma(+\frac{1}{2})} = \left[1 - 4\left(\frac{3}{5}\right)^{1/2} \frac{M_2}{M_1}\right] \frac{\rho(E_{-1/2})}{\rho(E_{+1/2})} \qquad \text{(XI-81)}$$

where $\rho(E)$ is a (known) energy dependent factor relevant to a decay of the multipolarity and energy considered. For the $A = 13$ decays under consideration, inserting the experimental data into eq. (XI-81) gives

$M_2/M_1 = -0.026 \pm 0.116$, setting an upper limit of the order 15% on the relative isotensor amplitude. A similar consideration of the more complicated case of γ-decay from the same $T = \frac{3}{2}$ states to the $T = \frac{1}{2}$ 3.51 MeV and 3.56 MeV states in ^{13}N and the 3.68 MeV and 3.85 MeV states in ^{13}C leads to a rather lower limit of the order 7%.

As with $\Delta T = 2$ decays, the analysis should take into account the effects of isospin mixing due to Coulomb and other charge dependent forces. Here Warburton and Weneser (1969) argue that such effects should be small and Blin-Stoyle (1969) in a more quantitative treatment estimates an upper limit of the order 3% in the amplitude ratio.

This approach then seems to lead to rather better limits being imposed on the isotensor amplitude than the study of $\Delta T = 2$ transitions. Further, choice of inhibited decays (i.e. M_1 small) could lead to appreciable amplification of the effects of an isotensor component. This is not the case with the example given here since the γ-widths are of the order of the single particle Weisskopf estimate (e.g. Preston (1962)). The study of more suitable decays could be rewarding *. Further and deeper theoretical analysis is also needed from the point of view of relating the matrix element M_2 (which is that of a two-nucleon electromagnetic interaction) to the actual isotensor component of the fundamental electromagnetic interaction.

XI-6.3. E1 *transitions in conjugate and self-conjugate nuclei*

There have been a number of measurements on corresponding E1 transitions in conjugate nuclei (e.g. Robinson *et al.* (1968) and Riess *et al.* (1968) on ^{13}C and ^{13}N respectively; Sowerby and McCallum (1968) on ^{25}Mg and ^{25}Al; Warburton (1966) on ^{15}N and ^{15}O). In all cases there are large differences between the transitions but these can be accounted for in terms of nuclear wave function differences linked with corresponding Thomas–Ehrman shifts (Thomas (1951, 1952); Ehrman (1951)). Because of these nuclear complexities the experimental data is unable to throw any light on the structure of the electromagnetic interaction at this stage.

Similarly with E1 transitions in self-conjugate nuclei, although this field has been extensively studied during the last decade or so starting with the work of Wilkinson (1958b, 1960), the results are too dependent on nuclear structure uncertainties to allow the fine analysis necessary for our purposes (see Warburton and Weneser (1969) for a general review). Suffice it to say that the E1 transitions studied are generally inhibited by an order of magni-

* Adelberger and Balamuth (1971) have pointed out the value of related studies of $T = 2 \rightarrow T = 1$ γ-transitions in $T_3 = 0, \pm 1$ nuclei.

tude or more but that because of nuclear complications no more than semi-quantitative interpretations of the implicit isospin mixing can be given. These studies nevertheless do throw light on different aspects of the nuclear structure problem.

XI-7. Isospin in nuclear reactions

Isospin impurities and charge dependent effects in general certainly play a role in nuclear reactions processes. However, whilst significant information about these matters can be obtained from a study of weak and electromagnetic interaction processes this is no longer the case when we are dealing with what are essentially perturbations in strong interaction processes. In the former cases the relevant operators are known with some precision, particularly as far as their isospin properties are concerned, and can generally be treated by perturbation theory. In the latter, however, the reactions are complex many body processes for which simple perturbation theory is usually not appropriate. Since the fundamental effects we are concerned with (e.g. charge dependence of nuclear forces, isotensor component of the electromagnetic interaction) have a relative strength of the order 1% or less, the complexity just referred to makes it virtually impossible to disentangle them.

The main approach would seem to be the study of isospin forbidden reactions which can certainly give information about isospin impurities. Soper (1969) points out that there are two major types of processes to be considered in this connection.

Firstly, there are processes in which a particle of isospin t_0 is captured or emitted by a state of isospin T_1 to form a state of spin T_2 where $|T_1 - T_2| > t_0$ (e.g. neutron emission from a $T = \frac{3}{2}$, $T_3 = \frac{1}{2}$ state to form a $T = 0$, $T_3 = 0$ final state). Obviously isospin is not conserved in such a process. Unfortunately highly excited (and therefore complex) states are inevitably involved making detailed analysis extremely difficult.

Secondly, there are nuclear reactions of the type A(a, b)B in which the particles a and b have zero isospin and final states studied in B have different isospin from A. Such reactions are (α, α)(d, d), $(\alpha$, d) and (d, $\alpha)$. However, even with a process of this kind since T_3 is conserved at least one state with $T > T_3$ is involved which will in general have high excitation energy. Soper (1969) suggests that low-lying $I = 0^+$, $T = 1$ states in odd–odd nuclei with $T_3 = 0$ would be most suitable for study since their excitation energy is not too high. The most useful reactions would then be (d, α) and $(\alpha$, d) on even–even nuclei. It is further important not to use reactions involving long lived compound states since, as pointed out by Wilkinson (1956a) and confirmed by Browne (1966), severe breakdowns in isospin selection rules can take place related solely to the complexity and longevity of such states. This situation does not occur in the case of "direct" reactions and information about the isospin purity of initial or final states might be obtained. So far, however, no significant information about the charge dependent properties of fundamental interactions has arisen from studies of this kind.

Nevertheless, in time with better experimental accuracy and theoretical understanding some useful information might be forthcoming and in conclusion we mention some recent work of Cocke and Adloff (1971) *. Using the reaction $^7Li(^3He, \alpha)\, ^6Li^*$ (5.36 MeV) (d) ^4He these authors study the isospin forbidden decay of the 5.36 MeV ($I = 2^+$, $T = 1$) state of ^6Li into α + d ($T = 0$). Their work sets an upper limit of the order 2% on the branching ratio of this state to α + d. Assuming a $(1p)^2$ configuration, this state is $^1D_2$ ($T = 1$) in nature and isospin mixing of the nearby $^3D_2$ ($T = 0$; 4.5 MeV) state is expected and from their data the authors conclude that the intensity of this mixing is less than about $\frac{1}{2}$%. This is a much better limit than the one previously set by Debevec *et al.* (1971) of $\lesssim 6.8\%$. Unfortunately, on the simple $(1p)^2$ model it is clear that the two particle isovector part of the Coulomb force or charge dependent nuclear force leads to zero mixing since both operators are proportional to T_3 $(= t_3^{(1)} + t_3^{(2)})$ (see eq. (XI-12) and, e.g., eq. (XI-33)) which has zero matrix element between two particle $T = 0$ and $T = 1$ states. This means that small isospin mixing is to be expected in any case. Nevertheless, similar studies in more suitable cases might be useful from a fundamental point of view quite apart from their importance in achieving greater understanding of nuclear structure and reaction mechanisms.

Another nuclear reaction approach to studying isospin purity and implicitly, therefore, the charge dependence of nuclear forces, was first suggested by Barshay and Temmer (1964). They consider a nuclear reaction of the form

$$A + B \rightarrow C + C'$$

where C and C′ are members of the same isospin multiplet and T_A *or* $T_B = 0$. If isospin is strictly conserved and C and C′ are *exactly* connected by a rotation in isospin space, then the differential cross section of the reaction products will be symmetrical about 90° in the centre-of-mass system, independently of the reaction mechanism. This result follows since the identity

* Other recent studies of this kind are by Adelburger *et al.* (1969, 1971), McDonald *et al.* (1970) and McGrath *et al.* (1970).

(apart from T_3) of C and C$'$, the fact that only one value of isospin contributes and the required symmetry (or antisymmetry) of the wave function for C and C$'$ restricts the orbital angular momenta occurring in the final state to be all even or all odd. There are therefore no odd–even interference terms and only even powers of $\cos\theta$ can occur in the angular distribution.

A number of experimental tests of this effect have been carried out and further related theorems demonstrated (see Simonius (1972) and references contained therein). Although significant deviations ($\approx$ 20%) are found in some cases, because of nuclear structure and reaction complexities the interpretation of the results is so difficult that it is not possible to derive any information about the charge dependence of the nuclear force itself and the results can broadly be attributed to Coulomb effects.

CHAPTER XII

STRONG INTERACTIONS IN THE NUCLEUS

XII-1. Introduction

Strong interactions manifest themselves in the nucleus primarily through the two-body internucleon potential and are responsible for the major features of nuclear wave functions. It therefore follows that, in principle, from a study of nuclear properties it should be possible to derive considerable information about the form of these interactions. Such, indeed, was the early history of nuclear physics which enabled the rough strength, range, shape and some aspects of the spin dependence of the internucleon potential to be determined. However, because of the difficulties associated with the many body problem more precise information was not forthcoming. In any case, with the development of particle accelerators and particle detectors it has been possible to make detailed studies of nucleon–nucleon scattering itself. The resulting data, linked with that on the bound two-nucleon system (^2H), has then been analysed with as much precision as was thought to be necessary in terms of different possible phenomenological forms for the internucleon potential (see Signell (1969) for a review). Parallel with such analyses runs a long history of work aimed at calculating the form of the internucleon potential in terms of the more fundamental aspects of the strong interactions and testing the extent to which these forms match up with phenomenological potentials and can account for the two-nucleon data. Recently the most profitable approach in this direction has been the calculation of potentials in terms of the exchange of different bosons (e.g. Signell (1969)). It has also been pointed out by Moravcsik (1967) that precise knowledge of the internucleon potential can throw light on more general aspects of elementary particle theories (e.g. group theoretical schemes, general conservation laws, analyticity of the S-matrix).

The form of the interaction between two-nucleons is not determined uniquely by scattering experiments and it is found possible to fit the data in significantly different ways. For example, Hamada and Johnston (1962) used a local potential including hard cores and a quadratic spin-orbit term; Reid

(1968) used a similar potential but which was different in each state of distinct isospin, total spin and total angular momentum. Other authors have introduced velocity dependent (i.e. non-local) potentials (e.g. Tabakin and Davies (1966); Nestor *et al.* (1968)). All account for the main features of the nucleon–nucleon scattering data and, therefore, lead to the same asymptotic forms for the wave functions. They do not, however, lead to the same short range behaviour and it is difficult for information on this point to be extracted from the data although some help can be gained from the study of inelastic processes such as ed scattering or nucleon–nucleon bremsstrahlung (see e.g. Signell (1969)).

A second and related lack of information about the internucleon potential provided by scattering data is about off-the-energy-shell matrix elements. Of necessity the two-nucleons involved in a scattering process are on-shell and so one has the situation that two potentials having virtually identical on-shell but quite different off-shell behaviour would be indistinguishable in the nucleon–nucleon scattering situation (Haftel and Tabakin (1971)).

Finally, nucleon–nucleon scattering can obviously give no direct information about 3- or higher-body forces. Here one must deal at least with a three-body system in order to derive information on this point.

For these reasons among others, attention during the last few years has again focussed on the nucleus as a possible agent for distinguishing between different potential forms and for clarifying the role of (in particular) three-body forces. As has been remarked earlier and referred to at other points, many-body complexities in general make the task a very difficult one. Nevertheless, at the two extremes of nuclear matter and three-body nuclei some progress has been made and it is with these two aspects of the situation that this chapter is concerned. Both areas of activity are extremely complicated in their different ways and no attempt will be made here to go into the details of the various calculations that have been performed. Rather an outline of the information which has so far come out of this work will be given.

XII-2. Studies in nuclear matter

Nuclear matter is a hypothetical idealised state consisting of an infinite homogeneous fermion gas composed of equal numbers of neutrons and protons interacting through the internucleon potential but ignoring Coulomb effects. Because of these simplifying assumptions it is easier to deal with theoretically than a real nucleus where surface, Coulomb effects and unequal numbers of protons and neutrons are major complications. On the other hand, it is expected to bear some asymptotic relation to the state of matter at the centre of a heavy nucleus and in particular the volume binding energy per nucleon should be that given by the Weizsäcker semi-empirical mass formula (Weizsäcker (1935); Bethe and Bacher (1936)).

This formula expresses the total nuclear binding energy B for a finite nucleus in the following form

$$B = b_{\text{vol}}A - b_{\text{surf}}A^{2/3} - \frac{1}{2}b_{\text{sym}}\frac{(N-Z)^2}{A} - \frac{3}{5}\frac{Z^2e^2}{R_c} - \delta\ . \qquad \text{(XII-1)}$$

The first term is the volume binding energy; the second term reflects surface effects and allows for the fact that nucleons in the surface have fewer neighbours; the third term represents the tendency for nuclear stability to be achieved when $N = Z$; the fourth term accounts for Coulomb effects corresponding to a uniformly charged sphere of radius R_c; the fifth term δ describes the effects of pairing. The above formula agrees with the general trend of nuclear binding energies if the different parameters are chosen as follows (e.g. Bohr and Mottelson (1969); De Benedetti (1964))

$$\begin{aligned}
b_{\text{vol}} &\approx 16\ \text{MeV}\\
b_{\text{surf}} &\approx 17\ \text{MeV}\\
b_{\text{sym}} &\approx 50\ \text{MeV} \qquad \text{(XII-2)}\\
R_c &\approx 1.24\, A^{1/3}\ \text{fm}\\
\delta &\approx \begin{cases} -33\, A^{-3/4}\ \text{MeV} & \text{for even-even nuclei}\\ 0 & \text{for even-odd nuclei}\\ +33\, A^{-3/4}\ \text{MeV} & \text{for odd-odd nuclei}\ . \end{cases}
\end{aligned}$$

In the nuclear matter limit ($A \to \infty$, $N = Z$, $e = 0$) we have for the volume binding energy per nucleon

$$B/A \to b_{\text{vol}} \approx 16\ \text{MeV} \qquad \text{(XII-3)}$$

and any effective nuclear matter calculation should produce this result. Correspondingly the claculation should lead to an equilibrium density ρ, in agreement with that in the interior of a nucleus. On this point Brandow (1964) concludes that

$$\rho = 0.17 \text{ nucleon fm}^{-3} . \tag{XII-4}$$

In terms of a simple Fermi-gas model of the nucleus this corresponds to a wavenumber (k_F) at the Fermi surface of *

$$k_F = [3\pi^2(\tfrac{1}{2}\rho)]^{1/3} = 1.36 \text{ fm}^{-1} \tag{XII-5}$$

Correspondingly, the Fermi energy (i.e. the maximum kinetic energy in the Fermi gas) is

$$\epsilon_F = \frac{(\hbar k_F)^2}{2M} \approx 37 \text{ MeV} \tag{XII-6}$$

and the average kinetic energy per nucleon, obtained by summing over all occupied orbits, is

$$\bar{\epsilon}_{kin} = \tfrac{3}{5}\epsilon_F \approx 22 \text{ MeV} . \tag{XII-7}$$

The simple Fermi gas treatment of the nucleus is of course unsatisfactory since it takes no account of correlations between nucleons due to the inter-nucleon potential. Earlier improvements on the basic theory (e.g. Euler (1937)) used non-singular potentials with exchange mixtures so chosen as to give saturation. This approach enabled straightforward perturbation theory to be employed. However, nucleon–nucleon scattering data makes it clear that the internucleon potential is singular in character (e.g. has hard-core properties) and so new calculational methods had to be developed. Work of this kind was pioneered by Brueckner and his collaborators (Brueckner (1959); see Brown (1967), Day (1967) or Rajaraman and Bethe (1967) for accounts of this and related work) and highly sophisticated and elaborate approaches to this complex many-body problem have been developed. These methods will not be described here. Suffice it to say that although there is semi-quantitative agreement between the theoretical and experimental values for the binding energy per nucleon and the equilibrium density, there are still many uncertainties (see e.g. Bethe (1968)). Some of these certainly stem from the calculational methods employed but others arise from the different possible potential models already referred to which agree with the nucleon–nucleon scattering data and also from possible contributions from 3- and higher body forces. We shall now briefly review progress in these latter areas.

* The factor $\frac{1}{2}$ arises in eq. (XII-5) since we are assuming in nuclear matter that $N = Z$ so that the neutron or proton densities separately are taken to be $\frac{1}{2}\rho$.

XII-2.1. *Detailed form of the nucleon–nucleon potential*

Some recent work by Haftel and Tabakin (1970) (see also equivalent work by Haftel *et al.* (1972) in ^{16}O) has taken as its task "to study the relationship between nuclear saturation and smoothness of the two-nucleon interaction". This is by far the most detailed study of its kind and clearly illustrates the possibility of nuclear properties being able to distinguish between potentials which are equivalent from the scattering point of view. For those interested in many body technicalities it can be said that the authors solve the Brueckner equation for infinite nuclear matter using numerical matrix inversion in momentum space. They also assume that the angle averaged Pauli operator can be used, that the effective mass approximation is valid for nucleons below the Fermi surface and that there is zero single particle potential for nucleons above the Fermi surface. The following different potentials were used in these calculations

Potential R [Reid (1968)]
This potential is local and has a Yukawa core.

Potential BS [Bryan and Scott (1969)]
This is a one-boson exchange potential (exchange of $\rho, \omega, \pi, \eta, \sigma, \sigma_0$). It is non-local and has a quadratic velocity dependence.

Potential AP [Appel (1969)]
This has a Yukawa form but includes a non-local tensor force.

Potentials A and B [Haftel and Tabakin (1970)]
These are phenomenological potentials characterised by very weak non-local tensor forces of two different and non-equivalent kinds.

Each potential is able to account for the nucleon–nucleon scattering data but gives significantly different results for the total binding energy per nucleon (b_{vol}) and equilibrium density. This is illustrated in fig. XII-1 where the arrows indicate the experimentally determined values for b_{vol} and k_F. Equilibrium, of course, corresponds to the minima in the different curves.

Haftel and Tabakin (1970) classify the different potentials according to their "smoothness" where the latter is measured by the magnitude, κ, of the "wound" integral in the internucleon wave function in nuclear matter. Precisely, κ is given by the following expression (Brandow (1966))

$$\kappa = \rho \int |\xi(r)|^2 \, \mathrm{d}r \qquad \text{(XII-8)}$$

where ρ is the nucleon density and $\xi = \phi - \psi_{BG}$ is the difference (wound)

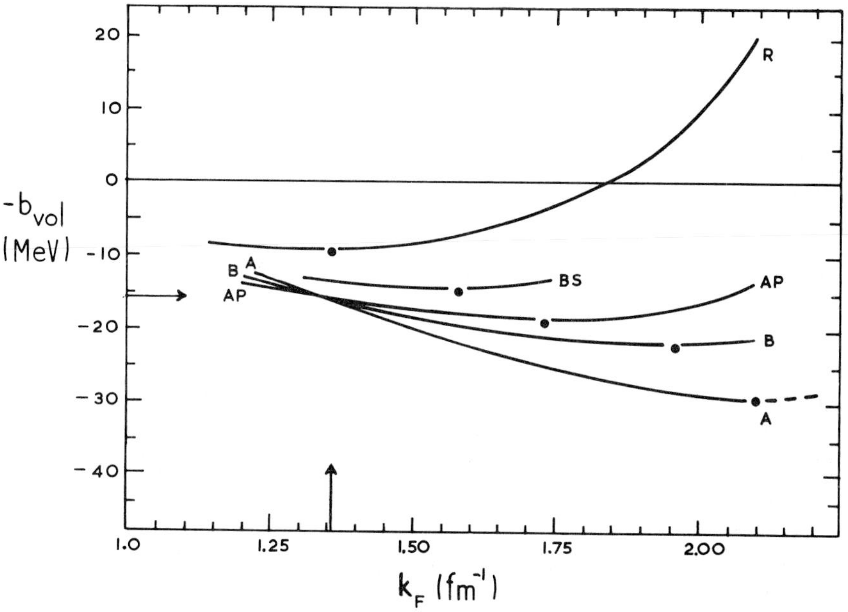

Fig. XII-1. Total binding energy per nucleon as a function of k_F.

between the unperturbed relative wave function for two nucleons and the correlated (Bethe and Goldstone (1957)) wave function calculated for an "average pair" in the Fermion sea. Clearly the higher the value of κ the less smooth is the corresponding potential. For the potentials considered here the values of κ are as follows:

$$R(\kappa = 0.1589);\quad BS(\kappa = 0.1044);\quad AP(\kappa = 0.0329);$$

$$B(\kappa = 0.0242);\quad A(\kappa = 0.0137)\,. \tag{XII-9}$$

In the light of these values of κ and the results given in fig. XII-1 it can be concluded that smooth non-local potentials (e.g. A and B) lead to overbinding and collapse and that only strong potentials with strong tensor forces yield proper saturation. These conclusions agree with the earlier results of a number of authors (e.g. Moszkowski (1963); Preston and Bhaduri (1963); Wong (1964, 1965); Bhargava and Sprung (1967); Ingber (1968); Kallio and Day (1969); Sprung *et al.* (1970); Siemens (1970)) and clearly demonstrate the possibility of nuclear matter calculations distinguishing between different equivalent internucleon potentials.

XII-2.2. *Three-body forces in nuclear matter*

That three-body forces should exist is beyond question within the framework of current theories of strong interactions. What is in considerable doubt is their detailed form and strength and the magnitude of their contribution to the binding energy of nuclei and, especially nuclear matter. Calculations of these forces from meson theory have been carried out in the past by various authors (e.g. Klein (1953); Drell and Huang (1953); Wentzel (1953); Gelbard (1955); Fujita *et al.* (1962); Miyazawa (1964); Harrington (1966); Loiseau and Nogami (1967)) and recently work by Brown *et al.* (1968) restimulated considerable interest in them and their role in nuclear matter (McKellar and Rajaraman (1968); Brown and Green (1969); Bhaduri *et al.* (1970); Loiseau *et al.* (1971); Green and Shucan (1971)). The calculations have been aimed at calculating the two-pion exchange contribution to the three-body potential together with some higher order corrections. Typical contributing diagrams are shown in fig. XII-2. These correspond to emission of a virtual meson by one nucleon, scattering by a second and absorption by a third and are responsible for the long range part of the force. It must be remembered here that the first diagram in the expansion is already included in the iterated two-body potential in any nuclear matter calculation. Taking only diagrams of the type illustrated in fig. XII-2 into account Brown *et al.* (1968) showed, by relating the strength of the potential to the pion–nucleon forward scattering amplitude, that because of the consistency condition imposed by PCAC theory (e.g. Adler (1965d); Hamilton (1967); Weinberg (1966, 1967); see also Chapt. IV) this amplitude is essentially zero *. The three-body force, it was then argued, must also be small. However, McKellar and Rajaraman (1968) pointed out that in nuclear matter the two-nucleon force can excite nucleons to states having high momenta so that, correspondingly, the virtual pions will have higher four-momenta than assumed by Brown *et al.* (1968). This situation is illustrated in fig. XII-3 where g signifies the two-body reaction matrix. The authors also raised other queries about the calculation.

In a subsequent paper, Brown and Green (1969) considered these issues further. They took into account higher order effects and made a further study

* More precisely the consistency condition requires this amplitude to be zero when one or both of the pion four-momenta are zero. Brown *et al.* (1968) argue that in nuclear matter $q^2 \approx -m_\pi^2$ where q is the pion four momentum and that the properties of the amplitude are such that on extrapolating from $q^2 = 0$ to $q^2 = -m_\pi^2$ there is little change.

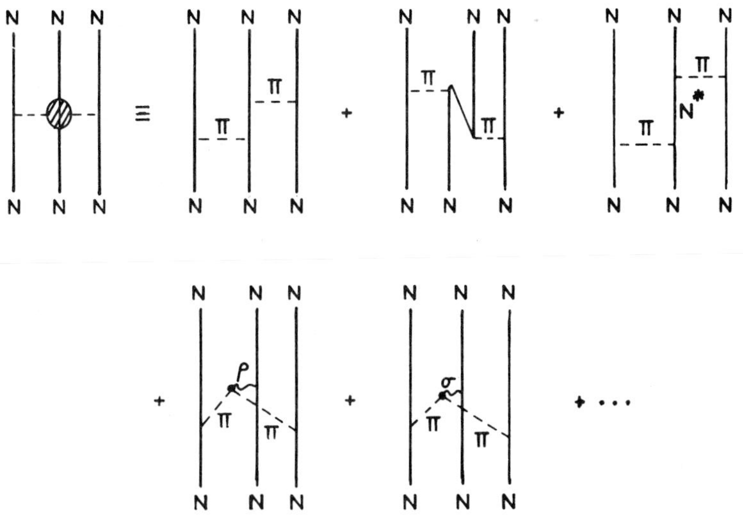

Fig. XII-2. Diagrams contributing to a three-body nuclear potential.

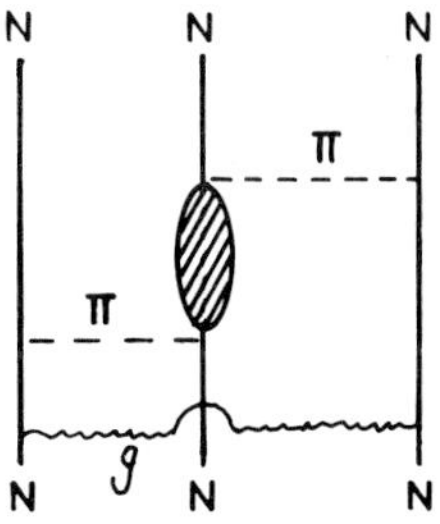

Fig. XII-3. Higher order three-body force contribution.

of the procedures for extrapolating the π–N scattering amplitude off the mass shell to values of $q^2 \approx -10\,m_\pi^2$ as required in calculating the higher order effects. Although there are a number of uncertainties they expressed confidence in their conclusion that the binding energy per nucleon stemming from three-body forces in nuclear matter is $\approx$ 2.5 MeV. They also pointed out that the contribution from the lowest order diagrams in fig. XII-2 can be viewed as a Pauli correction to part of the two-body force. This arises because in nuclear matter intermediate nucleons are limited in the momentum states they can occupy due to the presence of other nucleons *.

A further study by Bhaduri *et al.* (1970) sought to clarify the relationship between three-body effects as calculated by Brown and Green (1969), in which no explicit form for the three-body force in configuration space is derived and, those calculated, for example, by Loiseau and Nogami (1967) in which a conventional potential approach is used. In their calculations they also took into account the effect of introducing cut-offs in the nucleon–nucleon interparticle distances to simulate correlation effects and also uncertainty about the short range behaviour of the 3-body force. Naturally the results were sensitive to the form of cut-off and the authors concluded only that the net binding energy effect lay in the range 4 to 8 MeV/nucleon (attraction) in nuclear matter. A further calculation along the same lines by Loiseau *et al.* (1971), however, leads to a smaller binding energy contribution $\approx$ 2 MeV/nucleon (see also McKellar and Rajaraman (1971) who obtain a similar result).

The situation is then still unclear and Loiseau *et al.* (1971) do stress that their conclusions are subject to "restrictions and ambiguities". But clearly more precise calculations of the 2-body contribution to binding in nuclear matter can help with our understanding of 3-body forces.

XII-3. Studies in the A = 3 system

At the other extreme from nuclear matter are the smallest non-trivial nuclei namely ^{3}H and ^{3}He. These nuclei have been the subject of extensive study over the years (see Delves and Phillips (1969) for a review) and gradually very precise knowledge of the wave functions obtained by using realistic potentials are becoming available.

Recent work of this kind is by Delves and Hennell (1971) who carry out extremely accurate and complicated variational calculations for ^{3}H and ^{3}He using the Hamada–Johnston (1962) potential. Their results for the binding energies (B_{HJ}) of the two nuclei are as follows

$$B_{\mathrm{HJ}}(^3\mathrm{H}) = 6.5 \pm 0.2 \text{ MeV}$$

$$B_{\mathrm{HJ}}(^3\mathrm{He}) = 5.95 \pm 0.2 \text{ MeV} \tag{XII-10}$$

* Note that a similar situation arises, for example, in the case of electromagnetic exchange effects (see §XI-4).

to be compared with the experimental values (Mattauch *et al.* (1965))

$$B_{exp}(^3H) = 8.482 \text{ MeV}$$

$$B_{exp}(^3He) = 7.718 \text{ MeV} . \tag{XII-11}$$

There is, therefore, an underbinding of the order 2 MeV. This can certainly be accounted for by appealing to a contribution from three-body forces. Indeed an earlier calculation by Pask (1967) using two forms of the three-body force derived from a simple form of meson theory leads to a contribution $\gtrsim 1$ MeV to the binding energy. A similar conclusion was also arrived at by Loiseau and Nogami (1967) and Yip *et al.* (1971). However, as Delves and Hennell (1971) point out, although their results point to the existence of three-body force effects of the above magnitude, the "evidence cannot be considered strong until calculations have been performed with alternative interactions which also fit the two-body data". In this connection the work of Kok *et al.* (1968) is to be noted. These authors carried out model calculations for a system of three identical spinless particles acting pairwise through (local or non-local) central interactions. They found that a non-local separable Yamaguchi (1954) potential and an equivalent Bargman (Newton (1960)) potential gave binding energies differing by $\approx 14\%$ inspite of the fact that both potentials lead to identical S wave scattering phase shifts. Delves and Phillips (1969) comment that "the unknown behaviour of the two-body t-matrix off shell leads to an uncertainty ~ 1.2 MeV in the triton binding energy".

Delves and Hennell (1971) then go on to explore the sensitivity of their overall results to the inclusion of a simple central, spin independent three-body force of the form

$$V_3 = -V_3^{(0)} \exp\left[-\alpha(r_{12} + r_{13} + r_{23})\right] . \tag{XII-12}$$

They use values for the range α corresponding to 1, 2 or 3 pion masses and adjust the strength parameter $V_3^{(0)}$ to give the observed binding energy. In general inclusion of V_3 does not seriously disturb other physical quantities calculated and the fit with experiment is quite good.

This work, then, can be taken as supporting evidence for a significant binding energy contribution from three-body forces. Clearly, though, as with nuclear matter it is very important to carry out calculations on the $A = 3$ system using two-nucleon potentials other than the Hamada–Johnston potential and which are able to account for the nucleon–nucleon scattering data. In this connection some work of Hadjimichael and Jackson (1972) should be noted which uses a series of two-nucleon potentials generated by Haftel and Tabakin (1971) by applying a unitary transformation to the Reid (1968) potential (see §XII-2.1). These different potentials all account for the usual two-body observables as calculated by the Reid potential but have a short range non-locality. Hadjimichael and Jackson use a variational approach in a basis of oscillator states and find that the triton bound state observables depend sensitively on the off-the-energy-shell properties of the potentials. In particular they find that there can be significant contributions to the binding energy discrepancy just discussed. They conclude that the triton is a very suitable nucleus for testing the short range properties of the internucleon potential. In addition, there is the further possibility that relativistic corrections could be important. Gupta *et al.* (1965) have studied this point (see also Primakoff (1947)) using second order relativistic corrections to a two-body potential given by Shirokov (1959). Although the results should be regarded as preliminary they too indicate that a further contribution to the $A = 3$ binding energy $\approx \frac{1}{2}$ MeV could arise. There is obviously then considerably more theoretical work to be done before this evidence for the size of the three-body potential can be regarded as strong.

APPENDIX A

SOME NOTATION AND CONVENTIONS

A-1. Space-time coordinates

$x, y, z, ict \equiv x_1, x_2, x_3, x_4 \equiv x_\mu$ ($\mu = 1, 2, 3, 4$) (Greek indices)
x signifies the four-vector whose components are x_μ
$x, y, z \equiv x_1, x_2, x_3 \equiv x_k$ ($k = 1, 2, 3$) (Roman indices)
$\boldsymbol{x}$ signifies the three-vector whose components are x_k.

A-2. Momentum-energy coordinates

$p_x, p_y, p_z, \frac{iE}{c} \equiv p_1, p_2, p_3, p_4 \equiv p_\mu$ ($\mu = 1, 2, 3, 4$) (Greek indices)
p signifies the four-vector whose components are p_μ
$p_x, p_y, p_z \equiv p_1, p_2, p_3 \equiv p_k$ ($k = 1, 2, 3$) (Roman indices)
$\boldsymbol{p}$ signifies the three-vector whose components are p_k.

A free particle of mass m has values of $\boldsymbol{p}$ and E which satisfy

$$E^2 = \boldsymbol{p}^2 c^2 + m^2 c^4 .$$

A-3. Scalar products

e.g. $$p \cdot x \equiv p_\mu x_\mu = \boldsymbol{p} \cdot \boldsymbol{x} - Et$$

$$\boldsymbol{p} \cdot \boldsymbol{x} = p_k x_k$$

where the dummy suffix convention is used in which repeated indices are summed over. Further examples are

$$x^2 = x_\mu x_\mu = \boldsymbol{x}^2 - c^2 t^2$$

$$p^2 = p_\mu p_\mu = \boldsymbol{p}^2 - \frac{E^2}{c^2} \; (= -m^2c^2 \text{ for a free particle}) \, .$$

A-4. Differentials

$$\frac{\partial}{\partial x_\mu} \equiv \partial_\mu \equiv \left(\frac{\partial}{\partial x_k}, -\frac{i}{c}\frac{\partial}{\partial t} \right)$$

$$\Box^2 \equiv \nabla^2 - \frac{1}{c^2}\frac{\partial}{\partial t^2} \, .$$

A-5. Quantum mechanical energy-momentum operator

Let P_μ denote the energy-momentum four-vector operator, i.e.

$$P_\mu = -\,\mathrm{i}\hbar \frac{\partial}{\partial x_\mu} = (-\,\mathrm{i}\hbar\nabla, \frac{-\hbar}{c}\frac{\partial}{\partial t}) \, .$$

For any operator $F(x)$

$$[P_\mu, F(x)] = -\,\mathrm{i}\hbar \frac{\partial F(x)}{\partial x_\mu}$$

and $F(x) = \mathrm{e}^{-\mathrm{i}P\cdot x} F(0)\, \mathrm{e}^{\mathrm{i}P\cdot x}$.

A-6. General symbols

(i) $a^\times$ – complex conjugate.

The suffix $\times$ signifies that the complex conjugate of a (whether c-number, matrix or operator) is taken.

(ii) $\widetilde{A}$ – transpose.

The suffix $\sim$ transposes a finite matrix (e.g. Dirac γ-matrices) but leaves Hilbert space operators and c-numbers unchanged.

(iii) $A^\dagger \equiv \widetilde{A}^\times$ – adjoint.

The suffix $\dagger$ transposes a finite matrix and takes its complex conjugate. For a c-number it takes the complex conjugate and for a field operator it acts as the Hermitian conjugate. Thus, for a Dirac spinor operator

$$\psi^\dagger \equiv (\psi_1^\dagger \, \psi_2^\dagger \, \psi_3^\dagger \, \psi_4^\dagger) \, .$$

(iv) A_λ^* – adjoint of four-vector quantities.
The suffix $*$ is defined for four-vector quantities as follows:

$$A_\lambda^* \equiv A_1^\dagger, A_2^\dagger, A_3^\dagger, -A_4^\dagger \, .$$

The need for this symbol arises from the imaginary nature of the fourth component of four-vectors.

A-7. Units

In general, but not invariably, units are used for which $\hbar = c = 1$.

APPENDIX B

NON-RELATIVISTIC QUANTUM MECHANICS

B-1. **The Schrödinger equation**

The Schrödinger equation can be written as follows for a particle of mass m in a potential V

$$i\hbar\frac{\partial\psi}{\partial t} = -\frac{\hbar^2}{2m}\nabla^2\psi + V\psi \,. \tag{B1}$$

The corresponding probability and current densities are

$$\rho = \psi^\times\psi \tag{B2}$$

$$j = \frac{\hbar}{2mi}(\psi^\times\nabla\psi - \psi\nabla\psi^\times) \tag{B3}$$

and satisfy

$$\operatorname{div} j + \frac{\partial\rho}{\partial t} = 0 \,. \tag{B4}$$

B-2. **Transformation properties under space reflection**

The space reflection operation is defined by

$$\boldsymbol{x} \to \boldsymbol{x}' = -\boldsymbol{x}\,; \qquad t \to t' = t \,.$$

The corresponding operator (the parity operator) P has the following properties

$$P\boldsymbol{x}P^{-1} = -\boldsymbol{x}\,; \quad P\boldsymbol{p}P^{-1} = -\boldsymbol{p}\,; \quad P\boldsymbol{J}P^{-1} = \boldsymbol{J} \tag{B5}$$

where $\boldsymbol{x}, \boldsymbol{p}$ and $\boldsymbol{J}$ are space, linear momentum and angular momentum operators respectively.

B-3. Transformation properties under time reversal

The time reversal operation is defined by

$$\boldsymbol{x} \to \boldsymbol{x}' = \boldsymbol{x} ; \quad t \to t' = -t .$$

The corresponding operator T is antiunitary (Wigner (1932)) and can be written in the form

$$T = U_T K \tag{B6}$$

where U_T is unitary and K is the complex conjugation operator. T is antilinear, i.e.

$$T(a\psi_1 + b\psi_2) = a^\times T\psi_1 + b^\times T\psi_2 . \tag{B7}$$

If A is some operator with corresponding time reversed operator A^T, where

$$A^T = TAT^{-1} = U_T A^\times U_T^{-1} \tag{B8}$$

then for two states $\psi_1(t)$ and $\psi_2(t)$

$$\langle T\psi_1(t)|A^T|T\psi_2(t)\rangle = \langle \psi_1(-t)|A|\psi_2(-t)\rangle^\times . \tag{B9}$$

For an angular momentum state $|J,M\rangle$ we use the phase convention (Coester (1953))

$$T|J,M\rangle = (-1)^{J-M}|J,-M\rangle . \tag{B10}$$

APPENDIX C

RELATIVISTIC QUANTUM MECHANICS

C-1. Non-covariant form of the Klein–Gordon equation

The equation can be written as follows:

$$\frac{\partial^2 \phi}{\partial t^2} = \nabla^2 \phi - m^2 \phi \tag{C1}$$

in units for which $\hbar = c = 1$. ϕ is a function of $\boldsymbol{x}$ and t, and m is the mass of the particle.

The corresponding probability and current densities are

$$\rho = \frac{i}{2m}\left(\phi^\times \frac{\partial \phi}{\partial t} - \phi \frac{\partial \phi^\times}{\partial t}\right) \tag{C2}$$

$$\boldsymbol{j} = \frac{1}{2mi}(\phi^\times \boldsymbol{\nabla}\phi - \phi \boldsymbol{\nabla}\phi^\times) \tag{C3}$$

and satisfy

$$\operatorname{div} \boldsymbol{j} + \frac{\partial \rho}{\partial t} = 0\,. \tag{C4}$$

C-2. Covariant form of the Klein–Gordon equation

Writing $x_\mu = (\boldsymbol{x}, it)$; the Klein–Gordon equation can be written

$$(\Box - m^2)\phi = 0 \tag{C5}$$

where $\Box \equiv \partial^2/\partial x_\mu^2$.

The continuity equation becomes

$$\frac{\partial j_\mu}{\partial x_\mu} = 0 \tag{C6}$$

where

$$j_\mu = (\boldsymbol{j}, i\rho) = \frac{1}{2mi}\left(\phi^\times \frac{\partial \phi}{\partial x_\mu} - \phi \frac{\partial \phi^\times}{\partial x_\mu}\right). \tag{C7}$$

C-3. Non-covariant form of the Dirac equation

The equation can be written in the following form:

$$\frac{\partial \psi}{\partial t} + \boldsymbol{\alpha} \cdot \nabla \psi + im\beta\psi = 0 \tag{C8}$$

in units for which $c = \hbar = 1$. ψ is a four component spinor function of $\boldsymbol{x}$ and t; m is the mass of the Dirac particle. $\boldsymbol{\alpha}$ and β are 4×4 Hermitian matrices conventionally represented as follows

$$\boldsymbol{\alpha} = \begin{Vmatrix} 0 & \boldsymbol{\sigma} \\ \boldsymbol{\sigma} & 0 \end{Vmatrix}, \qquad \beta = \begin{Vmatrix} 1 & 0 \\ 0 & -1 \end{Vmatrix} \tag{C9}$$

where $\boldsymbol{\sigma}$, 1 and 0 signify the Pauli spin matrices, the 2×2 unit matrix and the 2×2 null matrix respectively. $\boldsymbol{\alpha}$ and β satisfy the following relations:

$$\begin{aligned} &\boldsymbol{\alpha}\beta + \beta\boldsymbol{\alpha} = 0 \\ &\alpha_i\alpha_j + \alpha_j\alpha_i = 2\delta_{ij} \qquad (i,j = 1,2,3) \\ &\beta^2 = 1\,. \end{aligned} \tag{C10}$$

The corresponding probability and current densities are respectively

$$\begin{aligned} &\rho = \psi^\dagger \psi \\ &\boldsymbol{j} = \psi^\dagger \boldsymbol{\alpha} \psi \end{aligned} \tag{C11}$$

and satisfy the continuity equation

$$\operatorname{div} \boldsymbol{j} + \frac{\partial \rho}{\partial t} = 0\,. \tag{C12}$$

Eq. (C8) can be written in Hamiltonian form, viz:

$$H\psi = \mathrm{i}\frac{\partial \psi}{\partial t} \tag{C13}$$

where $H = \beta m - \mathrm{i}\,\boldsymbol{\alpha}\cdot\boldsymbol{\nabla} = \beta m + \boldsymbol{\alpha}\cdot\boldsymbol{p}$ and $\boldsymbol{p}$ is the 3-momentum operator.

C-4. **Covariant form of the Dirac equation**

To go to a covariant form the following definitions are made

$$\begin{aligned}
&\gamma_\mu = (\boldsymbol{\gamma}, \gamma_4) = (\gamma_k, \gamma_4) \\
&\gamma_k = -\,\mathrm{i}\beta\alpha_k \quad (k = 1, 2, 3) \\
&\gamma_4 = \beta\,.
\end{aligned} \tag{C14}$$

The Dirac equation can then be written

$$\gamma_\mu \frac{\partial \psi}{\partial x_\mu} + m\psi = 0 \tag{C15}$$

or

$$(\gamma_\mu p_\mu - \mathrm{i}m)\psi = (\not{p} - \mathrm{i}m)\psi = 0$$

where ψ is a function of x_μ, p_μ is the four-momentum and $\not{p}$ signifies $\gamma_\mu p_\mu$. The matrices γ_μ are all Hermitian and satisfy the relation

$$\gamma_\mu\gamma_\nu + \gamma_\nu\gamma_\mu = 2\delta_{\mu\nu}\,. \tag{C16}$$

In the conventional representation they can be written

$$\gamma_k = \begin{Vmatrix} 0 & -\mathrm{i}\sigma_k \\ \mathrm{i}\sigma_k & 0 \end{Vmatrix}, \qquad \gamma_4 = \begin{Vmatrix} 1 & 0 \\ 0 & -1 \end{Vmatrix}. \tag{C17}$$

It is also convenient to define the matrix γ_5, given by

$$\gamma_5 = \gamma_1\gamma_2\gamma_3\gamma_4 \tag{C18}$$

which satisfies the relation

$$\gamma_\mu \gamma_5 + \gamma_5 \gamma_\mu = 0\,. \tag{C19}$$

γ_5 has the conventional representation

$$\gamma_5 = \begin{Vmatrix} 0 & -1 \\ -1 & 0 \end{Vmatrix}. \tag{C20}$$

The adjoint spinor to ψ is defined by

$$\bar{\psi} = \psi^\dagger \gamma_4 \tag{C21}$$

and satisfies the adjoint equation

$$\frac{\partial \bar{\psi}}{\partial x_\mu} \gamma_\mu - m\bar{\psi} = 0 \tag{C22}$$

or

$$\bar{\psi}(\not{p} - \mathrm{i}m) = 0\,.$$

Corresponding to eq. (C11) the four component probability current can be written

$$j_\mu = (\boldsymbol{j}, \mathrm{i}\rho) = \mathrm{i}(\bar{\psi}\gamma_\mu\psi) \tag{C23}$$

and satisfies the continuity equation (cf. eq. (C12))

$$\frac{\partial j_\mu}{\partial x_\mu} = 0\,. \tag{C24}$$

In the presence of an electromagnetic field (four-vector potential A_μ), the Dirac equation and its adjoint are modified as follows

$$\gamma_\mu \left(\frac{\partial}{\partial x_\mu} - \mathrm{i}eA_\mu \right) \psi + m\psi = 0$$

$$\left(\frac{\partial}{\partial x_\mu} + \mathrm{i}eA_\mu \right) \bar{\psi}\gamma_\mu - m\bar{\psi} = 0 \tag{C25}$$

where e is the charge of the Dirac particle.

C-5. Transformation properties of Dirac spinors

C-5.1. *Lorentz transformation*

Under a Lorentz transformation

$$x_\mu \to x'_\mu = a_{\mu\nu} x_\nu \tag{C26}$$

the Dirac spinors $\psi(x)$ and $\bar{\psi}(x)$ transform as follows:

$$\begin{aligned} \psi(x) &\to \psi'(x') = S\psi(x) \\ \bar{\psi}(x) &\to \psi'(x') = \bar{\psi}(x) S^{-1} . \end{aligned} \tag{C27}$$

The matrix S (which is not unitary) satisfies the following relations

$$\begin{aligned} S^{-1}\gamma_\mu S &= a_{\mu\nu}\gamma_\nu \\ S^\dagger \gamma_4 &= \gamma_4 S^{-1} . \end{aligned} \tag{C28}$$

Under an infinitesimal Lorentz transformation

$$a_{\mu\nu} = \delta_{\mu\nu} + \epsilon_{\mu\nu} \tag{C29}$$

where $\epsilon_{\mu\nu} + \epsilon_{\nu\mu} = 0$, S can be written

$$S = \tfrac{1}{8}\epsilon_{\mu\nu}(\gamma_\mu\gamma_\nu - \gamma_\nu\gamma_\mu) . \tag{C30}$$

C-5.2. *Charge conjugation*

The charge conjugate spinors ψ^c and $\bar{\psi}^c$ are given by

$$\begin{aligned} \psi^c(x) &= C\tilde{\bar{\psi}}(x) \\ \bar{\psi}^c(x) &= -\tilde{\psi}(x) C^{-1} \end{aligned} \tag{C31}$$

where the matrix C satisfies the relations

$$\begin{aligned} C^{-1}\gamma_\mu C &= -\tilde{\gamma}_\mu \\ C^{-1}\gamma_5 C &= +\tilde{\gamma}_5 \\ C^{-1} &= C^\dagger \\ \tilde{C} &= -C . \end{aligned} \tag{C32}$$

In the usual representation $C = \gamma_4\gamma_2$. The charge conjugate spinors ψ^c and $\bar{\psi}^c$ satisfy the Dirac equations:

$$\gamma_\mu \left(\frac{\partial}{\partial x_\mu} + \mathrm{i}eA_\mu\right) \psi^c + m\psi^c = 0$$

$$\left(\frac{\partial}{\partial x_\mu} - \mathrm{i}eA_\mu\right) \bar{\psi}^c\gamma_\mu - m\bar{\psi}^c = 0 \tag{C33}$$

(cf. eq. (C25)) and refer to a particle having charge $-e$.

C-5.3. *Space reflection*

Under space reflection

$$x_k \to x_k' = -x_k\,; \qquad x_4 \to x_4' = x_4 \tag{C34}$$

the Dirac spinors $\psi(x)$ and $\bar{\psi}(x)$ transform as follows

$$\psi(x) \to \psi^P(x') = \lambda\gamma_4\psi(x)$$

$$\bar{\psi}(x) \to \bar{\psi}^P(x') = \lambda^\times\bar{\psi}(x)\gamma_4 \tag{C35}$$

where $\lambda = \pm 1, \pm\mathrm{i}$. Note that if ψ and ψ^c are to have exactly the same transformation properties then only $\lambda = \pm\mathrm{i}$ is allowed.

C-5.4. *Time reversal*

Under time reversal

$$x_k \to x_k' = x_k\,; \qquad x_4 \to x_4' = -x_4 \tag{C36}$$

the Dirac spinors $\psi(x)$ and $\bar{\psi}(x)$ transform as follows:

$$\psi(x) \to \psi^t(x') = \eta B\psi^\times(x)$$

$$\bar{\psi}(x) \to \bar{\psi}^t(x') = \eta^\times\bar{\psi}^\times B^{-1} \tag{C37}$$

where $\eta^2 = \pm 1$ and the matrix B satisfies

$$B^{-1}\gamma_\mu B = \tilde{\gamma}_\mu$$
$$B^{-1}\gamma_5 B = \tilde{\gamma}_5$$
$$B^\dagger = B^{-1}$$
$$\tilde{B} = -B\,. \tag{C38}$$

B is related to C by $B = -\gamma_5 C$. Again if ψ and ψ^c are to have the same transformation properties, then only $\eta = \pm 1$ is allowed.

C-6. Bilinear forms of Dirac spinors

Five bilinear covariant quantities can be constructed from the spinors ψ and $\bar{\psi}$ and the γ-matrices as follows

$$S = \bar{\psi}\psi$$
$$V_\lambda = \mathrm{i}\,\bar{\psi}\gamma_\lambda\psi$$
$$T_{\lambda\nu} = \mathrm{i}\,\bar{\psi}\gamma_\lambda\gamma_\nu\psi(\lambda \neq \nu)$$
$$= 0 \qquad (\lambda = \nu)$$
$$A_\lambda = \mathrm{i}\bar{\psi}\gamma_5\gamma_\lambda\psi$$
$$P = \mathrm{i}\,\bar{\psi}\gamma_5\psi \tag{C39}$$

which transform as a scalar (S), polar vector (V_λ), second rank antisymmetric tensor ($T_{\lambda\nu}$), axial vector (A_λ) and pseudoscalar (P) respectively.

C-7. Plane wave solutions of the Dirac equation

For a given four-momentum p ($\equiv(\boldsymbol{p}, \mathrm{i}E)$) there are four linearly independent solutions of the Dirac equation

$$\psi_r = \begin{cases} u_+^{(r)}(p)\,\mathrm{e}^{\mathrm{i}px} & r = 1, 2 \\ u_-^{(r)}(-p)\,\mathrm{e}^{-\mathrm{i}px} & r = 1, 2 \end{cases} \tag{C40}$$

where the u_+ solutions correspond to a particle with four-momentum p (i.e. positive energy solutions) and the u_- solutions correspond to a particle with four-momentum $-p$ (i.e. negative energy solutions).

The spinors $u_+^{(r)}, u_-^{(r)}$ satisfy the equations

$$(\not{p} - \mathrm{i}m)u_+^{(r)}(p) = 0$$

$$(\not{p} + \mathrm{i}m)u_-^{(r)}(-p) = 0 \tag{C41}$$

and are given explicitly by

$$u_+^{(r)}(p) = \sqrt{\frac{E+m}{2E}} \left\| \begin{matrix} \xi^{(r)} \\ \dfrac{\boldsymbol{\sigma}\cdot\boldsymbol{p}}{E+m}\xi^{(r)} \end{matrix} \right\|$$

$$u_-^{(r)}(-p) = \sqrt{\frac{E+m}{2E}} \left\| \begin{matrix} \dfrac{-\boldsymbol{\sigma}\cdot\boldsymbol{p}}{E+m}\xi^{(r)} \\ \xi^{(r)} \end{matrix} \right\| \tag{C42}$$

where $\xi^{(r)}$ is a column vector given by

$$\xi^{(1)} = \begin{vmatrix} 1 \\ 0 \end{vmatrix}; \quad \xi^{(2)} = \begin{vmatrix} 0 \\ 1 \end{vmatrix}. \tag{C43}$$

The different spinors are orthogonal to one another and are invariantly normalised to one particle in a volume $mc^2/|E|$ so that

$$u^\dagger u = |E|/mc^2 . \tag{C44}$$

This implies that

$$\bar{u}_+^{(r)} u_+^{(r')} = \delta_{rr'}$$

$$\bar{u}_-^{(r)} u_-^{(r')} = \delta_{rr'} . \tag{C45}$$

In addition

$$\bar{u}_+^{(r)} u_-^{(r')} = \bar{u}_-^{(r)} u_+^{(r')} = 0 .$$

To clarify notation we also use spinors w and v defined as follows:

$$w^{(r)}(p) = u_+^{(r)}(p)\,; \qquad r = 1, 2$$

$$w^{(r+2)}(p) = v^{(r)}(p) = u_-^{(r)}(-p)\,; \quad r = 1, 2\,. \tag{C46}$$

Thus

$$\bar{w}^{(s)}(p)w^{(s')}(p) = \epsilon_s \delta_{ss'} \qquad (s, s' = 1, 2, 3, 4) \tag{C47}$$

where

$$\epsilon_{1,2} = 1\,, \quad \epsilon_{3,4} = -1\,.$$

C-8. Zero mass solutions of the Dirac equation

For a massless particle, the Dirac equation has the form

$$\gamma_\mu \frac{\partial \psi}{\partial x_\mu} = 0\,. \tag{C48}$$

Of particular interest are the two-component solutions of this equation (Weyl (1929); Pauli (1931)) which are eigenfunctions of γ_5 namely

$$\psi_+ = \tfrac{1}{2}(1 + \gamma_5)\psi = \begin{Vmatrix} \varphi_+ \\ -\varphi_+ \end{Vmatrix}$$

and

$$\psi_- = \tfrac{1}{2}(1 - \gamma_5)\psi = \begin{Vmatrix} \varphi_- \\ \varphi_- \end{Vmatrix}\,. \tag{C49}$$

These functions satisfy the chirality equations $\gamma_5 \psi_+ = \psi_+$, $\gamma_5 \psi_- = -\psi_-$. φ_+ and φ_- are two component functions and satisfy

$$i\hbar \frac{\partial \varphi_+}{\partial t} = E\varphi_+ = -\boldsymbol{\sigma} \cdot \boldsymbol{p}\varphi_+$$

$$i\hbar \frac{\partial \varphi_-}{\partial t} = E\varphi_- = \boldsymbol{\sigma} \cdot \boldsymbol{p}\varphi_- \tag{C50}$$

where $\boldsymbol{\sigma}$ is the Pauli spin operator and $\boldsymbol{p}$ and E are the momentum operator and energy respectively of the particle.

$\boldsymbol{\sigma} \cdot \hat{\boldsymbol{p}}$ (where $\hat{\boldsymbol{p}}$ is the unit vector in the direction of $\boldsymbol{p}$; i.e. $\boldsymbol{p} = p\hat{\boldsymbol{p}}$) is known as the helicity operator and since for a massless particle $p = E$ it follows that

$$\begin{aligned} \boldsymbol{\sigma} \cdot \hat{\boldsymbol{p}}\varphi_+ &= -\varphi_+ \\ \boldsymbol{\sigma} \cdot \hat{\boldsymbol{p}}\varphi_- &= \varphi_- \ . \end{aligned} \tag{C51}$$

Thus φ_+ has helicity $h = -1$ and φ_- has $h = +1$ in the case of positive energy solutions. The converse holds in the case of the corresponding charge conjugate (i.e. antiparticle) functions.

From the definition of the helicity operator it is clear that for positive energy solutions the function φ_+ describes a particle spinning in a left-handed fashion about its directions of motion. Similarly φ_- refers to right-handed spin. For this reason an alternative notation is sometimes used, namely,

$$\begin{aligned} \varphi_L &\equiv \varphi_+ \\ \varphi_R &\equiv \varphi_- \ . \end{aligned} \tag{C52}$$

APPENDIX D

QUANTISED FIELDS

We now consider the situation that φ, ψ etc. are field operators.

D-1. Quantisation of a free scalar or pseudoscalar field

In the case of a charged particle the scalar (or pseudoscalar) field $\varphi(x)$ satisfies the Klein–Gordon equation and is complex. It can be expanded as follows:

$$\varphi(x) = \frac{1}{\sqrt{V}} \sum_{k} \frac{1}{\sqrt{2E(k)}} [a(\boldsymbol{k})\, \mathrm{e}^{\mathrm{i}k\cdot x} + b^{\dagger}(\boldsymbol{k})\, \mathrm{e}^{-\mathrm{i}k\cdot x}]$$

$$\varphi^{\dagger}(x) = \frac{1}{\sqrt{V}} \sum_{k} \frac{1}{\sqrt{2E(k)}} [b(\boldsymbol{k})\, \mathrm{e}^{\mathrm{i}k\cdot x} + a^{\dagger}(\boldsymbol{k})\, \mathrm{e}^{-\mathrm{i}k\cdot x}] \qquad \text{(D1)}$$

where $E(\boldsymbol{k}) = -\mathrm{i}k_4 = \sqrt{\boldsymbol{k}^2 + m^2}$ and k_μ is the four-momentum of the particle. V denotes a (large) normalisation volume. The operators $a(\boldsymbol{k})$, $b(\boldsymbol{k})$ etc. satisfy the communication relations

$$[a(\boldsymbol{k}), a^{\dagger}(\boldsymbol{k}')] = [b(\boldsymbol{k}), b^{\dagger}(\boldsymbol{k}')] = \delta_{\boldsymbol{k}\boldsymbol{k}'} \,. \qquad \text{(D2)}$$

All other possible commutators vanish. The Hamiltonian for the system is

$$H = \sum_{\boldsymbol{k}} E(\boldsymbol{k}) (a^{\dagger}(\boldsymbol{k}) a(\boldsymbol{k}) + b^{\dagger}(\boldsymbol{k}) b(\boldsymbol{k})) \,. \qquad \text{(D3)}$$

The operators $a^{\dagger}(\boldsymbol{k})$ and $a(\boldsymbol{k})$ create and annihilate respectively the positively charged particle whilst $b^{\dagger}(\boldsymbol{k})$ and $b(\boldsymbol{k})$ create and annihilate the negatively charged particle.

In the case of a neutral field, $\varphi(x) = \varphi^{\dagger}(x)$ so that $a(\boldsymbol{k}) \equiv b(\boldsymbol{k})$ and only one type of particle is involved. Also

$$H = \sum_{k} E(k) a^{\dagger}(k) a(k) . \tag{D4}$$

D-2. **Transformation properties * of a free scalar or pseudoscalar field under charge conjugation, space reflection and time reversal**

D-2.1. *Charge conjugation*

The charge conjugate field $\phi^{C}(x)$ and its adjoint are given by

$$\phi^{C}(x) = U_{C}\phi(x)U_{C}^{-1} = \eta_{C}\phi^{\dagger}(x)$$

$$\phi^{\dagger C}(x) = U_{C}\phi^{\dagger}(x)U_{C}^{-1} = \eta_{C}^{\times}\phi(x) \tag{D5}$$

where $|\eta_{C}|^2 = 1$. For a Hermitian (neutral) field $\eta_{C} = \pm 1$ and the one particle state is an eigenstate of the charge conjugation operator.

D-2.2. *Space reflection*

The space reflected field $\phi^{P}(x)$ and its adjoint are given by

$$\phi^{P}(x) = U_{P}\phi(x)U_{P}^{-1} = \eta_{P}\phi(-\boldsymbol{x}, x_4)$$

$$\phi^{\dagger P}(x) = U_{P}\phi^{\dagger}(x)U_{P}^{-1} = \eta_{P}^{\times}\phi^{\dagger}(-\boldsymbol{x}, x_4) \tag{D6}$$

where η_{P} is a phase factor (the intrinsic parity) having values + 1 (scalar field) and – 1 (pseudoscalar field).

D-2.3. *Time reversal (Weak, Wigner)*

The time reversed field $\phi^{T}(x)$ and its adjoint are given by

$$\phi^{T}(x) = U_{T}\phi(x)U_{T}^{-1} = \eta_{T}\phi^{\times}(\boldsymbol{x}, -x_4)$$

$$\phi^{\dagger T}(x) = U_{T}\phi^{\dagger}(x)U_{T}^{-1} = \eta_{T}^{\times}\phi^{\dagger\times}(\boldsymbol{x}, -x_4) . \tag{D7}$$

This transformation is non-linear and $|\eta_{T}|^2 = 1$.

* See, for example, Roman (1964); Schweber (1964); Källen (1964).

D-3. Quantisation of a free Dirac field

The Dirac field $\psi(x)$ and its adjoint $\bar{\psi}(x)(\bar{\psi}(x) = \psi^\dagger(x)\gamma_4)$ are expanded in terms of plane wave solutions of the Dirac equation as follows

$$\psi(x) = \frac{1}{\sqrt{V}} \sum_{\boldsymbol{k}} \left(\frac{m}{E(\boldsymbol{k})}\right)^{1/2} \sum_{r=1,2} [b_r(\boldsymbol{k}) w^{(r)}(\boldsymbol{k})\, \mathrm{e}^{\mathrm{i}k\cdot x} + d_r^\dagger(\boldsymbol{k}) v^{(r)}(\boldsymbol{k})\, \mathrm{e}^{-\mathrm{i}k\cdot x}]$$

$$\bar{\psi}(x) = \frac{1}{\sqrt{V}} \sum_{\boldsymbol{k}} \left(\frac{m}{E(\boldsymbol{k})}\right)^{1/2} \sum_{r=1,2} [b_r^\dagger(\boldsymbol{k}) \bar{w}^{(r)}(\boldsymbol{k})\, \mathrm{e}^{-\mathrm{i}k\cdot x} + d_r(\boldsymbol{k}) \bar{v}^{(r)}(\boldsymbol{k})\, \mathrm{e}^{\mathrm{i}k\cdot x}] \tag{D8}$$

where V is the normalisation volume and $w^{(r)}$ and $v^{(r)}$ are the spinors defined in eq. (C46). The operators $b_r^\dagger(\boldsymbol{k})$ and $b_r(\boldsymbol{k})$ are creation and annihilation operators respectively for a particle of momentum $\boldsymbol{k}$ and spin label r. Similarly $d_r^\dagger(\boldsymbol{k})$ and $d_r(\boldsymbol{k})$ are creation and annihilation operators for the corresponding anti-particle. These different operators satisfy the anti-commutation relations

$$\{b_r(\boldsymbol{k}), b_{r'}^\dagger(\boldsymbol{k}')\} = \{d_r(\boldsymbol{k}), d_{r'}^\dagger(\boldsymbol{k}')\} = \delta_{r,r'}\delta_{\boldsymbol{k},\boldsymbol{k}'} \tag{D9}$$

whilst all other possible anti-commutators vanish. The Hamiltonian for the system is

$$H = \sum_{\boldsymbol{k},r} E(\boldsymbol{k})[b_r^\dagger(\boldsymbol{k})\, b_r(\boldsymbol{k}) + d_r^\dagger(\boldsymbol{k})\, d_r(\boldsymbol{k})] \tag{D10}$$

and the current density operator is written

$$J_\lambda(x) = \frac{\mathrm{i}}{2} [\bar{\psi}(x), \gamma_\lambda \psi(x)]$$

$$= \frac{\mathrm{i}}{2} [\bar{\psi}_\alpha(x)(\gamma_\lambda)_{\alpha\beta}\psi_\beta(x) - (\gamma_\lambda)_{\alpha\beta}\psi_\beta(x)\bar{\psi}_\alpha(x)]\,. \tag{D11}$$

D-4. Transformation properties * of a Dirac field under charge conjugation, space reflection and time reversal

D-4.1. *Charge conjugation*

The charge conjugate field ψ^{C} and its adjoint are given by

$$\psi^{\mathrm{C}}(x) = U_{\mathrm{C}}\psi(x)U_{\mathrm{C}}^{-1} = \eta_{\mathrm{C}}C\tilde{\bar{\psi}}(x)$$

$$\bar{\psi}^{\mathrm{C}}(x) = U_{\mathrm{C}}\bar{\psi}(x)U_{\mathrm{C}}^{-1} = -\eta_{\mathrm{C}}^{\times}\tilde{\psi}(x)C^{-1} \qquad \text{(D12)}$$

where $|\eta_{\mathrm{C}}|^2 = 1$ and C is the matrix defined in eq. (C32).

D-4.2. *Space reflection*

The space reflected field ψ^{P} and its adjoint are given by

$$\psi^{\mathrm{P}}(x) = U_{\mathrm{P}}\psi(x)U_{\mathrm{P}}^{-1} = \eta_{\mathrm{P}}\gamma_4\psi(-\boldsymbol{x}, x_4)$$

$$\bar{\psi}^{\mathrm{P}}(x) = U_{\mathrm{P}}\bar{\psi}(x)U_{\mathrm{P}}^{-1} = \eta_{\mathrm{P}}^{\times}\bar{\psi}(-\boldsymbol{x}, x_4)\gamma_4 \qquad \text{(D13)}$$

where $\eta_{\mathrm{P}}^2 = \pm 1$. In order for the charge conjugate field to have the same properties as ψ, η_{P} is restricted to be $\pm\,\mathrm{i}$.

D-4.3. *Time reversal (Weak, Wigner)*

The time reversed field ψ^{T} and its adjoint are given by

$$\psi^{\mathrm{T}}(x) = U_{\mathrm{T}}\psi(x)U_{\mathrm{T}}^{-1} = \eta_{\mathrm{T}}B\psi^{\times}(\boldsymbol{x}, -x_4)$$

$$\bar{\psi}^{\mathrm{T}}(x) = U_{\mathrm{T}}\bar{\psi}(x)U_{\mathrm{T}}^{-1} = \eta_{\mathrm{T}}^{\times}\bar{\psi}^{\times}(\boldsymbol{x}, -x_4)B^{-1} \qquad \text{(D14)}$$

where $|\eta_{\mathrm{T}}|^2 = 1$ and B is the matrix defined in eq. (C38).

* See, for example, Roman (1964); Schweber (1964); Källen (1964).

APPENDIX E

ISOSPIN, CHARGE SYMMETRY AND *G*-PARITY

E-1. Isospin

We use the isospin convention common to elementary particle physics, namely

$$\tau_3|\mathrm{p}\rangle = |\mathrm{p}\rangle\ ; \qquad \tau_3|\mathrm{n}\rangle = -|\mathrm{n}\rangle \tag{E1}$$

where $\boldsymbol{\tau}$ is the usual Pauli operator and $|\mathrm{p}\rangle$ and $|\mathrm{n}\rangle$ refer to proton and neutron states respectively.

The nucleon isospin operator $\boldsymbol{t}$ is then defined as

$$\boldsymbol{t} = \tfrac{1}{2}\boldsymbol{\tau} \tag{E2}$$

and frequent use is made of the displacement operators

$$t_+ = \tfrac{1}{2}(\tau_1 + \mathrm{i}\tau_2)\ ; \qquad t_- = \tfrac{1}{2}(\tau_1 - \mathrm{i}\tau_2) \tag{E3}$$

which effect the following operations

$$t_+|\mathrm{n}\rangle = |\mathrm{p}\rangle\ ; \qquad t_-|\mathrm{p}\rangle = |\mathrm{n}\rangle\ . \tag{E4}$$

Similarly for pion states

$$T_3|\pi^+\rangle = |\pi^+\rangle\ ; \qquad T_3|\pi^0\rangle = 0\ ; \qquad T_3|\pi^-\rangle = -|\pi^-\rangle \tag{E5}$$

where $\boldsymbol{T}$ is the isospin operator.

E-2. Charge symmetry

The charge symmetry operation U_S is one in which a system is rotated through 180° about the 2-axis in isospin space. For an operator O it is defined by the following relation

$$O \to U_S O U_S^{-1} . \qquad \text{(E6)}$$

An isospinor function ψ transforms as follows

$$U_S \psi U_S^{-1} = e^{i\pi\tau_2/2} \psi = i\tau_2 \psi \qquad \text{(E7)}$$

and, for nucleons:

$$\psi_p \to \psi_n , \qquad \psi_n \to -\psi_p . \qquad \text{(E8)}$$

Similarly an isovector function ϕ transforms as follows

$$U_S \phi U_S^{-1} \to e^{i\pi T_2} \phi \qquad \text{(E9)}$$

and, for pions:

$$\phi^+ \to -\phi^- ; \qquad \phi^- \to -\phi^+ ; \qquad \phi^0 \to -\phi^0 . \qquad \text{(E10)}$$

E-3. *G*-parity

The G-operation is defined as the product of the charge symmetry operation and charge conjugation. Thus

$$G = U_S U_C . \qquad \text{(E11)}$$

It then follows that for nucleons under the G-operation

$$\psi_p \to \psi_{\bar{n}} ; \qquad \psi_n \to -\psi_{\bar{p}} \qquad \text{(E12)}$$

and, for pions,

$$\phi^i \to -\phi^i \qquad \text{(E13)}$$

where i = +, 0, –. The pion field thus has negative *G*-parity.

This result illustrates the value of *G*-parity which enables the concept of charge conjugation eigenstates (cf. App. D-2.1) to be extended to charged states.

APPENDIX F

CURRENT MATRIX ELEMENTS

At various points we have to deal with the matrix element between nucleon states of either the electromagnetic current J_λ or the weak polar vector and axial vector currents $\mathcal{V}_\lambda$ and $\mathcal{A}_\lambda$ respectively. We deal in this appendix with the general form of such matrix elements when the nucleons are on the mass shell.

Consider first the matrix element $\langle \mathrm{p}|\mathcal{V}_\lambda(x)|\mathrm{n}\rangle$ of $\mathcal{V}_\lambda$ taken between neutron and proton states having four-momenta n and p respectively. Because of translational invariance, we can write

$$\mathcal{V}_\lambda(x) = \mathrm{e}^{-\mathrm{i}P\cdot x}\, \mathcal{V}_\lambda(0)\, \mathrm{e}^{\mathrm{i}P\cdot x} \tag{F1}$$

where P is the total energy momentum operator. Thus, we have

$$\begin{aligned}\langle \mathrm{p}|\mathcal{V}_\lambda(x)|\mathrm{n}\rangle &= \langle \mathrm{p}|\mathrm{e}^{-\mathrm{i}P\cdot x}\, \mathcal{V}_\lambda(0)\, \mathrm{e}^{\mathrm{i}P\cdot x}|\mathrm{n}\rangle \\ &= \mathrm{e}^{\mathrm{i}(n-p)\cdot x}\langle \mathrm{p}|\mathcal{V}_\lambda(0)|\mathrm{n}\rangle\,.\end{aligned} \tag{F2}$$

Further, the matrix element $\langle \mathrm{p}|\mathcal{V}_\lambda(0)|\mathrm{n}\rangle$ must have the general form

$$\langle \mathrm{p}|\mathcal{V}_\lambda(0)|\mathrm{n}\rangle = (\bar{u}_\mathrm{p}|O_\lambda^{(V)}|u_\mathrm{n}) \tag{F3}$$

where u_p and u_n are Dirac spinors for the proton and neutron corresponding to four momenta p and n respectively and $O_\lambda^{(V)}$ is an operator of such a form that $(\bar{u}_\mathrm{p}|O_\lambda^{(V)}|u_\mathrm{n})$ transforms as a polar vector under a Lorentz transformation. $O_\lambda^{(V)}$ can be constructed only from Dirac γ matrices and the four momenta p and n, these latter being the only dynamical variables in the problem.

Now p and n satisfy the relativistic conditions $p^2 = -M_\mathrm{p}^2$, $n^2 = -M_\mathrm{n}^2$ where M_p and M_n are the masses of the proton and neutron respectively. Thus only one scalar variable can be constructed, and for convenience this is chosen to be $k^2 = (p-n)^2$. Further, rather than expressing $O_\lambda^{(V)}$ in terms of p and n it is more convenient to use $k = p - n$ and $K = p + n$. It is then straightforward to write down the most general form for $O_\lambda^{(V)}$, namely:

$$O_\lambda^{(V)} = a\gamma_\lambda + b\sigma_{\lambda\nu}k_\nu + ck_\lambda + d\sigma_{\lambda\nu}K_\nu + eK_\lambda \tag{F4}$$

where each of the coefficients a, ..., e is an arbitrary function of the scalar k^2 and $\sigma_{\lambda\nu} = \frac{1}{2i}(\gamma_\lambda\gamma_\nu - \gamma_\nu\gamma_\lambda)$. However, the matrix elements with respect to u_p and u_n of the different terms in eq. (F4) are not all linearly independent. For example, using the Dirac equations

$$(\not{p} - \mathrm{i}M)u = 0$$

$$\bar{u}(\not{p} - \mathrm{i}M) = 0 \tag{F5}$$

for a spinor u corresponding to a particle of mass M, it is easy to show that

$$(\bar{u}_\mathrm{p}|\sigma_{\lambda\nu}K_\nu|u_\mathrm{n}) = -\mathrm{i}(\bar{u}_\mathrm{p}|k_\lambda|u_\mathrm{n}) + (M_\mathrm{n} - M_\mathrm{p})(\bar{u}_\mathrm{p}|\gamma_\lambda|u_\mathrm{n})$$

$$(\bar{u}_\mathrm{p}|K_\lambda|u_\mathrm{n}) = \mathrm{i}(M_\mathrm{n} + M_\mathrm{p})(\bar{u}_\mathrm{p}|\gamma_\lambda|u_\mathrm{n}) + \mathrm{i}(\bar{u}_\mathrm{p}|\sigma_{\lambda\nu}k_\nu|u_\mathrm{n})\,. \tag{F6}$$

The most general form of $(\bar{u}_\mathrm{p}|O_\lambda^{(V)}|u_\mathrm{n})$ can therefore be written:

$$\langle \mathrm{p}|\mathcal{V}_\lambda(0)|\mathrm{n}\rangle = (\bar{u}_\mathrm{p}|O_\lambda^{(V)}|u_\mathrm{n}) = \mathrm{i}(\bar{u}_\mathrm{p}|f_\mathrm{V}\gamma_\lambda + f_\mathrm{W}\sigma_{\lambda\nu}k_\nu + \mathrm{i}f_\mathrm{S}k_\lambda|u_\mathrm{n}) \tag{F7}$$

where the form factors f_V, f_W and f_S are functions of k^2.

An exactly analogous discussion for the axial vector current leads to the result

$$\langle \mathrm{p}|\mathcal{A}_\lambda(0)|\mathrm{n}\rangle = (\bar{u}_\mathrm{p}|O_\lambda^{(A)}|u_\mathrm{n}) = \mathrm{i}(\bar{u}_\mathrm{p}|-f_\mathrm{A}\gamma_\lambda\gamma_5 + \mathrm{i}f_\mathrm{P}\gamma_5 k_\lambda + f_\mathrm{T}\sigma_{\lambda\nu}k_\nu\gamma_5|u_\mathrm{n}) \tag{F8}$$

where, again, the form factors f_A, f_P and f_T are functions of k^2.

In the case of the electromagnetic current J_λ which is a polar vector we have equivalently to eq. (F7) and in the usual notation

$$\langle \mathrm{p}'|J_\lambda|\mathrm{p}\rangle = \mathrm{i}(\bar{u}_{\mathrm{p}'}|F_1^{(\mathrm{p})}\gamma_\lambda - F_2^{(\mathrm{p})}\sigma_{\lambda\nu}k_\nu + \mathrm{i}F_3^{(\mathrm{p})}k_\lambda|u_\mathrm{p})\,. \tag{F9}$$

However, J_λ is conserved, i.e.

$$\partial_\lambda J_\lambda = 0\,. \tag{F10}$$

Taking the matrix element of this equation then gives

$$\langle \mathrm{p}' | \partial_\lambda J_\lambda | \mathrm{p} \rangle = -\langle \mathrm{p}' | \mathrm{i}\, [P_\lambda, J_\lambda] \, | \mathrm{p} \rangle = -\mathrm{i} k_\lambda \langle \mathrm{p}' | J_\lambda | \mathrm{p} \rangle = 0 \, . \tag{F11}$$

Substituting for $\langle \mathrm{p}' | J_\lambda | \mathrm{p} \rangle$ from eq. (F9) gives

$$(\bar{u}_{\mathrm{p}'} | F_1^{(\mathrm{p})} k_\lambda \gamma_\lambda - F_2^{(\mathrm{p})} \sigma_{\lambda\nu} k_\nu k_\lambda + \mathrm{i} F_3^{(\mathrm{p})} k^2 | u_{\mathrm{p}}) = 0 \, . \tag{F12}$$

The term in F_1 vanishes since $\bar{u}_{\mathrm{p}'}$ and u_{p} both satisfy the Dirac eq. (F5); the term in F_2 vanishes since $\sigma_{\lambda\nu}$ is antisymmetric and so we must have $F_3^{(\mathrm{p})} = 0$. Thus for the electromagnetic current

$$\langle \mathrm{p}' | J_\lambda | \mathrm{p} \rangle = \mathrm{i} \, (\bar{u}_{\mathrm{p}'} | F_1^{(\mathrm{p})} \gamma_\lambda - F_2^{(\mathrm{p})} \sigma_{\lambda\nu} k_\nu | u_{\mathrm{p}}) \, . \tag{F13}$$

A similar expression arises for the case of neutron states. Equivalently, if $\mathcal{V}_\lambda$ is conserved, the form factor f_{s} in eq. (F7) is zero.

REFERENCES

Abers, E.S., R.E. Norton and D.A. Dicus, 1967, Phys. Rev. Letters 18, 676.

Abers, E.S., D.A. Dicus, R.E. Norton and M.R. Quinn, 1968, Phys. Rev. Letters 167, 1461.

Abov, Yu.G., O.N. Yermakov, A.D. Gulko, P.A. Krupchitsky and S.S. Troshin, 1962, Nucl. Phys. 34, 505.

Abov, Yu.G., P.A. Krupchitsky and Yu.A. Oratovsky, 1965, Sov. J. Nucl. Phys. 1, 341.

Abov, Yu.G., P.A. Krupchitsky, M.I. Bulgakov, O.N. Yermakov and I.L. Karpikhin, 1968, Phys. Letters 27B, 16.

Abov, Yu.G., P.A. Krupchitsky, M.I. Bulgakov, O.N. Yermakov and I.L. Karpikhin, 1970, Sov. J. Nucl. Phys. 10, 320.

Adams, J.B., 1967, Phys. Rev. 156, 1611.

Adelberger, E.G., and D.P. Balamuth, 1971, Phys. Rev. Letters 27, 1597.

Adelberger, E.G., and A.B. McDonald, 1967, Phys. Letters 24B, 270.

Adelberger, E.G., C.L. Cocke, C.N. Davids and A.B. McDonald, 1969, Phys. Rev. Letters 22, 352.

Adelberger, E.G., A.B. McDonald, H.B. Mak, A.P. Shukla and A.V. Nero, 1971, Bull. Am. Phys. Soc. 16, 829.

Adler, S.L., 1965a, Phys. Rev. 139, B1638.

Adler, S.L., 1965b, Phys. Rev. Letters 14, 1051.

Adler, S.L., 1965c, Phys. Rev. 140, B736.

Adler, S.L., 1965d, Phys. Rev. 137, B1022.

Adler, S.L., 1967, Phys. Rev. Letters 18, 519, 1036 (E).

Adler, S.L., 1968, Ann. Phys. (N.Y.) 50, 189.

Adler, S.L., and R.F. Dashen, 1968, *Current Algebras and Applications to Particle Physics* (W.A. Benjamin, Inc.).

Adler, S.L., and Y. Dothan, 1966, Phys. Rev. 151, 1267.

Ahrens, T., and E. Feenberg, 1952, Phys. Rev. 86, 64.

Akimova, M., L. Blokhinhev and Dolinsky, 1961, Sov. Phys. - JETP 12, 1260.

Alberi, J.L., R. Wilson and I.G. Schroder, 1972, Phys. Rev. Letters 29, 518.

Albright, C.H., and R.J. Oakes, 1970, Phys. Rev. D3, 1270.

Alburger, D.E., 1972, Phys. Rev. C5, 274.

Alburger, D.E., and D.H. Wilkinson, 1971, Phys. Rev. C3, 1957.

Alburger, D.E., R.E. Pixley, D.H. Wilkinson and P. Donovan, 1961, Phil. Mag. 6, 171.

Alder, K., B. Stech and A. Winther, 1957, Phys. Rev. 107, 728.

Altman, A., and W.M. MacDonald, 1962, Nucl. Phys. 35, 593.

Ambler, E., R.W. Hayward, D.D. Hoppes and R.P. Hudson, 1957a, Phys. Rev. 106, 1361.

Ambler, E., R.W. Hayward, D.D. Hoppes and R.P. Hudson, 1957b, Phys. Rev. 108, 503.

Anderson, J.D., C. Wong and J.W. McClure, 1965, Phys. Rev. 138, B615.

Anderson, R.L., R. Prepost and B.H. Wilk, 1969, Phys. Rev. Letters 22, 651.

Appel, H., 1959, Z. Phys. 155, 580.

Appel, H., and H. Schopper, 1957, Z. Phys. 149, 103.
Appel, H., H. Schopper and S.D. Bloom, 1958, Phys. Rev. 109, 2211.
Appel, H., R. Blatter and H. Schopper, 1962, Nucl. Phys. 30, 688.
Appel, K., 1969, Z. Phys. 219, 447.
Arenhövel, H., and M. Danos, 1968, Phys. Letters 28B, 299.
Arenhövel, H., M. Danos and H.T. Williams, 1970, Phys. Letters 31B, 109.
Arenhövel, H., M. Danos and H.T. Williams, 1971, Nucl. Phys. A162, 12.
Arzubov, B.A., and A.T. Filippov, 1968, Sov. Phys. - JETP Letters 8, 302.
Astbury, A., J.H. Bartley, I.M. Blair, M.A.R. Kemp, H. Muirhead and T. Woodhead, 1962, Proc. Phys. Soc. (London) 79. 1011.
Astbury, A., L.B. Auerbach, D. Cutts, R.J. Esterling, D.A. Jenkins, N.H. Lipman and R.E. Schafer, 1964, Nuovo Cimento 33, 1020.
Atac, M., B. Christman, P. Debrunner and H. Frauenfelder, 1968, Phys. Rev. Letters 20, 691.
Atkinson, J., L.G. Mann, K.G. Tirsell and S.D. Bloom, 1968, Nucl. Phys. A114, 143.
Auerbach, N., and A. Lev, 1972, Nucl. Phys. A180, 337.
Auerbach, L.B., R.J. Esterling, R.E. Hill, D.A. Jenkins, J.T. Lach and N.H. Lipman, 1965, Phys. Rev. 138, B127.
Auerbach, N., J. Hufner, A.K. Kerman and C.M. Shakin, 1969a, Phys. Rev. Letters 23, 484.
Auerbach, E.H., S. Kahana and J. Weneser, 1969b, Phys. Rev. Letters 23, 1253.
Auerbach, E.H., S. Kahana, C.K. Scott and J. Weneser, 1969c, Phys. Rev. 188, 1747.
Auerbach, N., A. Lev and E. Kashy, 1971, Phys. Letters 36B, 453.
Austern, N., and E. Rost, 1959, Phys. Rev. 117, 1506.
Austern, N., and R.G. Sachs, 1951, Phys. Rev. 81, 710.

Backenstoss, G., H. Daniel, H. Koch, U. Lynen, Ch. Von der Malsburg, G. Poelz, H.P. Povel, H. Schmitt, K. Springer and L. Tauscher, 1971, Phys. Letters 36B, 403.
Baba, K., 1970, Prog. Theor. Phys. 43, 390.
Bahcall, J.N., 1963a, Phys. Rev. 129, 2683.
Bahcall, J.N., 1963b, Phys. Rev. 131, 1756.
Bahcall, J.N., 1963c, Phys. Rev. 132, 362.
Bahcall, J.N., 1966, Nucl. Phys. 75, 10.
Bailin, D., 1964, Phys. Rev. 135, B166.
Bailin, D., 1970, Private communication.
Bailin, D., 1972, Rep. Prog. in Phys. 34, 491.
Baird, J.K., P.D. Miller, W.B. Dress and N.F. Ramsey, 1969, Phys. Rev. 179, 1285.
Baker, K.D., and W.D. Hamilton, 1970, Phys. Letters 31B, 557.
Baker, K.D., and W.D. Hamilton, 1971, Nucl. Phys. A175, 350.
Baker, W.F., and C. Rubbia, 1959, Phys. Rev. Letters 3, 179.
Banerjee, P., and H.D. Zeh, 1960, Z. Phys. 159, 170.
Bardin, R.K., C.A. Barnes, W.A. Fowler and P.A. Seeger, 1962, Phys. Rev. 127, 583.
Bardin, R.K., P.J. Gollon, J.D. Ullman and C.S. Wu, 1970, Nucl. Phys. A158, 337.
Bargov, N.A., and G.A. Lobov, 1967, Sov. Phys. - JETP 25, 344.
Barrett, B., and G. Barton, 1964, Phys. Rev. 133, B466.
Barshay, S., 1965, Phys. Letters 17, 78.
Barshay, S., 1966, Phys. Rev. Letters 17, 49.
Barshay, S., and G.M. Temmer, 1964, Phys. Rev. Letters 12, 728.

Bartlett, D.F., C.E. Friedberg, K. Goulianos, I.S. Hammerman and D.P. Hutchinson, 1969, Phys. Rev. Letters 23, 893, 1205 (E).
Bartlett, D.F., C.E. Friedberg, P.E. Goldhagen and K. Goulianos, 1971, Phys. Rev. Letters 27, 881.
Barton, G., 1961, Nuovo Cimento 19, 512.
Barton, G., and E.D. White, 1969, Phys. Rev. 184, 1660.
Bassi, P., B. Ferretti, G. Venturini, G.C. Bertolini, F. Cappellani, V. Mandl, G.B. Restelli and A. Rota, 1962, Nuovo Cimento 24, 560.
Bassi, P., B. Ferretti, G. Venturini, G.C. Bertolini, F. Cappellani, V. Mandl, G.B. Restelli and A. Rota, 1963, Nuovo Cimento 28, 1049.
Beck, E., and H. Daniel, 1967, *Symposium on Nuclear β-decay and Weak Interactions Zagreb, Yugoslavia.*
Beg, M.A.B., and J. Bernstein, 1972, Phys. Rev. D5, 714.
Beg, M.A.B., J. Bernstein and A. Sirlin, 1969, Phys. Rev. Letters 23, 270.
Behrens, H., 1966, Dissertation. Karlsruhe. KFK 434.
Behrens, H., 1967, Z. Phys. 201, 153.
Behrens, H., and W. Bühring, 1968, Nucl. Phys. A106, 433.
Behrens, H., and W. Bühring, 1970, Nucl. Phys. A150, 481.
Behrens, H., and W. Bühring, 1971, Nucl. Phys. A162, 111.
Behrens, H., and W. Bühring, 1972, Nucl. Phys. A179, 297.
Behrens, H., and J. Jänecke, 1969, *Numerical tables for beta decay and electron capture*, Landolt–Börnstein, new series vol. I/4 (Springer, Berlin).
Bell, J.S., and R.J. Blin-Stoyle, 1957, Nucl. Phys. 6, 87.
Bell, J.S., and J. Lovseth, 1964, Nuovo Cimento 17, 408.
Bell, J.S., and F. Mandl, 1958, Proc. Phys. Soc. (London) A71, 272.
Berenyi, D., 1968, Rev. Mod. Phys. 40, 390.
Berge, J., 1966, Proc. *Thirteenth Ann. Intern. Conf. on High Energy Physics, Berkeley* (University of California Press, Berkeley, 1967) p. 46.
Bergkvist, K.E., 1969, *Proc. Topical Conf. on Weak Interactions, CERN, Geneva,* p. 91.
Bergmann, E.E., 1968, Phys. Rev. 172, 1441.
Berman, S.M., 1958, Phys. Rev. 112, 267.
Berman, S.M., and A. Sirlin, 1962, Ann. of Phys. (N.Y.) 20, 20.
Berovic, N., 1971, Phys. Letters 35B, 475.
Bernabeu, J., and P. Pascual, 1969, Phys. Letters 29B, 555.
Bernardini, G., J.K. Bieulein, G. Von Dardel, H. Faissner, F. Ferrero, J.M. Gaillard, H.J. Gerber, B. Hahn, V. Kaftanov, F. Frienen, C. Manfredotti, M. Reinharz and R.A. Salmenon, 1966, Phys. Letters 13, 86.
Bernstein, J., 1959, Phys. Rev. 115, 694.
Bernstein, J., and R.R. Lewis, 1958, Phys. Rev. 112, 232.
Bernstein, J., T.D. Lee, C.N. Yang and H. Primakoff, 1958, Phys. Rev. 111, 313.
Bernstein, J., M. Gell-Mann and L. Michel, 1960a, Nuovo Cimento 16, 560.
Bernstein, J., S. Fabini, M. Gell-Mann and W. Thirring, 1960b, Nuovo Cimento 17, 757.
Bernstein, J., G. Feinberg and T.D. Lee, 1965, Phys. Rev. 139, B1650.
Berovic, N., 1971, Phys. Letters 35B, 475.
Berthier, J., 1962, Thesis, University of Paris (unpublished).
Bertolini, E., A. Citron, G. Gianienella, S. Focardi, A. Mukhin, C. Rubbia and S. Saporetti, *Proc. 1962 Annual Intern. Conf. on High Energy Physics at Geneva* (CERN, Geneva) p. 421.

Bethe, H.A., 1968, *Proc. Intern. Conf. on Nuclear Structure*, J. Phys. Soc., Japan, 24 Suppl. p. 56.
Bethe, H.A., and R.F. Bacher, 1936, Rev. Mod. Phys. 8, 82.
Bethe, H.A., and J. Goldstone, 1957, Proc. Roy. Soc. A238, 551.
Bethe, H.A., and E.E. Salpeter, 1957, *Handbuch der Physik* 35 (Springer-Verlag, Berlin) sect. 12.
Bethe, H.A., and P.J. Siemens, 1968, Phys. Letters 27B, 549.
Bhaduri, R.K., Y. Nogami and C.K. Ross, 1970, Phys. Rev. C2, 2082.
Bhalla, C.P., 1964, NBS Mongraph 82.
Bhalla, C.P., 1966, Phys. Letters 19, 691.
Bhalla, C.P., and M.E. Rose, 1960a, Oak Ridge National Laboratory Report 2954.
Bhalla, C.P., and M.E. Rose, 1960b, Phys. Rev. 120, 1415.
Bhalla, C.P., and M.E. Rose, 1961, Oak Ridge National Laboratory Report 3207.
Bhargava, P.C., and D.W. Sprung, 1967, Ann. Phys. 42, 222.
Bhattacherjee, S.K., S.K. Mitra and H.C. Padhi, 1967, Nucl. Phys. A96, 81.
Biedenharn, L.C., and M.E. Rose, 1953, Rev. Mod. Phys. 25, 729.
Bienlein, H., R. Fleischmann and H. Wegener, 1958, Z. Phys. 150, 80.
Bietti, A., 1965, Nuovo Cimento 37, 337.
Bincer, A.M., 1958, Phys. Rev. 112, 244.
Bisi, A., A. Fasana and L. Zappa, 1960, Nuovo Cimento 16, 1374.
Bisi, A., A. Fasana and L. Zappa, 1963, Nuovo Cimento 45, 405.
Bjorken, J.D., 1966, Phys. Rev. 148, 1467.
Blatt, J.M., 1952, Phys. Rev. 88, 945.
Blatt, J.M., and L.M. Delves, 1964, Phys. Rev. Letters 12, 544.
Blatt, J.M., and V.F. Weisskopf, 1952, *Theoretical Nuclear Physics* (John Wiley, N.Y.).
Bleser, E., L. Lederman, J. Rosen, J. Rothberg and E. Zavattini, 1952, Phys. Rev. Letters 8, 288.
Blin-Stoyle, R.J., 1952, Proc. Phys. Soc. (London) A65, 452.
Blin-Stoyle, R.J., 1960a, Phys. Rev. 118, 1605.
Blin-Stoyle, R.J., 1960b, Phys. Rev. 120, 181.
Blin-Stoyle, R.J., 1964a, Nucl. Phys. 57, 232.
Blin-Stoyle, R.J., 1964b, Phys. Rev. Letters 13, 55.
Blin-Stoyle, R.J., 1969a, Phys. Letters 29B, 12.
Blin-Stoyle, R.J., 1969b, *Isospin in Nuclear Physics*, Ed. D.H. Wilkinson (North-Holland Publ. Comp., Amsterdam) p. 115.
Blin-Stoyle, R.J., 1969c, Phys. Rev. 188, 1540.
Blin-Stoyle, R.J., 1969d, Phys. Rev. Letters 23, 535.
Blin-Stoyle, R.J., and H. Feshbach, 1961, Nucl. Phys. 27, 395.
Blin-Stoyle, R.J., and J. Freeman, 1970, Nucl. Phys. A150, 369.
Blin-Stoyle, R.J., and P. Herczeg, 1966, Phys. Letters 23, 376.
Blin-Stoyle, R.J., and P. Herczeg, 1968, Nucl. Phys. B5, 291.
Blin-Stoyle, R.J., and M.J. Kearsley, 1960, Proc. Phys. Soc. (London) 75, 147.
Blin-Stoyle, R.J., and J. Le Tourneux, 1961, Phys. Rev. 123, 627.
Blin-Stoyle, R.J., and J. Le Tourneux, 1962, Ann. of Phys. 18, 12.
Blin-Stoyle, R.J., and E. Maqueda, 1966, Nucl. Phys. A91, 460.
Blin-Stoyle, R.J., and S.C.K. Nair, 1963, Phys. Letters 7, 161.
Blin-Stoyle, R.J., and S.C.K. Nair, 1966, Adv. Phys. 15, 493.
Blin-Stoyle, R.J., and S.C.K. Nair, 1967, Nucl. Phys. A105, 640.

Blin-Stoyle, R.J., and L. Novakovic, 1964, Nucl. Phys. 51, 133.
Blin-Stoyle, R.J., and S. Papageorgiou, 1965a, Nucl. Phys. 64, 1.
Blin-Stoyle, R.J., and S. Papageorgiou, 1965b, Phys. Letters 14, 343.
Blin-Stoyle, R.J., and M. Rosina, 1965, Nucl. Phys. 70, 321.
Blin-Stoyle, R.J., and R.M. Spector, 1961, Phys. Rev. 124, 1199.
Blin-Stoyle, R.J., and M. Tint, 1967, Phys. Rev. 160, 803.
Blin-Stoyle, R.J., and C. Yalçin, 1965, Phys. Letters 15, 258.
Blin-Stoyle, R.J., and C.T. Yap, 1966, Nucl. Phys. 79, 561.
Blin-Stoyle, R.J., J.A. Evans and A.M. Khan, 1971, Phys. Letters 36B, 202.
Blin-Stoyle, R.J., V. Gupta and H. Primakoff, 1959, Nucl. Phys. 11, 444.
Blin-Stoyle, R.J., V. Gupta and J.S. Thompson, 1959/60, Nucl. Phys. 14, 685.
Blin-Stoyle, R.J., S.C.K. Nair and S. Papageorgiou, 1965, Proc. Phys. Soc. (London) 85, 477.
Bloch, R., R.E. Pixley and P. Truol, 1967, Phys. Letters 25B, 215.
Block, M.M., H. Burmeister, D.C. Cundy, B. Eiben, C. Franzinetti, J. Keren, R. Møllerud, G. Myatt, M. Nikolic, A. Orkin-Lecourtois, M. Paty, D.H. Perkins, C.A. Ramm, K. Schultze, H. Sletten, K. Soop, R. Stump, W. Venus and H. Yoshiki, 1966, Phys. Letters 12, 281.
Blomqvist, J., 1970, Phys. Letters 32B, 1.
Blomqvist, J., 1971, Phys. Letters 35B, 375.
Bloom, S.D., 1964, Nuovo Cimento 32, 1023.
Bloom, S.D., 1966, *Isobaric Spin in Nuclear Physics* (Academic Press Inc., New York) p. 123.
Bloom, S.D., L.G. Mann, R. Polichar, J.R. Richardson and A. Scott, 1964, Phys. Rev. 134, B481.
Bludman, S.A., 1963, Nuovo Cimento 27, 751.
Bock, P., 1969, Thesis, Karlsruhe.
Bock, P., and B. Jenschke, 1971, Nucl. Phys. A160, 550.
Bodansky, D., W.J. Braithwaite, D.C. Shreve, D.W. Storm and W.G. Weitkamp, 1968, Phys. Rev. Letters 17, 589.
Bodenstedt, E., L. Ley, H.O. Schlenz and U. Wehmann, 1969a, Phys. Letters 29B, 165.
Bodenstedt, E., L. Ley, H.O. Schlenz and U. Wehmann, 1969b, Nucl. Phys. A137, 33.
Boehm, F., 1968, *Hyperfine Structure and Nuclear Radiations* (North-Holland).
Boehm, F., and E. Kankeleit, 1968, Nucl. Phys. A109, 457.
Boehm, F., and A.H. Wapstra, 1957a, Phys. Rev. 106, 1364.
Boehm, F., and A.H. Wapstra, 1957b, Phys. Rev. 107, 1202.
Boehm, F., and A.H. Wapstra, 1958, Phys. Rev. 109, 458.
Bogdan, Jr., A., 1968, Nucl. Phys. B5, 431.
Bogdan, Jr., A., 1969a, Phys. Rev. Letters 22, 71.
Bogdan, Jr., A., 1969b, Nucl. Phys. B12, 89.
Bogdan, D., and A. Vata, 1968a, Nucl. Phys. A109, 347.
Bogdan, D., C. Protop and I. Vata, 1968b, Nucl. Phys. A119, 113.
Bohr, A., and B.R. Mottelson, 1969, *Nuclear Structure, Vol. 1* (W.A. Benjamin, Inc.).
Bohr, A., J. Damgaard and B.R. Mottelson, 1967, *Nuclear Structure Lectures given at The International Seminar on Low Energy Nuclear Physics Dacca* (North-Holland Publ. Comp., Amsterdam) p. 1–10.
Bonar, D.C., C.W. Drake, R.D. Headrick and V.W. Hughes, 1968, Phys. Rev. 174, 1200.
Borchi, E., and S. De Gennaro, 1970, Phys. Rev. C2, 1012.

Borchi, E., and R. Gatto, 1964, Nuovo Cimento 33, 1472.

Botterill, D.R., R.M. Brown, A.B. Clegg, I.F. Corbett, G. Culligan, J.M. McL. Emmerson, R.C. Field, J. Garvey, P.B. Jones, N. Middlemas, D. Newton, T.W. Quirk, G.L. Salmon, P. Steinberg and W.C.S. Williams, 1968, Phys. Rev. 174, 1661.

Bouchiat, C.C., 1959, Phys. Rev. Letters 3, 516.

Bouchiat, C.C., 1960, Phys. Rev. 118, 540.

Bouyssy, A., and N. Vinh Mau, 1972, Nucl. Phys. A185, 32.

Bouyssy, A., M. Castaignet and N. Vinh Mau, 1967, Phys. Letters 25B, 533.

Boyd, D.P., P.F. Donovan, B. Marsh, D.E. Alburger, D.H. Wilkinson, P. Assimakopoulos and E. Beardsworth, 1968, Bull. Am. Phys. Soc. 13, 1424.

Brandow, B.H., 1964, Ph. D. Thesis, Cornell University.

Brandow, B.H., 1966, Phys. Rev. 152, 863.

Breit, G., 1962, Rev. Mod. Phys. 34, 766.

Breit, G., and I. Bloch, 1947, Phys. Rev. 72, 135.

Breit, G., and M.L. Rustgi, 1971, Nucl. Phys. A161, 337.

Brene, N., M. Roos and A. Sirlin, 1968, Nucl. Phys. B6, 255.

Brink, D.M., 1965, *Nuclear Forces* (Pergamon Press, Oxford).

Brink, D.M., and G.R. Satchler, 1962, *Angular Momentum* (Oxford Library of the Physical Sciences, Clarendon Press, Oxford).

Broadhurst, D.J., 1970, Nucl. Phys. B20, 603.

Broadhurst, D.J., 1972, Phys. Rev. D5, 1228.

Brodine, J.C., 1970, Phys. Rev. D1, 100.

Brodine, J.C., 1971, Phys. Rev. D2, 2090.

Bromley, D.A., H.E. Grove, J.A. Kuehner, A.E. Litherland and E. Almqvist, 1959, Phys. Rev. 114, 758.

Brosi, A.R., A.I. Galansky, B.H. Ketelle and H.B. Willard, 1962, Nucl. Phys. 33, 353.

Brown, G.E., 1967, *Unified Theory of Nuclear Models*, 2nd Edition (North-Holland Publ. Comp., Amsterdam).

Brown, G.E., and A.M. Green, 1969, Nucl. Phys. A137, 1.

Brown, G.E., A.M. Green and W.J. Gerace, 1968, Nucl. Phys. A115, 435.

Brown, L.S., 1964, Phys. Rev. 136, B314.

Browne, C.P., 1966, *Isobaric Spin in Nuclear Physics* Eds. J.D. Fox and D. Robson (Academic Press Inc).

Brueckner, K.A., 1959, *The Many Body Problem*. Ecole d'Eté de Physique Théorique, Les Houches, 1958 (Wiley).

Bryan, R.A., and B.L. Scott, 1969, Phys. Rev. 177, 1435.

Bühring, W., 1963a, Nucl. Phys. 40, 472.

Bühring, W., 1963b, Nucl. Phys. 49, 190.

Bühring, W., 1965, Nucl. Phys. 61, 110.

Bühring, W., 1967, Preprint (Heidelberg).

Bühring, W., and L. Schulke, 1965, Nucl. Phys. 65, 369.

Bulgakov, M.I., A.D. Gulko, G.V. Danilyan, I.L. Karpikhin, P.A. Krupchitsky, V.V. Novitsky, V.S. Pavlov, Yu.A. Oratovsky, E.I. Tarkovsky and S.S. Trostin, 1972, Preprint.

Bunyatan, G.G., 1966, Sov. J. Nucl. Phys. 3, 613.

Buon, J., V. Gracco, J. Lefrançois, P. Lehmann, B. Mephel and Ph. Roy, 1968, Phys. Letters 26B, 595.

Burgov, N.A., and G.A. Lobov, 1967, Zh. Eksp. i Teor. Fiz. 52, 527.

Cabibbo, N., 1963, Phys. Rev. Letters 10, 531.
Cabibbo, N., 1964, Phys. Letters 12, 137.
Cabibbo, N., 1965, Phys. Rev. Letters 14, 965.
Cabibbo, N., and R. Gatto, 1960, Phys. Rev. Letters 5, 114.
Cabibbo, N., and R. Gatto, 1961, Nuovo Cimento 19, 612.
Cabibbo, N., L. Maiani and G. Preparata, 1967a, Phys. Letters 25B, 31.
Cabibbo, N., L. Maiani and G. Preparata, 1967b, Phys. Letters 25B, 132.
Calaprice, F.P., E.D. Commins, H.M. Gibbs, G.L. Wick and D.A. Dobson, 1967, Phys. Rev. Letters 18, 918.
Calaprice, F.P., E.D. Commins, H.M. Gibbs, G.L. Wick and D.A. Dobson, 1969, Phys. Rev. 184, 1117.
Callan, Jr., C.G., and S.B. Treiman, 1967, Phys. Rev. 162, 1494.
Camiz, P., and N. Vinh Mau, 1964, J. Phys. (Paris) 25, 371.
Camp, D.C., L.G. Mann and S.D. Bloom, 1965, Nucl. Phys. 73, 174.
Cantwell, R.M., 1956, Ph. D. Thesis, Washington University.
Carhart, R.A., 1967, Phys. Rev. 153, 1077.
Carruthers, P.A., 1966, *Introduction to Unitary Symmetry* (Interscience Publishers).
Castle, R.T., and R.W. Finlay, 1964, Phys. Rev. 134, B929.
Cavanagh, P.E., J.F. Turner, C.F. Coleman, G.A. Gard and B.W. Ridley, 1957, Phil. Mag. 2, 1105.
Celenza, L.S., R.M. Dreizler, A. Klein and G.J. Dreiss, 1966, Phys. Letters 23, 241.
Cerny, J., 1968, Ann. Rev. Nucl. Sci., 18, 27.
Cerny, J., R.H. Pehl and G.T. Garvey, 1964, Phys. Letters 12, 234.
Chang, L.N., and Y.B. Dai, 1961, Scientia Sinica 10, 420.
Chemtob, M., 1969, Thesis "Les courants d'interaction nucleaires à deux corps", Université de Paris.
Chemtob, M., and M. Rho, 1969, Phys. Letters 29B, 540.
Chemtob, M., and M. Rho, 1971, Nucl. Phys. A163, 1.
Chen, F.S., L. Durand III, and I.J. McGee, 1966, Phys. Rev. 146, 638.
Chen, H.H., 1969a, Phys. Rev. 185, 2003.
Chen, H.H., 1969b, Phys. Rev. 185, 2007.
Cheng, W.K., 1966, University of Pennsylvania Thesis (unpublished).
Cheng, W.K., and E. Fischbach, 1969, Phys. Rev. 188, 1530.
Cheng, W.K., E. Fischbach, H. Primakoff, D. Tadić and K. Trabert, 1971, Phys. Rev. D3, 2289.
Chern, B., T.A. Halpern and L. Logue, 1967, Phys. Rev. 161, 1116.
Chew, G.F., 1966, *The Analytic S-matrix* (W.A. Benjamin, Inc.).
Christensen, C.J., A. Nielsen, A. Bahnsen, W.K. Brown and B.M. Rustad, 1967, Phys. Letters 26B, 11.
Christensen, C.J., V.E. Krohn and G.R. Ringo, 1969, Phys. Letters 28B, 411.
Christensen, C.J., A. Nielsen, A. Bahnsen, W.K. Brown and B.M. Rustad, 1972, Phys. Rev. D5, 1628.
Christenson, J.H., J.W. Cronin, V.L. Fitch and R. Turlay, 1964, Phys. Rev. Letters 13, 138.
Chu, W.T., I. Nadelhaft and J. Ashkin, 1963, Bull. Am. Phys. Soc. 8, 34.
Chu, W.T., I. Nadelhaft and J. Ashkin, 1965, Phys. Rev. 137, B352.
Clark, G.J., J.M. Freeman, D.C. Robinson, J.S. Ryder, W.E. Burcham and G.T.A. Squier, 1971, Phys. Letters 35B, 503.
Clay, D.R., J.W. Keuffel, R.L. Wayner, Jr., and R.M. Edelstein, 1965, Phys. Rev. 140, B587.

Clement, C.F., and L. Heller, 1971, Phys. Rev. Letters 27, 545.
Cocke, C.L., and J.C. Adloff, 1971, Nucl. Phys. A172, 417.
Cocke, C.L., J.C. Adloff and P. Chevallier, 1968, Phys. Rev. 176, 1120.
Coester, F., 1953, Phys. Rev. 89, 619.
Cohen, R.C., S. Devons and A.D. Kanaris, 1963, Phys. Rev. Letters 11, 134.
Cohen, R.C., S. Devons and A.D. Kanaris, 1964, Nucl. Phys. 57, 255.
Cohen, V.W., R. Nathans, H.B. Silsbee, E. Lipworth and N.F. Ramsey, 1969, Phys. Rev. 177, 1942.
Collard, H., R. Hofstadter, E.B. Hughes, A. Johansson, M.R. Yearian, R.B. Day and R.T. Wayner, 1965, Phys. Rev. 138, B57.
Conforto, G., M. Conversi and L. Di Lella, 1962, Phys. Rev. Letters 9, 22.
Conversi, M., R. Diebold and L. Di Lella, 1964, Phys. Rev. 136, B1077.
Coussement, R., and L. Van Neste, 1967, Nucl. Phys. A102, 363.
Coutinho, F., 1970, Private communication.
Coutinho, F., and P. Ridley, 1971, (to be published).
Coutinho, F., 1972, D. Phil. thesis, University of Sussex.
Cox, A.E., S.A.R. Wynchank and C.H. Collie, 1965, Nucl. Phys. 4, 497.
Cramer, J.G., and N.F. Mangelson, 1968, Phys. Letters 27B, 1507.
Cronin, J.W., 1968, *14th Intern. Conf. on High Energy Physics, Vienna.*
Cruse, D.W., and W.D. Hamilton, 1969, Nucl. Phys. 125, 241.
Csonka, P.L., and M.J. Moravcsik, 1966, Phys. Rev. 152, 1310.
Cutkosky, R.E., 1957, Phys. Rev. 107, 330.

Dai, Y.B., D.C. San, T.H. Ho and H.Y. Tzu, 1959, Acta Physica Sinica 15, 262.
Dalitz, R.H., 1952, Proc. Phys. Soc. (London) A65, 175.
Dalitz, R.H., 1954, Phys. Rev. 95, 799.
Dalitz, R.H., and F. Von Hippel, 1964, Phys. Letters 10, 155.
Dal'karov, O.D., 1965, Sov. Phys. - JETP Letters 2, 197.
Damgaard, J., 1966, Nucl. Phys. 79, 374.
Damgarrd, J., 1969, Nucl. Phys. A130, 233.
Damgaard, J., and A. Winther, 1964, Nucl. Phys. 54, 615.
Damgaard, J., and A. Winther, 1965, Phys. Letters 23, 345.
Damgaard, J., C.K. Scott and E. Osnes, 1970, Nucl. Phys. A154, 12.
Daniel, H., 1958, Nucl. Phys. 8, 191.
Daniel, H., 1962, Nucl. Phys. 31, 293.
Daniel, H., 1968, Rev. Mod. Phys. 40, 659.
Daniel, H., and G.Th. Kaschl, 1966, Nucl. Phys. 76, 97.
Daniel, H., and A. Schmitt, 1965, Nucl. Phys. 65, 481.
Danilov, G.S., 1965, Phys. Letters 18, 40.
Danilov, G.S., 1970, Nucl. Phys. B24, 165.
Danilov, G.S., 1971, Phys. Letters 35B, 579.
Das, T., 1968, Phys. Rev. Letters 21, 409.
Dashen, R., S. Frautschi, M. Gell-Mann and Y. Hara, 1964, *The Eightfold Way* (W.A. Benjamin, Inc., New York) p. 254.
Davis, R., 1955, Phys. Rev. 97, 766.
Davis, R., 1958, *Proc. Intern. Conf. on Radioisotopes in Scientific Research, Paris* (Pergamon, London).

Davis, R., and C. St. Pierre, 1969, Nucl. Phys. A138, 545.
Day, B.D., 1967, Rev. Mod. Phys. 39, 719.
De Benedetti, S., 1964, *Nuclear Interactions* (John Wiley and Sons, Inc.).
Debevec, P.T., G.T. Jarvey and B.E. Hingerby, 1971, Phys. Letters 34B, 497.
Debrunner, P., and W. Kundig, 1957, Helv. Phys. Acta, 30, 261.
DeForest, T., 1965, Phys. Rev. 139, B1217.
De Groot, S.R., and H.A. Tolhoek, 1950, Physica 16, 456.
De Groot, S.R., H.A. Tolhoek and W.J. Huiskamp, 1965, *Alpha-, Beta- and Gamma-ray Spectroscopy*, Ed. K. Siegbahn (North-Holland Publ. Comp., Amsterdam) p. 1199.
Delorme, J., 1970, Nucl. Phys. B19, 573.
Delorme, J., and M. Ericson, 1970, Phys. Letters 32B, 443.
Delorme, J., and M. Rho, 1971a, Phys. Letters 34B, 239.
Delorme, J., and M. Rho, 1971b, Nucl. Phys. B34, 317.
Delves, L.M., 1964, Phys. Rev. 135, B1316.
Delves, L.M., and J.M. Blatt, 1967, Nucl. Phys. A98, 503.
Delves, L.M., and M.A. Hennell, 1971, Nucl. Phys. A168, 347.
Delves, L.M., and A.C. Phillips, 1969, Rev. Mod. Phys. 41, 497.
Delves, L.M., J.M. Blatt, C. Pask and B. Davies, 1969, Phys. Letters 28B, 472.
Der Mateosian, E., and M. Goldhaber, 1966, Phys. Rev. 146, 810.
De Raedt, J., 1968, *Proc. Beta Spectroscopy and Nuclear Structure, Groningen.*
Derenzo, S.E., 1969, Phys. Rev. 181, 1854.
De Sabbata, V., 1961, Nuovo Cimento 21, 659.
De Saintignon, P., and Chabre, 1970, Phys. Letters 33B, 463.
De Saintignon, P., J.J. Lucas, J.B. Viano, M. Chabre and P. Depommier, 1970, Nucl. Phys. A160, 53.
D'Espagnat, B., 1963, Phys. Letters 7, 209.
Desplanques, B., 1972, Phys. Rev. Letters 41B, 461.
Desplanques, B., and N. Vinh Mau, 1971, Phys. Letters 35B, 28.
De Toledo Piza, A.F.R., A.K. Kerman, S. Fallieros and R.H. Venter, 1966, Nucl. Phys. 89, 369.
Deutsch, J.P., and P. Lipnik, 1965, Nucl. Phys. 61, 97.
Deutsch, J.P., L. Grenacs, P. Igo-Kemenes, P. Lipnik and C.P. Macq, 1968a, Phys. Letters 26B, 315.
Deutsch, J.P., L. Grenacs, J. Lehmann, P. Lipnik and C.P. Macq, 1968b, Phys. Letters 28B, 178.
Deutsch, J.P., L. Grenacs, J. Lehmann, P. Lipnik and C.P. Macq, 1969, Phys. Letters 29B, 66.
Devanathan, V., and M.E. Rose, 1967, J. Math. and Phys. Sci. 1, 137.
De Vries, E., and J.E. Jonker, 1968, Nucl. Phys. B6, 213.
De Waard, H., and O.J. Poppema, 1957, Physica 23, 597.
De Wit, P., and C. van der Leun, 1969, Phys. Letters 30B, 639.
Dicus, D.A., and R.E. Norton, 1970, Phys. Rev. D1, 1360.
Diehl, H., G. Hopfensitz, E. Kankeleit and E. Kuphal, 1969, *High Energy Physics and Nuclear Structure*, Ed. S. Devons (Plenum Press, New York–London) p. 722.
Dietrich, F.S., M. Suffert, A.V. Nero and S.S. Hanna, 1968, Phys. Rev. 168, 1169.
Dihella, L., I. Hammerman and L.M. Rosenstein, 1971, Phys. Rev. Letters 27, 830.

Divakaran, P.P., V. Gupta and G. Rajasekaran, 1968, Phys. Rev. 166, 1792.
Dobrokhotov, E.I., V.R. Lazarenko and S.Yu. Luk'yanov, 1959, Sov. Phys. - JETP 9, 54.
Do Dang, G., 1972, Phys. Letters 38B, 397.
Dolinskii, E., and L. Blokhintsev, 1958, Sov. Phys. - JETP 8, 1040.
Dombey, N., and P.K. Kabir, 1966, Phys. Rev. Letters 17, 730.
Donovan, P.F., D.E. Alburger and D.H. Wilkinson, 1961, *Proc. Rutherford Jubilee Conf., Manchester*; Ed. J.B. Birks (Heywood and Co., London) p. 827.
Dorman, G., 1964, Nuovo Cimento 32, 1226.
Downs, B.W., and Y. Nogami, 1967, Nucl. Phys. B2, 459.
Drechsler, W., and B. Stech, 1964, Z. Phys. 178, 1.
Drell, S.D., and K. Huang, 1953, Phys. Rev. 91, 1527.
Drell, S.D., and J.D. Walecka, 1960, Phys. Rev. 120, 1069.
Dress, W.B., J.K. Baird, P.D. Miller and N.F. Ramsey, 1968, Phys. Rev. 170, 1200.
Durand III, L., 1964, Phys. Rev. 135, B310.
Dumitrescu, O., M. Gari, H. Kummel and J.G. Zabolitzky, 1971a,b, Preprints, Bochum.
Dydak, F., H.D. Polashegg, P. Riehs and P. Weinzierl, 1971, Phys. Letters 37B, 375.
Dzhelepov, B.S., and L.N. Zyrianova, 1956, *Influence of atomic electric fields on beta-decay* (Izdatel'stvo Akademica Nauk SSR, Moskva).

Eckhause, M., R. Siegel, R.E. Welsh and T.A. Filippas, 1966, Nucl. Phys. 81, 575.
Eden, R.J., P.V. Landshoff, O.I. Olive and J.C. Polkinghorne, 1966, *The Analytic S-Matrix* (Cambridge University Press).
Ehrman, J.B., 1951, Phys. Rev. 81, 412.
Eichler, J., 1963, Z. Phys. 171, 463.
Eichler, J., 1968, Nucl. Phys. 120, 535.
Eichler, J., 1969, Nucl. Phys. A127, 693.
Eichler, J., Tombrello, T.A. and J.H. Bahcall, 1964, Phys. Letters 13, 146.
Elliott, J.P., and B.H. Flowers, 1957, Proc. Roy. Soc. A242, 57.
Elton, L.R.B., 1967, Phys. Rev. 158, 970.
Eman, B., and D. Tadić, 1971, Phys. Rev. C4, 661.
Eman, B., F. Krmpotić, D. Tadić and A. Nielsen, 1967, Nucl. Phys. A104, 386.
Enz, C.P., 1957, Nuovo Cimento 6, 250.
Eramzhyan, R.A., V.N. Fetisov and Yu.A. Salganic, 1972, Nucl. Phys. B39, 216.
Ericson, T.E.O., 1963, Ann. Phys. 23, 390.
Ericson, T.E.O., 1966, Phys. Letters 23, 97.
Ericson, T., J.C. Sens and H.P.C. Rood, 1964, Nuovo Cimento 34, 51.
Erozolimsky, B.G., L.N. Bondarenko, Yu.A. Mostovoy, B.A. Obinyakov, Z.P. Zacharova and V.A. Titov, 1968, Phys. Letters 27B, 557.
Erozolimsky, B.G., L.N. Bondarenko, Yu.A. Mostovoy, B.A. Obinyakov, V.A. Titov, V.P. Zacharova and A.I. Frank, 1970, Phys. Letters 33B, 351.
Euler, H., 1937, Z. Physik 105, 553.
Evseev, V.S., V.S. Roganov, V.A. Chernogorova, Jun-wa Chang and M. Szymczak, 1967a, Sov. J. Nucl. Phys. 4, 245.
Evseev, V.S., F. Kil'binger, V.S. Roganov, V.A. Chernagorova and M. Szymczak, 1967b, Sov. J. Nucl. Phys. 4, 387.

Fairbairn, W.M., 1961, Proc. Phys. Soc. (London) 77, 599.
Fairbairn, W.M., 1963, Nucl. Phys. 45, 437.
Falomkin, A.I., A.I. Filippov, M.M. Kulyukin, B. Pontecorvo, Yu.A. Scherbakov, R.M. Sulyaev, V.M. Tsupko-Sitnikov and O.A. Zaimidoraga, 1963a, Phys. Letters 3, 229.
Falomkin, A.I., A.I. Filippov, M.M. Kulyukin, B. Pontevorvo, Yu.A. Scherbakov, R.M. Sulyaev, V.M. Tsupko-Sitnikov and O.A. Zaimidoraga, 1963b, Phys. Letters 6, 100.
Fayans, S.A., 1971, Phys. Letters 37B, 155.
Fayans, S.A., and V.A. Khodel, 1969, Phys. Letters 30B, 5.
Fearing, H.W., 1966, Phys. Rev. 146, 723.
Feuer, M., 1969, Ph.D. Thesis, Harvard.
Feynman, R.P., 1949, Phys. Rev. 76, 749.
Feynman, R.P., and M. Gell-Mann, 1958, Phys. Rev. 109, 193.
Fierz, M., 1937, Z. Phys. 104, 553.
Fink, M., M. Gari and J.G. Zabolitzky, 1972, Phys. Letters 38B, 189.
Fischbach, E., 1968, Phys. Rev. 170, 1398.
Fischbach, E., and D. Tadić, 1972, Phys. Reports, to be published.
Fischbach, E., and K. Trabert, 1968, Phys. Rev. 174, 1843.
Fischbach, E., D. Tadić and K. Trabert, 1969, Phys. Rev. 186, 1688.
Fischbach, E., D. Tadić and K. Trabert, 1970, *High Energy Physics and Nuclear Structure*; Ed. S. Devons (Plenum Press, New York–London) p. 742.
Fischbach, E., F. Iachello, A. Lande, M.M. Nieto and C.K. Scott, 1971a, Phys. Rev. Letters 26, 1200.
Fischbach, E., M.M. Nieto, H. Primakoff, C.K. Scott and J. Smith, 1971b, Phys. Rev. Letters 27, 1403.
Fischbach, E., M.M. Nieto and C.K. Scott, 1972a, preprint.
Fischbach, E., E.P. Harper, Y.E. Kim, A. Tubis and W.K. Cheng, 1972b, Phys. Letters 38B, 8.
Flowers, B.H., 1951, Proc. Roy. Soc. A212, 248.
Foldy, L.L., 1953, Phys. Rev. 92, 178.
Foldy, L.L., and R.H. Klein, 1967, Phys. Letters 24B, 540.
Foldy, L.L. and J.D. Walecka, 1964, Nuovo Cimento 34, 1026.
Foldy, L.L., and J.D. Walecka, 1965, Phys. Rev. 140, B1339.
Foldy, L.L., and S.A. Wouthuysen, 1950, Phys. Rev. 78, 29.
Frauenfelder, H., R. Bobone, E. von Goeler, N. Levine, H.R. Lewis, R.N. Peacock, A. Rossi and G. de Pasquali, 1957a, Phys. Rev. 106, 386.
Frauenfelder, H., N. Levine, A.O. Hanson, A. Rossi and G. de Pasquali, 1957b, Phys. Rev. 107, 643.
Frauenfelder, H., and R.M. Steffen, 1965a, *Alpha-, Beta- and Gamma-Ray Spectroscopy*; Ed. K. Siegbahn (North-Holland Publ. Comp., Amsterdam) p. 997.
Frauenfelder, H., and R.M. Steffen, 1965b, *Alpha-, Beta- and Gamma-Ray Spectroscopy*; Ed. K. Siegbahn (North-Holland Publ. Comp., Amsterdam) p. 1431.
Frazier, J., and C.W. Kim, 1969, Phys. Rev. 177, 2568.
Freeman, J.M., J.G. Jenkin and G. Murray, 1966a, Phys. Letters 22, 177.
Freeman, J.M., J.G. Jenkin, G. Murray and W.E. Burcham, 1966b, Phys. Rev. Letters 16, 959.
Freeman, J.M., J.G. Jenkin, D.C. Robinson, G. Murray and W.E. Burcham, 1968, Phys. Letters 27B, 156.
Freeman, J.M., J.G. Jenkin and G. Murray, 1969a, Nucl. Phys. A124, 393.

Freeman, J.M., J.G. Jenkin, G. Murray and W.E. Burcham, 1969b, Nucl. Phys. A132, 593.
Freeman, J.M., D.C. Robinson and G.L. Wick, 1969c, Phys. Letters 29B, 296.
Freeman, J.M., D.C. Robinson and G.L. Wick, 1969d, Phys. Letters 30B, 240.
Friar, J.L., 1966, Nucl. Phys. 87, 407.
Friar, J.L., 1970, Nucl. Phys. 156, 43.
Friedman, E., 1971, Phys. Letters 35B, 543.
Friedman, E., and B. Mandelbaum, 1969, Nucl. Phys. A135, 472.
Fryberger, D., 1968, Phys. Rev. 166, 1379.
Fubini, S., and S. Furlan, 1965, Physics 1, 229.
Fujii, A., 1960, Phys. Rev. 118, 870.
Fujii, A., and H. Primakoff, 1959, Nuovo Cimento 12, 327.
Fujii, A., and Y. Yamaguchi, 1964, Prog. Theor. Phys. 31, 107.
Fujii, A., M. Morita and H. Ohtsubo, 1968, Suppl. to Prog. Theor. Phys. p. 303.
Fujii, Y., and J.I. Fujita, 1965, Phys. Rev. 140, B329.
Fujita, I., M. Kawai and M. Tanifuji, 1962, Nucl. Phys. 29, 252.
Fujita, J.I., 1962a, Phys. Rev. 126, 202.
Fujita, J.I., 1962b, Prog. Theor. Phys. 28, 338.
Fujita, J.I., 1967, Phys. Letters 24B, 123.
Fujita, J.I., and A. Fujii, 1969, Phys. Rev. 185, 1475.
Fujita, J.I., and K. Ikeda, 1965, Nucl. Phys. 67, 145.
Fujita, J.I., and K. Ikeda, 1966a, Prog. Theor. Phys. 36, 288.
Fujita, J.I., and K. Ikeda, 1966b, Prog. Theor. Phys. 36, 530.
Fulton, T., 1958, Nucl. Phys. 6, 319.
Furlan, G., F.G. Lannoy, C. Rossetti and G. Segré, 1965, Nuovo Cimento 38, 1747.
Fuschini, E., V. Gadjokov, C. Maroni and P. Veronesi, 1964a, Nuovo Cimento 33, 709.
Fuschini, E., V. Gadjokov, C. Maroni and P. Veronesi, 1964b, Nuovo Cimento 33, 1309.

Galindo, A., and P. Pascual, 1968, Nucl. Phys. B4, 295.
Galindo, A., and P. Pascual, 1969, Nucl. Phys. B14, 37.
Gammel, J.L., and R.M. Thaler, 1960, *Progress in Elementary Particle and Cosmic Ray Physics* (North-Holland Publ. Comp., Amsterdam) Vol. V., Chapter 2.
Gari, M., 1969, Z. Phys. 231, 412.
Gari, M., 1970, Phys. Letters 31B, 627.
Gari, M., and A.H. Huffman, 1971, Phys. Letters 36B, 442.
Gari, M., and H. Kümmel, 1969, Phys. Rev. Letters 23, 26.
Gari, M., H. Kümmel and J.G. Zabolitzky, 1971a, Nucl. Phys. A161, 625.
Gari, M., O. Dumitrescu, J.G. Zabolitzky and H. Kümmel, 1971b, Phys. Letters 35B, 19.
Garrell, M., H. Frauenfelder, D. Ganek and D.C. Sutton, 1969, Phys. Rev. 187, 1410.
Garvey, G.T., 1969, Ann. Rev. Nucl. Sci. 19, 433.
Garwin, R.L., L.M. Lederman and M. Weinrich, 1957, Phys. Rev. 105, 1415.
Geiger, J.S., G.T. Ewan, R.L. Graham and D.R. MacKenzie, 1958, Phys. Rev. 112, 1684.
Gelbard, E.M., 1955, Phys. Rev. 100, 1530.
Gell-Mann, M., 1953, Phys. Rev. 92, 833.
Gell-Mann, M., 1958, Phys. Rev. 111, 362.
Gell-Mann, M., 1962, Phys. Rev. 125, 1067.
Gell-Mann, M., 1964a, Phys. Letters 8, 214.

Gell-Mann, M., 1964b, Physics 1, 63.
Gell-Mann, M., and S.M. Berman, 1959, Phys. Rev. Letters 3, 99.
Gell-Mann, M., and M. Levy, 1960, Nuovo Cimento 16, 705.
Gell-Mann, M., and Y. Ne'eman, 1964, *The Eightfold Way* (W.A. Benjamin, Inc.).
Gell-Mann, M., and V.L. Telegdi, 1953, Phys. Rev. 91, 169.
Gell-Mann, M., and F. Zachariasen, 1961, Phys. Rev. 124, 953.
Gerhart, J.B., 1958, Phys. Rev. 109, 897.
Gerhart, J.B., F.H. Schmidt, H. Bichsel and J.C. Hopkins, 1959, Phys. Rev. 114, 1095.
Gerling, E.K. Yu.A. Shukolyukov and G.Sh. Ashkimadze, 1968, Sov. J. Nucl. Phys. 6, 226.
Gershtein, S.S., 1958, Sov. Phys. - JETP 7, 318.
Gershtein, S.S., 1959, Sov. Phys. - JETP 9, 927.
Gerstenberger, R.V., and Y. Nogami, 1972, Phys. Rev. Letters 29, 233.
Gibson, B.F., 1965, Phys. Rev. 139, B1153.
Gillet, V., and D.A. Jenkins, 1965, Phys. Rev. 140, B32.
Gillet, V., and N. Vinh Mau, 1964, Nucl. Phys. 54, 321.
Ginsberg, E.S., 1966, Phys. Rev. 142, 1035.
Ginsberg, E.S., 1967, Phys. Rev. 162, 1570.
Girvin, D.C., 1972, Ph.D thesis (Berkeley).
Gittelman, B., and W. Schmidt, 1965, Phys. Rev. 175, 1998.
Glashow, S.L., 1965, Phys. Rev. Letters 14, 35.
Glashow, S.L., and A.N. Mitra, 1967, Phys. Letters 24B, 27.
Glashow, S.L., and S. Weinberg, 1968, Phys. Rev. Letters 20, 224.
Glass, N.W., and R.W. Peterson, 1963, Phys. Rev. 130, 299.
Goldberg, H., 1965, Nuovo Cimento 40, 243.
Goldberg, H., and Y. Ne'eman, 1963, Nuovo Cimento 27, 1.
Goldberger, M.C., and S.B. Treiman, 1958, Phys. Rev. 110, 1178.
Goldhaber, M., L. Grodzins and A.W. Sunyar, 1958, Phys. Rev. 109, 1015.
Goulard, B., G. Goulard and H. Primakoff, 1964, Phys. Rev. 133, B186.
Grabowski, J., 1970, Phys. Letters 33B, 268.
Green, A.M., and M. Rho, 1969, Nucl. Phys. A130, 112.
Green, A.M., and T.H. Schucan, 1972, Nucl. Phys. A188, 289.
Green, A.M., T.K. Dahlblom, A. Kallio and M. Rho, 1970, Phys. Letters 31B, 189.
Greenberg, J.S., D.P. Malone, R.L. Gluckstern and V.W. Hughes, 1960, Phys. Rev. 120, 1393.
Greenlees, G.W., G.J. Pyle and Y.C. Tang, 1968, Phys. Rev. 171, 1115.
Greuling, E., and R.C. Whitten, 1960, Ann. Phys. 11, 510.
Grishin, V.G., V.L. Lyuboshitz, V.I. Ogievetskii and M.I. Podgoretski, 1967, Sov. J. Nucl. Phys. 4, 90.
Grismore, R., 1968, Nucl. Phys. A108, 1.
Gross, E.E., J.J. Malanify, A. Van der Woude and A. Zucker, 1968, Phys. Rev. Letters 21, 1476.
Gruhle, E., K.H. Lauterjung and B. Schimmer, 1963, Nucl. Phys. 42, 321.
Gupta, V.K., and A.N. Mitra, 1967, Phys. Letters 24B, 27.
Gupta, V.K., B.S. Bhakar and N. Mitra, 1965, Phys. Rev. Letters 15, 974.
Gustafson, G., 1969, Nucl. Phys. B11, 213.

Haas, R., L.B. Leipuner and R.K. Adair, 1959, Phys. Rev. 116, 1221.
Haase, E.L., H.A. Hill and D.B. Knudsen, 1963, Phys. Letters 4, 338.
Hadjimichael, E., 1972, Private communication.
Hadjimichael, E., and E. Fischbach, 1971, Phys. Rev. D3, 755.
Hadjimichael, E., and A.D. Jackson, 1972, Nucl. Phys. A180, 217.
Hadjimichael, E., E. Harms and V. Newton, 1971, Phys. Rev. Letters 27, 1322.
Haftel, M.I., and F. Tabakin, 1970, Nucl. Phys. A158, 1.
Haftel, M.I., and F. Tabakin, 1971, Phys. Rev. C3, 921.
Haftel, M.I., E. Lambert and P.U. Sauer, 1972, Nucl. Phys. A192, 225.
Halpern, A., 1964a, Phys. Rev. Letters 13, 660.
Halpern, A., 1964b, Phys. Rev. 135, A34.
Halpern, T.A., 1970, Phys. Rev. C1, 1928.
Halpern, T.A., and B. Chern, 1968, Phys. Rev. 175, 1314.
Hamada, T., and I.D. Johnston, 1962, Nucl. Phys. 34, 382.
Hamilton, D.R., W.P. Alford and L. Gross, 1953, Phys. Rev. 92, 1521.
Hamilton, J., 1967, Nucl. Phys. B1, 449.
Hamilton, W.D., 1968, Prog. Nucl. Phys. 10, 1.
Hamilton, W.D., 1969, *High Energy Physics and Nuclear Structure*, Ed. S. Devons (Plenum Press, New York–London) p. 696.
Handler, R., S.C. Wright, L. Pondrom, P. Limon, S. Olsen and P. Kloeppel, 1967, Phys. Rev. Letters 19, 933.
Hanna, S.S., 1969, *Isospin in Nuclear Physics*, Ed. D.H. Wilkinson (North-Holland Publ. Comp., Amsterdam) p. 591.
Hannon, J.P., and G.T. Trammell, 1968, Phys. Rev. Letters 21, 726.
Hannon, J.P., 1971, Nucl. Phys. A177, 493.
Hardy, J.C., J.E. Esterl, R.G. Sextro and J. Cerny, 1971a, Phys. Rev. C3, 700.
Hardy, J.C., J.M. Loiseaux, J. Cerny and G.T. Garvey, 1971b, to be published.
Hardy, J.C., H. Schmeing, J.S. Geiger, R.L. Graham and I.S. Towner, 1972, Phys. Rev. Letters 29, 1027.
Harper, E.R., Y.E. Kim, A. Tubis and M. Rho, 1972, Phys. Letters 40B, 533.
Harrington, D., 1966, Phys. Rev. 141, 1494.
Harrison, G.E., P.G.H. Sandars and S.J. Wright, 1969, Phys. Rev. Letters 22, 1263.
Hartman, G., 1967, Thesis (MIT).
Hättig, H., K. Hünchen, P. Roth and H. Wäffler, 1969, Nucl. Phys. A137, 144.
Hättig, H., K. Hünchen and H. Wäffler, 1970, Phys. Rev. Letters 25, 941.
Heisenberg, W., 1936, Z. Phys. 101, 533.
Heller, L., P. Signell and N.R. Yodor, 1964, Phys. Rev. Letters 13, 577.
Henley, E.M., 1968, Phys. Letters 28B, 1.
Henley, E.M., 1969a, Ann. Rev. Nucl. Sci. 19, 367.
Henley, E.M., 1969b, *Isospin in Nuclear Physics*, Ed. D.H. Wilkinson (North-Holland Publ. Comp., Amsterdam) p. 15.
Henley, E.M., 1971, Phys. Rev. Letters 27, 542.
Henley, E.M., and A.H. Huffman, 1968, Phys. Rev. Letters 20, 1191.
Henley, E.M., and B.A. Jacobsohn, 1959, Phys. Rev. 113, 225.
Henley, E.M., and B.A. Jacobsohn, 1966, Phys. Rev. Letters 16, 706.
Henley, E.M., and T.E. Keliher, 1972, Nucl. Phys. A189, 632.
Henley, E.M., and C. Lacy, 1969, Phys. Rev. 184, 1228.
Henley, E.M., and L.K. Morrison, 1965, Phys. Rev. 141, 1489.
Henley, E.M., T.E. Keliher and D.U.L. Yu, 1969, Phys. Rev. Letters 23, 941.

Hennel, M.A., and L.M. Delves, 1972, Phys. Letters 40B, 20.
Herczeg, P., 1963, Nucl. Phys. 48, 263.
Herczeg, P., 1966, Nucl. Phys. 75, 655.
Herczeg, P., 1971, Phys. Rev. D4, 1239.
Herczeg, P., 1972, Phys. Rev. D6, 1934.
Hildebrand, R.H., 1962, Phys. Rev. Letters 8, 34.
Hirooka, H., T. Konishi, R. Morita, H. Narumi, M. Soga and M. Morita, 1968, Prog. Theor. Phys. 40, 808.
Hirooka, M., 1969, preprint.
Hocquenghem, J.C., and J. Berthier, 1968, Nucl. Phys. A155, 661.
Hodgson, P.E., 1971, *Nuclear Reactions and Nuclear Structure* (Clarendon Press, Oxford).
Höhler, G., and R. Strauss, 1967, Phys. Letters 24B, 1967.
Holmes, M.J., W.D. Hamilton and R.A. Fox, 1971, Phys. Letters 37B, 170.
Holstein, B.R., 1971a, Phys. Rev. C4, 764.
Holstein, B.R., 1971b, Phys. Rev. C4, 740.
Holstein, B.R., 1972, Phys. Rev. C5, 1947.
Holstein, B.R., and S.B. Treiman, 1971, Phys. Rev. C3, 1921.
Holstein, B.R., W. Shanahan and S.B. Treiman, 1971, to be published.
Hopkins, J.C., J.B. Gerhart, F.H. Schmidt and J.E. Stroth, 1961, Phys. Rev. 121, 1185.
Hrasko, P., 1969, Phys. Letters 28B, 470.
Huang, K., C.N. Yang and T.D. Lee, 1957, Phys. Rev. 108, 1340.
Huffaker, J.N., and E. Greuling, 1962, Trans. N.Y., Acad. Sci. 24, 591.
Huffaker, J.N., and E. Greuling, 1963, Phys. Rev. 132, 738.
Huffaker, J.N., and C.E. Laird, 1967, Nucl. Phys. A92, 584.
Huffman, A.H., 1968, Thesis, University of Washington (unpublished).
Huffman, A.H., 1970a, Phys. Rev. D1, 882.
Huffman, A.H., 1970b, Phys. Rev. D1, 890.
Humblet, J., and G. Lebon, 1967, Nucl. Phys. A96, 593.

Ignatenko, A.E., L.B. Egorov, B. Khalupa and D. Chultem, 1959, Sov. Phys. - JETP 8, 621.
Ingber, L., 1968, Phys. Rev. 174, 1250.
Ioffe, B., 1958, Sov. Phys. - JETP 6, 240.

Jackson, J.D., and J.M. Blatt, 1950, Rev. Mod. Phys. 22, 77.
Jackson, J.D., S.B. Treiman and H.W. Wyld, Jr., 1957a, Phys. Rev. 106, 517.
Jackson, J.D., S.B. Treiman and H.W. Wyld, Jr., 1957b, Nucl. Phys. 4, 206.
Jacobsohn, B.A., and E.M. Henley, 1959, Phys. Rev. 113, 234.
Jänecke, J., 1969a, Nucl. Phys. A128, 632.
Jänecke, J., 1969b, *Isospin in Nuclear Physics*, Ed. D.H. Wilkinson (North-Holland Publ. Comp., Amsterdam) p. 297.
Jarlskog, C., 1966, Nucl. Phys. 75, 659.
Jaus, W., 1968, Nucl. Phys. B8, 408.
Jaus, W., 1971a, Nucl. Phys. A162, 97.
Jaus, W., 1971b, Nucl. Phys. A177, 70.
Jaus, W., 1972, Phys. Letters 40B, 616.

Jaus, W., and G. Rasche, 1970, Nucl. Phys. A143, 202.
Jenschke, B., and P. Bock, 1970, Phys. Letters 31B, 65.
Johnson, K., F.E. Low and H. Suura, 1967, Phys. Rev. Letters 18, 1224.
Jones, C.M., J.L.C. Ford, Jr. and F.E. Obenshain, 1970, Phys. Rev. Abs. 1, No. 22, 15.
Jones, P.B., 1968, Phys. Rev. Letters 21, 1553.
Kabir, P.K., 1966, Z. Phys. 191, 447.
Kajfosz, J., J. Kopecky and J. Honzatko, 1966, Phys. Letters 20, 284.
Kajfosz, J., J. Kopecky and J. Honzatko, 1968, Nucl. Phys. A120, 225.
Kahana, S., 1972, Phys. Rev. C5, 63.
Kahana, S., and D.L. Pursey, 1957, Nuovo Cimento 6, 1469.
Källen, G., 1964, *Elementary Particle Physics* (Addison–Wesley).
Källen, G., 1967, Nucl. Phys. B1, 225.
Kallio, A., B.D. Day, 1969, Nucl. Phys. A124, 177.
Kankeleit, E., 1964, *Compt. Rend. du Congress Intern. de Phys. Nucl. Paris* 5/C293.
Karpman, G., R. Leonardi and F. Strocci, 1968, Phys. Rev. 174, 1957.
Kaufmann, W., and H. Wäffler, 1961, Nucl. Phys. 24, 62.
Kelly, P.S., and S.A. Moszkowski, 1960, Phys. 158, 304.
Kemmer, N., 1939, Proc. Roy. Soc. A173, 91.
Kerman, A.K., and L.S. Kisslinger, 1969, Phys. Rev. 180, 1483.
Khadkikar, S.B., and C.S. Warke, 1969, Nucl. Phys. A130, 577.
Khanna, F.C., 1967, Nucl. Phys. A97, 417.
Khodel, V.A., 1970, Phys. Letters 32B, 583.
Khrylin, B.A., 1968, Sov. J. of Nucl. Phys. 6, 691.
Kim, C.W., 1966, Phys. Rev. 146, 691.
Kim, C.W., 1971, Phys. Letters 34B, 383.
Kim, C.W. and T. Fulton, 1971, Phys. Rev. C4, 390.
Kim, C.W., and S.L. Mintz, 1970, Phys. Letters 31B, 503.
Kim, C.W., and S.L. Mintz, 1971, Nucl. Phys. B27, 621.
Kim, C.W., and H. Primakoff, 1965a, Phys. Rev. 139, B1447.
Kim, C.W., and H. Primakoff, 1965b, Phys. Rev. 140, B566.
Kim, C.W., and H. Primakoff, 1966, Phys. Rev. 147, 1034.
Kim, C.W., and H. Primakoff, 1969, Phys. Rev. 180, 1500.
Kim, C.W., and M. Ram, 1967, Phys. Rev. Letters 18, 327.
Kim, Y.E., and J.O. Rasmussen, 1963, Nucl. Phys. 47, 184.
Kinoshita, T., and A. Sirlin, 1959, Phys. Rev. 113, 1652.
Kirsten, T., W. Gentner and O.A. Schaeffer, 1967, Z. Physik 202, 273.
Kirsten, T., O.A. Schaeffer, E. Norton and R.W. Stoenner, 1968, Phys. Rev. Letters 20, 1300.
Kisslinger, L.S., 1969, Phys. Letters 29B, 211.
Kistner, O.C., 1967, Phys. Rev. Letters 19, 872.
Klein, A., 1953, Phys. Rev. 90, 1101.
Klein, R., 1966, Phys. Rev. 146, 756.
Klein, R., and L. Wolfenstein, 1962, Phys. Rev. Letters 9, 408.
Klein, R., T. Neal and L. Wolfenstein, 1965, Phys. Rev. 138, B86.
Kok, L.P., G. Erens and R. Van Wageningen, 1968, Nucl. Phys. A122, 684.
Konijn, J., B. van Nooijen, H.L. Hagedoorn and A.H. Wapstra, 1958, Nucl. Phys. 9, 296.
Konopinski, E.J., 1966, *The Theory of Beta Radioactivity* (Oxford).
Krane, K.S., C.E. Olsen, J.R. Sites and W.A. Steyert, 1971a, Phys. Rev. Letters 26, 1579.
Krane, K.S., C.E. Olsen, J.R. Sites and W.A. Steyert, 1971b, Phys. Rev. C4, 1906.

Krane, K.S., J.R. Sites and W.A. Steyert, 1972, Phys. Rev. C5, 1104.
Krmpotić, F., and D. Tadić, 1966, Phys. Letters 21, 680.
Krmpotić, F., and D. Tadić, 1969, Phys. Rev. 178, 1804.
Kroll, N.M., T.D. Lee and B. Zumino, 1967, Phys. Rev. 157, 1376.
Krueger, D.A., and A. Goldberg, 1964, Phys. Rev. 135, B934.
Krüger, J., and P. Van Leuven, 1969, Phys. Letters 28B, 623.
Krüger, L., 1959, Z. Physik 157, 369.
Krupchitsky, P.A., and G.A. Lobov, 1969, *Atomic Energy Review VII*, 91.
Kuan, H.M., D.W. Heikkinen, K.A. Snover, F. Riess and S.S. Hanna, 1967, Phys. Letters 25B, 217.
Kuebbing, R.A., and K.J. Casper, 1969, Nucl. Phys. A130, 672.
Kuphal, E., 1972, Z. Phys. 253, 314.
Kurath, D., 1960, Phys. Rev. Letters 4, 180.
Kuston, R.L., D.E. Lundquist, T.B. Novey, A. Yokosawa and F. Chilton, 1969, Phys. Rev. Letters 22, 1014.

Lacaze, R., 1968, Nucl. Phys. B4, 657.
Lane, A.M., amd J.M. Soper, 1962a, Nucl. Phys. 37, 506.
Lane, A.M., and J.M. Soper, 1962b, Nucl. Phys. 37, 663.
Landau, L., 1967, Nucl. Phys. 3, 127.
Langer, L.M., and R.J.D. Moffat, 1952, Phys. Rev. 88, 689.
Laverne, A., and G. DoDang, 1971, Nucl. Phys. A177, 665.
Lazarenko, V.R., 1967, Sov. Phys. Uspekhi, 9, 860.
Lazarenko, V.R., and S.Yu. Luk'yanov, 1966, Sov. Phys. - JETP 22, 521.
Lederman, L.M., and B.G. Pope, 1971, Phys. Rev. Letters 27, 765.
Ledingham, K.W.D., J.Y. Gourlay, J.L. Campbell, M.L. Fitzpatrick, J.G. Lynch and J. McDonald, 1971, Nucl. Phys. A170, 663.
Lee, T.D., 1957, *Proc. Rehovoth Conf. Nucl. Struct.*, Ed. H.J. Lipkin (North-Holland Publ. Comp.) p. 336.
Lee, T.D., 1962, Phys. Rev. 128, 899.
Lee, T.D., 1965a, Phys. Rev. 140, B959.
Lee, T.D., 1965b, Phys. Rev. 140, B967.
Lee, T.D., 1968a, Phys. Rev. 171, 1731.
Lee, T.D., 1968b, *Lecture given at the Symposium for Joint Dedication of the Centre for Theoretical Physics and the Centre for Advanced Visual Studies, Mass. Inst. of Technology* (unpublished).
Lee, T.D., 1971, Phys. Rev. Letters 26, 801.
Lee, T.D., and L. Wolfenstein, 1965, Phys. Rev. 138, B1490.
Lee, T.D., and C.S. Wu, 1965, Ann. Rev. Nucl. Sci. 15, 381.
Lee, T.D., and C.N. Yang, 1956, Phys. Rev. 104, 254.
Lee, T.D., and C.N. Yang, 1957, Phys. Rev. 105, 1671.
Lee, T.D., and C.N. Yang, 1962, Phys. Rev. 126, 2239.
Lee, T.D., S. Weinberg and B. Zumino, 1967, Phys. Rev. Letters 18, 1029.
Lee, Y.K., L. Mo and C.S. Wu, 1963, Phys. Rev. Letters 10, 253.
Leung, J.S., and Y. Nogami, 1968, Nucl. Phys. B7, 527.
Leutz, H., 1961, Z. Phys. 164, 78.
Leutz, H., and H. Wenninger, 1967, Nucl. Phys. A99, 55.

Levinger, J.S., and B.K. Srivastava, 1965, Phys. Rev. 137, B426.
Lewis, V.E., 1970, Nucl. Phys. A151, 120.
Lin, D.L., 1964, Nucl. Phys. 60, 192.
Lipkin, H.J., 1971a, Phys. Letters 34B, 202.
Lipkin, H.J., 1971b, Phys. Rev. Letters 27, 432.
Lipnik, P., and J.W. Sunier, 1964, Nucl. Phys. 53, 305.
Lipnik, P., J.P. Deutsch, L. Granacs and P.C. Macq, 1962, Nucl. Phys. 30, 312.
Lipson, E.D., F. Boehm and J.C. Vanderleeden, 1971a, Phys. Letters 35B, 307.
Lipson, E.D., F. Boehm and J.C. Vanderleeden, 1971b, to be published.
Lloyd, S.P., 1951, Phys. Rev. 81, 161.
Lobashov, V.M., and V.A. Nazarenko, 1962, Sov. Phys. - JETP 15, 257.
Lobashov, V.M., V.A. Nazarenko, L.F. Saenko, L.M. Smotritskii and G.I. Kharevich, 1966, Sov. Phys. - JETP Letters 3, 173.
Lobashov, V.M., V.A. Nazarenko, L.F. Saenko, L.M. Smotritskii and G.I. Kharevich, 1967, Sov. Phys. - JETP Letters 5, 59.
Lobashov, V.M., N.A. Lozovoy, V.A. Nazarenko, L.M. Smotritskii and G.I. Kharevich, 1969, Phys. Letters 30B, 39.
Lobashov, V.M., A.E. Egorov, D.M. Kaninker, V.A. Nazarenko, L.F. Saenko, L.M. Smotritskii, G.I. Kharevich and V.A. Knyaz'kov, 1970, Sov. Phys. - JETP Letters 11, 76.
Lobashov, V.M., D.M. Kaminker, G.I. Kharkevich, V.A. Kniazkov, N.A. Lozovoy, V.A. Nazarenko, L.F. Sayenko, L.M. Smotritsky and A.I. Yegorov, 1972, Nucl. Phys. A197, 241.
Lobov, G.A., 1963, Nucl. Phys. 43, 430.
Lobov, G.A., 1965, Sov. Phys. - JETP Letters 1, 157.
Lobov, G.A., and J.S. Shapiro, 1963, Sov. Phys. - JETP 16, 1286.
Lodder, A., and C.C. Jonker, 1965, Phys. Letters 15, 245.
Logue, L.J., and B. Chern, 1968, Phys. Rev. 175, 1367.
Loiseau, B.A., and Y. Nogami, 1967, Nucl. Phys. B2, 470.
Loiseau, B.A., Y. Nogami and C.K. Ross, 1971, Nucl. Phys. A165, 601.
Longo, M., 1969, Bull. Am. Phys. Soc. 14, 598.
Longuemare, C., and C.A. Piketty, 1972, Phys. Letters 38B, 125.
Lovitch, L., 1963, Nucl. Phys. 46, 353.
Lovitch, L., 1964, Nucl. Phys. 53, 477.
Lovitch, L., 1965, Nucl. Phys. 62, 653.
Lovitch, L., and S. Rosati, 1967, Nucl. Phys. B1, 369.
Lubkin, E., 1960, Ann. of Phys. (N.Y.) 11, 414.
Lüders, G., 1954, Kgl. Danske, Videnskab. Selskab, Mat. Fys. Medd. 28, No. 5.
Lüders, G., 1957, Ann. of Phys. (N.Y.) 3, 1.
Lüders, G., 1958, Nuovo Cimento 7, 171.
Lundby, A., A.P. Patro and J.P. Stroot, 1957, Nuovo Cimento 6, 745.
Luyten, J.R., and H.A. Tolhoek, 1965, Nucl. Phys. 70, 641.
Luyten, J.R., H.P.C. Rood and H.A. Tolhoek, 1963, Nucl. Phys. 41, 236.

MacDonald, W.M., 1955, Phys. Rev. 100, 51.
MacDonald, W.M., 1958, Phys. Rev. 110, 1420.
Mafethe, M.E., and P.E. Hodgson, 1966, Proc. Phys. Soc. (London) 87, 429.

Mahaux, C., and H.A. Weidenmüller, 1966, Phys. Letters 23, 100.
Maiani, L., 1968, Phys. Letters 26B, 538.
Maier, E.J., R.M. Edelstein and R.T. Siegel, 1964, Phys. Rev. 133, B663
Majorana, E., 1937, Nuovo Cimento 14, 171.
Manacher, G.K., 1961, Carnegie Inst. of Tech. Report NYO 9284.
Manacher, G.K., and L. Wolfenstein, 1959, Phys. Rev. 116, 782.
Mang, H.J., 1964, Ann. Rev. Nucl. Sci. 14, 1.
Mann, L.G., and S.D. Bloom, 1965, Phys. Rev. 139, B540.
Mann, L.G., S.D. Bloom and R.J. Nagle, 1962, Nucl. Phys. 30, 636.
Mann, L.G., D.C. Camp, J.A. Miskel and R.J. Nagle, 1965, Phys. Rev. 137, B1.
Mann, L.G., K.G. Tirsell and S.D. Bloom, 1967, Nucl. Phys. A97, 425.
Mannque, R.H.O., 1967, Phys. Rev. Letters 18, 671.
Maqueda, E., 1966, Phys. Letters 23, 571.
Maqueda, E., and R.J. Blin-Stoyle, 1966, Nucl. Phys. A91, 460.
Marshak, R.E., Riazuddin and C.P. Ryan, 1968, *Theory of Weak Interactions in Particle Physics* (Wiley–Interscience).
Matese, J., and W.R. Johnson, 1966, Phys. Rev. 150, 846.
Matsson, L., 1969, Nucl. Phys. B12, 647.
Mayer-Kuckuk, T., and F.C. Michel, 1962, Phys. Rev. 127, 545.
McDonald, A.B., E.G. Adelberger, H.B. Mak, D. Ashery, A.P. Shukla, C.L. Cocke and C.N. Davids, 1970, Phys. Letters 31B, 119.
McGrath, R.L., J.C. Hardy and J. Cerny, 1968, Phys. Letters 27B, 443.
McGrath, R.L., J. Cerny, J.C. Hardy, G. Goth and A. Arima, 1970, Phys. Rev. C1, 184.
McKellar, B.H.J., 1967, Phys. Letters 26B, 107.
McKellar, B.H.J., 1968a, Phys. Rev. Letters 20, 1542.
McKellar, B.H.J., 1968b, Phys. Rev. Letters 21, 1822.
McKellar, B.H.J., 1969a, Phys. Rev. 178, 2160.
McKellar, B.H.J., 1969b, Phys. Rev. Letters 21, 1822.
McKellar, B.H.J., 1970, *High Energy Physics and Nuclear Structure*, Ed. S. Devons (Plenum Press, New York–London) p. 682.
McKellar, B.H.J., 1972, Phys. Letters 38B, 401.
McKellar, B.H.J., and P. Pick, 1971, preprint.
McKellar, B.H.J., and P. Pick, 1972, preprint.
McKellar, B.H.J., and R. Rajaraman, 1968, Phys. Rev. 21, 450.
McKellar, B.H.J., and R. Rajaraman, 1971, Phys. Rev. C3, 1877.
Mendelson, R., G.J. Wozniak, A.D. Bacher, J.M. Loiseaux and J. Cerny, 1970, Phys. Rev. Letters 25, 533.
Michel, F.C., 1964, Phys. Rev. 133, B329.
Migdal, A.B., 1964, Nucl. Phys. 57, 29.
Migdal, A.B., 1966, *Proc. Intern. School of Physics "Enrico Fermi"* (Academic Press).
Migdal, A.B., 1967, *Theory of finite Fermi-systems* (John Wiley, New York).
Migdal, A.B., and A.I. Larkin, 1964, Nucl. Phys. 51, 561.
Miller, P.D., W.B. Dress, J.K. Baird and N.F. Ramsey, 1967, Phys. Rev. Letters 19, 381.
Miller, P.D., W.B. Dress, J.K. Baird and N.F. Ramsey, 1968, *Proc. 14th Intern. Conf. on High Energy Physics, Vienna* p. 281 (as reported by J. Cronin).
Miyazawa, M., 1964, J. Phys. Soc. Japan 19, 1764.
Moldauer, P.A., 1968a, Phys. Rev. 165, 1136.
Moldauer, P.A., 1968b, Phys. Letters 26B, 713.

Moline, A., J. Morris, P. Dyer and C.A. Barnes, 1970, to be published.
Moravcsik, M.J., 1964, Phys. Rev. 136, 624.
Moravcsik, M.J., 1967, Rev. Mod. Phys. 39, 670.
Morita, M., 1959, Phys. Rev. 113, 1584.
Morita, M., 1963, Suppl. Progr. Theor. Phys. 26, 1.
Morita, M., and A. Fujii, 1960, Phys. Rev. 118, 606.
Morita, M., and R.S. Morita, 1957, Phys. Rev. 107, 1316.
Morita, M., and R.S. Morita, 1958, Phys. Rev. 110, 461.
Morita, M., and R. Morita, 1964, J. Phys. Soc. Japan 19, 1759.
Morita, M., R.S. Morita and M. Yamada, 1958, Phys. Rev. 111, 237.
Morita, M., R. Morita and T. Shirafuji, 1965, Suppl. Progr. Theor. Phys. (Commemoration issue) p. 96.
Morpurgo, G., 1958, Phys. Rev. 110, 721.
Morpurgo, G., 1959, Phys. Rev. 114, 1075.
Morrison, L.K., 1967, Thesis, University of Washington.
Morrison, L.K., 1968, Ann. of Phys. (N.Y.) 50, 6.
Mosher, J.M., R.W. Kavanagh and T.A. Tombrello, 1971, Phys. Rev. C3, 438.
Moskalev, A.N., 1968, Sov. J. Nucl. Phys. 8, 672.
Moskalev, A.N., 1969, Sov. J. Nucl. Phys. 9, 163.
Moszkowski, S.A., 1963, Phys. Rev. 129, 1901.
Moszkowski, S.A., 1965, *Alpha- Beta- and Gammay-ray Spectroscopy*, Ed. K. Siegbahn (North-Holland Publ. Comp., Amsterdam) p. 863.
Mukhopadhyay, N.C., and M.H. Macfarlane, 1971, Phys. Rev. Letters 27, 1823.
Murthy, A.S.V., and M.K. Ramaswamy, 1964, Ind. Journ. Pure and App. Phys. 2, 101.
Murthy, M., 1969, J. Phys. A (Gen. Phys.) 2, 672.

Nagel, B., and H. Snellman, 1971, Phys. Rev. Letters 27, 761.
Nakano, T., and K. Nishijima, 1953, Prog. Theor. Phys. 10, 581.
Nambu, Y., and M. Yoshimura, 1970, Phys. Rev. Letters 24, 25.
Negele, J.W., 1971, Nucl. Phys. A165, 305.
Nemirovskii, P.E., 1967, Sov. J. Nucl. Phys. 4, 334.
Nestor, C.W., K.T.R. Davies, S.J. Krieger and M. Baranger, 1968, Nucl. Phys. A113, 14.
Newby, Jr., N., and E.J. Konopinski, 1959, Phys. Rev. 115, 434.
Newton, R.G., 1960, J. Math. Phys. 1, 319.
Nilsson, S.G., 1955, Kgl. Danske Videnskab. Selskab. Mat. Fys. Medd. 29, No. 16.
Nishijima, K., and L.J. Schwank, 1966, Phys. Rev. 146, 1161.
Nishijima, K., and L.J. Schwank, 1967a, Nucl. Phys. B3, 553.
Nishijima, K., and L.J. Schwank, 1967b, Nucl. Phys. B3, 565.
Nolen, Jr., J.A., J.P. Schiffer, N. Williams and D. von Ehrenstein, 1967, Phys. Rev. Letters 18, 1140.
Nolen, Jr., J.A., and J.P. Schiffer, 1969a, Ann. Rev. Nucl. Sci. 19, 471.
Nolen, Jr., J.A., and J.P. Schiffer, 1969b, Phys. Letters 29B, 396.
Nordberg, M.E., F.B. Moringo and C.A. Barnes, 1960, Phys. Rev. Letters 5, 321.
Nordberg, M.E., F.B. Moringo and C.A. Barnes, 1962, Phys. Rev. 125, 321.
Noyes, H.P., 1965, Nucl. Phys. 74, 508.
Noyes, H.P., and H.M. Lipinski, 1971, Phys. Rev. C4, 995.
Nunberg, P., D. Properi and E. Pace, 1972, Phys. Letters 40B, 529.

Nyman, E.M., 1967, Nucl. Phys. B1, 535.
Nyman, E.M., 1970, Nucl. Phys. B17, 599.

Oakes, R.J., 1964, Phys. Rev. 136, B1848.
Oakes, R.J., 1968, Phys. Rev. Letters 20, 1539.
Ohtsubo, H., 1966, Phys. Letters 22, 480.
Ohtsubo, H., and A. Fujii, 1966, Nuovo Cimento 42, 109.
Okamoto, K., 1964, Phys. Letters 11, 150.
Okamoto, K., 1965, Prog. Theor. Phys. 34, 326.
Okamoto, K., 1966, *Isobaric spin in nuclear physics*, Eds. J.D. Fox and D. Robson (Academic Press, New York) p. 569.
Okamoto, K., and C. Lucas, 1967a, Nuovo Cimento 48A, 233.
Okamoto, K., and C. Lucas, 1967b, Nucl. Phys. B2, 347.
Okamoto, K., and C. Lucas, 1968, Phys. Letters 26B, 188.
Okamoto, K., and C. Pask, 1971a, Phys. Letters 36B, 317.
Okamoto, K., and C. Pask, 1971b, Ann. of Phys. (N.Y.) 68, 18.
Okubo, S., 1968a, Ann. Phys. 49, 219.
Okubo, S., 1968b, Nuovo Cimento 54A, 491.
Okun, L.B. *Weak Interactions of Elementary Particles* (Pergamon).
Oksen, P., and J.A. Rao, 1969, Phys. Letters 29B, 233.
Opat, G.I., 1964, Phys. Rev. 134, B428.
Osborne, R.K., and L.L. Foldy, 1950, Phys. Rev. 79, 795.
Oziewicz, Z., and A. Pikulski, 1967, Acta Physica Polonica 32, 873.

Padgett, D.W., W.M. Frank and J.G. Brennan, 1965a, Nucl. Phys. 73, 424.
Padgett, D.W., W.M. Frank and J.G. Brennan, 1965b, Nucl. Phys. 73, 445.
Palathingal, J.C., 1970, Phys. Rev. Letters 24, 524.
Pappademos, J.N., 1963, Nucl. Phys. 42, 122.
Pappademos, J.N., 1964, Nucl. Phys. 56, 351.
Particle Data Group, 1972, Phys. Letters 39B, 1.
Partovi, F., 1964, Ann. of Phys. (N.Y.) 27, 114.
Pascual, P., 1969, Preprint.
Pask, C., 1967, Phys. Letters 25B, 78.
Pauli, W., 1931, *Handbuch der Physik* (Springer Verlag) Vol. 24/1, p. 226.
Pauli, W., 1955, *Niels Bohr and the Development of Physics* (Pergamon).
Pauli, W., 1957, Nuovo Cimento 6, 204.
Paul, H., 1970, Nucl. Phys. A154, 160.
Peachey, S., 1969, D. Phil. Thesis (University of Sussex).
Perey, F.G., and J.P. Schiffer, 1966, Phys. Rev. Letters 17, 324.
Perkins, D.H., 1969, *Proc. Topical Conf. on Weak Interaction, CERN, Geneva.*
Perkins, D.H., and E.T. Ritter, 1968, Phys. Rev. 174, 1426.
Peterson, E.A., 1968, Phys. Rev. 167, 971.
Peterson, G.A., 1967, Phys. Letters 25B, 549.
Pignon, D., 1971, Phys. Letters 35B, 163.
Piketty, C.A., and J. Procureur, 1971, Nucl. Phys. B26, 390.
Pontecorvo, B., 1968, Phys. Letters 26B, 630.
Porter, F.T., 1959, Phys. Rev. 115, 450.

Postma, H., W.J. Huiskamp. A.R. Miedema, M.J. Steenland, H.A. Tolhoek and C.J. Gorter, 1957, Physica 23, 259.

Povel, H.P., 1968, Thesis, Karlsruhe.

Pratt, Jr., W.P., R.I. Schermer, J.R. Sites and W.A. Steyert, 1970, Phys. Rev. C2, 1499.

Prentki, J., and M. Veltman, 1965, Phys. Letters 15, 88.

Preston, M.A., 1962, *Physics of the Nucleus* (Addison Wesley).

Preston, M.A., and R.K. Bhaduri, 1963, Phys. Letters 6, 193.

Primakoff, H., 1952, Phys. Rev. 85, 888.

Primakoff, H., 1959, Rev. Mod. Phys. 31, 802.

Primakoff, H., 1963, *Lecture Notes on Weak Interactions and Topics in Dispersion Physics from the Second Bergen International School of Physics* (Benjamin).

Primakoff, H., 1964, *Lecture Notes of International School of Physics "Enrico Fermi" Varenna (Como) Italy* (Academic Press).

Primakoff, H., and S.P. Rosen, 1959, Rep. Prog. Phys. 22, 121.

Primakoff, H., and S.P. Rosen, 1961, Proc. Phys. Soc. (London) 78, 464.

Primakoff, H., and S.P. Rosen, 1969, Phys. Rev. 184, 1925.

Primakoff, H., and D.H. Sharp, 1969, Phys. Rev. Letters 23, 501.

Pursey, D.L., 1951, Phil. Mag. 42, 1193.

Pursey, D.L., 1957, Nuovo Cimento 6, 266.

Quaranta, A.A., A. Bertin, G. Matone, F. Palmonari, A. Placci, P. Dalpiaz, G. Torelli and E. Zavattini, 1967, Nuovo Cimento 47B, 72.

Quaranta, A.A., A. Bertin, G. Matone, F. Palmonari, G. Torelli, P. Dalpiaz, A. Placci and E. Zavattini, 1969, Phys. Rev. 177, 2118.

Radicati, L.A., 1952, Phys. Rev. 87, 521.

Rajaraman, R., and H.A. Bethe, 1967, Rev. Mod. Phys. 39, 745.

Ram, M., 1970, Phys. Rev. Letters 25, 391.

Ramaswamy, M.K., 1959, Ind. J. Phys. 33, 285.

Rapahel, R., H. Überall and C. Werntz, 1967, Phys. Letters 24B, 15.

Reid, R.V., 1968, Ann. of Phys. (N.Y.) 50, 411.

Reines, F., and C.L. Cowan, 1959, Phys. Rev. 113, 272.

Reitz, J.R., 1950, Phys. Rev. 77, 10.

Rho, M., 1965, Phys. Letters 16, 161.

Rho, M., 1967a, *Muon-Capture in Nuclei and Migdal Theory* (Lectures delivered at the Summer Institute of Nuclear and Particle Physics, McGill University, Montreal, Canada).

Rho, M., 1967b, Phys. Rev. Letters 18, 671.

Rho, M., 1967c, Phys. Letters 24B, 81.

Rho, M., 1967d, Phys. Rev. Letters 19, 248.

Rho, M., 1969, *Third Intern. Conf. on High Energy Physics and Nuclear Structure, Columbia University.*

Rho, M., 1970, *Lecture given at the Colloquium on Interaction of Elementary Particles with Nuclei, Valencia, Spain.*

Riazuddin, 1958, Nucl. Phys. 7, 217 (erratum 10, 96).

Riess, F., W.J. O'Connell, D.W. Heikkinen, H.M. Kuan and S.S. Hanna, 1967, Phys. Rev. Letters 19, 367.

Riess, F., P. Paul, J.B. Thomas and S.S. Hanna, 1968, Phys. Rev. 176, 1140.
Riska, D.O., and G.E. Brown, 1970, Phys. Letters 32B, 662.
Riska, D.O., and G.E. Brown, 1972, Phys. Letters 38B, 193.
Robinson, S.W., C.P. Swann and V.K. Rasmussen, 1968, Phys. Letters B26, 298.
Robinson, D.C., J.M. Freeman and T.T. Thwaites, 1972, Nucl. Phys. A181, 645.
Robson, D., 1968, Phys. Letters 26B, 117.
Roman, P., 1964, *Theory of Elementary Particles* (North-Holland Publ. Comp., Amsterdam).
Rood, H.P.C., and H.A. Tolhoek, 1963, Phys. Letters 6, 121.
Rood, H.P.C., and H.A. Tolhoek, 1965, Nucl. Phys. 70, 658.
Rood, H.P.C., and A.F. Yano, 1971, Phys. Letters 35B, 59.
Roos, M., and A. Sirlin, 1971, Nucl. Phys. B29, 296.
Rosati, S., and M. Barbi, 1966, Phys. Rev. 147, 730.
Rose, H.J., and D.M. Brink, 1967, Rev. Mod. Phys. 39, 306.
Rose, M.E., 1965, *Multipole Fields* (John Wiley and Sons Inc., New York).
Rose, M.E., and R.K. Osborn, 1954, Phys. Rev. 93, 1315.
Rosen, S.P., 1957, D. Phil. Thesis, Oxford.
Rosen, S.P, 1959, Proc. Phys. Soc. (London) 74, 350.
Rosen, S.P., 1971, to be published.
Rosen, S.P., and H. Primakoff, 1966, *Alpha-, Beta-, and Gamma-Ray Spectroscopy*, Ed. K. Siegbahn (North-Holland Publ. Comp.) p. 1499.
Ross, M., 1952, Phys. Rev. 88, 935.
Rothberg, J.E., E.W. Anderson, E.J. Bleser, L.M. Lederman, S.L. Meyer, J.L. Rosen and I.T. Wang, 1963, Phys. Rev. 132, 2664.
Russek, A., and L. Spruch, 1952, Phys. Rev. 87, 1111.

Sachs, R.G., 1953, *Nuclear Theory* (Addison–Wesley).
Sachs, R.G., and N. Austern, 1951, Phys. Rev. 81, 705.
Salam, A., 1957, Nuovo Cimento 5, 299.
Salam, A., and J.C. Ward, 1959, Nuovo Cimento 11, 568.
Salgo, R.C., and H.H. Staub, 1969, Nucl. Phys. A138, 417.
Salpeter, E.E., 1953, Phys. Rev. 91, 994.
Samaranayake, V.K., and G. Wilk, 1972, Preprint.
Sanda, A.I., and G. Shaw, 1970, Phys. Rev. Letters 24, 131.
Sanda, A.I., and G. Shaw, 1971a, Phys. Rev. D3, 243.
Sanda, A.I., and G. Shaw, 1971b, Phys. Rev. Letters 26, 1057.
Sandars, P.G.H., 1965, Phys. Letters 14, 194.
Sandars, P.G.H., 1966, Phys. Letters 22, 290.
Sandars, P.G.H., 1967, Phys. Rev. Letters 19, 1396.
Satchler, G.R., 1958, Nucl. Phys. 8, 65.
Schneider, R.E., and R.M. Thaler, 1965, Phys. Rev. 137, B874.
Schiffer, J.P., J.A. Nolen and N. Williams, 1969, Phys. Letters 29B, 399.
Schopper, H., 1957, Phil. Mag. 2, 710.
Schopper, H., 1966, *Weak Interactions and Nuclear Beta Decay* (North-Holland Publ. Comp., Amsterdam).
Schopper, H., and H. Müller, 1959, Nuovo Cimento 13, 1026.
Schopper, H., H. Müller, W. Jüngst, J. Görres, H. Behrens and H. Appel, 1964, Proc. *Intern. Conf. on Nuclear Physics, Paris*, Vol. II, p. 1201.

Schrock, B.L., R.P. Haddock, J.A. Helland, M.J. Longo, S.S. Wilson, K.K. Young, D. Cheng and V. Perez-Mendez, 1971, Phys. Rev. Letters 26, 1659.
Schülke, L., 1964, Z. Phys. 179, 331.
Schülke, L., 1972, Nucl. Phys. B40, 386.
Schweber, S.S., 1964, *An Introduction to Relativistic Quantum Field Theory* (Harper and Row, John Weatherill, N.Y. and Tokyo).
Schwela, D., 1971, Nucl. Phys. B26, 525.
Schwinger, J., 1950, Phys. Rev. 78, 135.
Schwinger, J., 1953, Phys. Rev. 91, 713.
Schwinger, J., 1957, Ann. of Phys. (N.Y.) 2, 407.
Schwinger, J., 1959, Phys. Rev. Letters 3, 296.
Schwinger, J., 1967, Phys. Rev. Letters 19, 1154.
Scobie, J., and G.M. Lewis, 1957, Phil. Mag. 2, 1089.
Segel, R.E., J.V. Kane and D.H. Wilkinson, 1958, Phil. Mag. 3, 204.
Segel, R.E., J.W. Olness and E.L. Sprenkel, 1961a, Phil. Mag. 6, 163.
Segel, R.E., J.W. Olness and E.L. Sprenkel, 1961b, Phys. Rev. 123, 1332.
Segré, G., 1968, Phys. Rev. 173, 1730.
Sengupta, S., 1961, Nucl. Phys. 21, 542.
Sengupta, S., 1962, Nucl. Phys. 30, 300.
Shaffer, R.A., 1962, Phys. Rev. 128, 1452.
Shaffer, R.A., 1963, Phys. Rev. 131, 2203.
Shann, R.T., 1971, Nuovo Cimento 5A, 591.
Shapiro, I.S., E.I. Dolinsky and L. Blokhintsev, 1957, Nucl. Phys. 4, 273.
Shapiro, M.H., S. Frankel, S. Koiski, W.D. Wales and G.T. Wood, 1967, Phys. Rev. 154, 1050.
Shaw, G., 1967, Nucl. Phys. B3, 338.
Sherr, R., B.F. Bayman, E. Rost, M.E. Rickey and C.G. Hoot, 1965, Phys. Rev. 139, B1272.
Shirokov, Yu.M., 1959, Sov. Phys. - JETP 9, 333.
Shrum, E.Y., and K.O.H. Ziock, 1971, Phys. Letters 37B, 114.
Shull, C.G., and R. Nathans, 1967, Phys. Rev. Letters 19, 384.
Siegbahn, K., 1965, *Alpha-, Beta-, and Gamma-Ray Spectroscopy* (North-Holland Publ. Company, Amsterdam).
Siegert, A.F.J., 1937, Phys. Rev. 52, 787.
Siemens, P.J., 1970, Nucl. Phys. A141, 225.
Signell, P., 1969, Adv. Nucl. Phys. 2, 223.
Silbar, R.R., 1964, Phys. Rev. 134, B542.
Simms, P.C., 1965, Phys. Rev. 138, B784.
Simonius, M., 1972, Phys. Letters 37B, 446.
Sirlin, A., 1967a, Phys. Rev. Letters 19, 877.
Sirlin, A., 1967b, Phys. Rev. 164, 1767.
Sirlin, A., 1968a, *Lecture given at the VII Internationale Universitäts Wochen für Kernphysik Schladming, Austria.*
Sirlin, A., 1968b, *Proc. 14th Intern. Conf. on High Energy Phys., Vienna*, p. 321.
Sirlin, A., 1969, *Proc. Topical Conf. on Weak Interactions, CERN, Geneva.*
Sirlin, A., 1972, Phys. Rev. 5D, 436.
Smith, J.H., E.M. Purcell, N.F. Ramsey, 1957, Phys. Rev. 108, 120.
Snover, K.A., F. Riess and S.S. Hanna, 1968, Private communication.

Snover, K.A., D.W. Heikkinen, F. Riess, H.M. Kuan and S.S. Hanna, 1969, Phys. Rev. Letters 22, 239.
Sober, D.I., D.G. Cassel, A.J. Sadoff, K.W. Chen and P.A. Crean, 1969, Phys. Rev. Letters 22, 430.
Sodemann, J., and A. Winther, 1965, Nucl. Phys. 69, 369.
Soergel, Rieselberg and Strocka, 1965, *Proc. Conf. on Beta-decay, Heidelberg.*
Soper, J.M., 1969, *Isospin in Nuclear Physics*, Ed. D.H. Wilkinson (North-Holland Publ. Comp., Amsterdam) p. 229.
Sosnovskii, A.N., P.E. Spivak, Yu.A. Prokof'ev, I.E. Kutinov and Yu.P. Dobrynin, 1959, Sov. Phys. - JETP 8, 739.
Sowerby, B.D., and G.J. McCallum, 1968, Nucl. Phys. A112, 453.
Spector, R.M., 1963a, Nucl. Phys. 39, 464.
Spector, R.M., 1963b, Nucl. Phys. 40, 338.
Spector, R.M., and R.J. Blin-Stoyle, 1962, Phys. Letters 1, 118.
Sprenkel-Segel, E.L., R.B. Segel and R.H. Siemsen, 1970, *High Energy Physics and Nuclear Structure*, Ed. S. Devons (Plenum Press, New York–London) p. 763.
Sprung, D.W.L., P.K. Banerjee, A.M. Jopko and M.K. Srivastava, 1970, Nucl. Phys. A144, 245.
Srivastava, B.K., 1965, Nucl. Phys. 67, 236.
Srivastava, P.P., 1968, CERN Report No. Th. 893.
Stech, B., and L. Schülke, 1964, Z. Phys. 179, 314.
Steffen, R.M., 1969, Phys. Rev. 115, 980.
Stevens, M., J. St., 1965, Phys. Letters 19, 499.
Stichel, P., 1958, Z. Phys. 150, 264.
Strubbe, H.J., and D.K. Callebaut, 1970, Nucl. Phys. A143, 537.
Sugawara, H., 1965a, Phys. Rev. Letters 15, 986.
Sugawara, H., 1965b, Phys. Rev. Letters 15, 997.
Sundelin, R.M., R.M. Edelstein, A. Suzuki and K. Takahashi, 1968a, Phys. Rev. Letters 20, 1198.
Sundelin, R.M., R.M. Edelstein, A. Suzuki and K. Takahashi, 1968b, Phys. Rev. Letters 20, 1201.
Suslov, Yu.P., 1967, Sov. J. Nucl. Phys. 4, 854.
Suzuki, M., 1965, Phys. Rev. Letters 15, 986.
Swamy, N.V.V.J., and A.E.S. Green, 1958, Phys. Rev. 112, 1719.
Szymanski, Z., 1966, Nucl. Phys. 76, 539.
Szymanski, Z., 1968, Nucl. Phys. A113, 385.

Tabakin, F., and K.T.R. Davies, 1966, Phys. Rev. 150, 793.
Tadić, D., 1968, Phys. Rev. 174, 1694.
Takaoka, N., and G. Ogata, 1966, Z. Naturforsch. 21A, 84.
Tang, Y.C., and R.C. Herndon, 1965, Phys. Letters 18, 42.
Tanner, N., 1957, Phys. Rev. 107, 1203.
Tarburton, R.M., and K.T.R. Davies, 1968, Nucl. Phys. A120, 1.
Taylor, B.N., W.H. Parker and D.N. Langenberg, 1969, Rev. Mod. Phys. 41, 375.
Taylor, J.C., 1958, Phys. Rev. 110, 1216.
Tee, C.S., and C.T. Yap, 1971a, Nucl. Phys. A170, 445.
Tee, C.S., and C.T. Yap, 1971b, Nucl. Phys. A174, 221.

Telegdi, V.L., 1960, *Proc. Ann. Intern. Conf. on High Energy Physics, Rochester*, p. 713.
Telegdi, V.L., 1963, *Weak Interactions*, Ed. Frandal (Benjamin, Inc., New York) pp. 16 and 26.
Thomas, R.G., 1951, Phys. Rev. 81, 148.
Thomas, R.G., 1952, Phys. Rev. 88, 1109.
Thornton, S.T., C.M. Jones, J.K. Bair, M.D. Mancus and H.B. Willard, 1968, Phys. Rev. Letters 21, 447.
Thornton, S.T., C.M. Jones, J.K. Bair, M.D. Mancus and H.B. Willard, 1971, Phys. Rev. C3, 1065.
Tolhoek, H.A., 1959, Nucl. Phys. 10, 606.
Tolhoek, H.A., 1963, *Selected Topics in Nuclear Theory* (Intern. Atomic Energy Agency, Vienna) p. 343.
Tolhoek, H.A., and J.R. Luyten, 1957, Nucl. Phys. 3, 679.
Touschek, B.Z., 1957, Nuovo Cimento 5, 754.
Towner, I.S., and J.C. Hardy, 1972, to be published.
Trainor, L.E.H., 1952, Phys. Rev. 85, 962.
Treiman, S.B., 1958, Phys. Rev. 110, 448.
Trentelman, G.F., and I.D. Proctor, 1971, Phys. Letters 35B, 570.
Trentelman, G.F., B.M. Preedom and E. Kashy, 1970, Phys. Rev. Letters 25, 530.

Überall, H., 1957, Nuovo Cimento 6, 593.
Ullman, J.D., H. Frauenfelder, H.J. Lipkin and A. Rossi, 1961, Phys. Rev. 122, 536.
Ullman, J.D., 1962, Phys. Letters 1, 339.

Vanderleeden, J.C., and F. Boehm, 1959, Phys. Letters 30B, 467.
Vanderleeden, J.C., and F. Boehm, 1970, Phys. Rev. C2, 748.
Vanderleeden, J.C., F. Boehm and E.D. Lipson, 1971, Phys. Rev. C4, 2218.
Van Giai Nguygen, D. Vautheriu and M. Veneroni, 1971, Phys. Letters 35B, 135.
Van Neste, L., R. Coussement and J.P. Deutsch, 1966, Phys. Letters 23, 122.
Van Neste, L., R. Coussement and J.P. Deutsch, 1967, Nucl. Phys. A98, 585.
Van Rooijen, J.J., P. Pronk, S.U. Ottevangers and J. Blok, 1967, Physica 37, 32.
Vatai, E., 1971, Phys. Letters 34B, 395.
Vatai, E., D. Varga and J. Uchrin, 1968, Nucl. Phys. A116, 637.
Veltman, M., 1967, Phys. Letters 24B, 587.
Verde, M., 1950, Helv. Phys. Acta 23, 453.
Villard, F., 1947, Phys. Rev. 72, 256.
Vinh Mau, N., and A.M. Bruneau, 1969, Phys. Letters 29B, 408.
Vishnevsky, M.E., V.K. Grigoriev, V.A. Ergakov, S.J. Nikitin, E.V. Pushkin and Yu.V. Trebukhovsky, 1957, Nucl. Phys. 4, 271.
Vogel, P., 1971, Bull. Am. Phys. Soc. 16, 603.
Von Wimmersperg, U., G. Kernel, B.W. Allardyce, W.M. Mason and N.W. Tanner, 1970, Phys. Letters 33B, 291.
Von Witsch, W., A. Richter and P. von Brentano, 1966, Phys. Letters 22, 631.
Von Witsch, W., A. Richter and P. von Brentano, 1967, Phys. Rev. Letters 19, 524.
Von Witsch, W., A. Richter and P. von Brentano, 1968, Phys. Rev. 169, 923.

Wäffler, H., 1970, Private communication to M. Gari.

Wahlborn, S., 1965, Phys. Rev. 138, B530.

Walker, G.E., 1967, Phys. Rev. 157, 845.

Walker, G.E., 1968, Phys. Rev. 174, 1290.

Wambach, U.M., M. Gari and H. Kümmel, 1970, Phys. Letters 33B, 253.

Wang, I.T., 1965, Phys. Rev. 139, B1544.

Warburton, E.K., 1966, *Isobaric Spin in Nuclear Physics*, Eds. J.D. Fox and D. Robson (Academic Press Inc.) p. 90.

Warburton, E.K., and J. Weneser, 1969, *Isospin in Nuclear Physics*, Ed. D.H. Wilkinson (North-Holland Publ. Comp., Amsterdam) p. 173.

Warming, E., 1969, Phys. Letters 29B, 564.

Warming, E., F. Stecher-Rasmussen, W. Ratynski and J. Kopechy, 1967, Phys. Letters 25B, 200.

Weber, G., 1967, *Proc. 1967 Intern. symp. on Electron and Photon Interactions at High Energy, Stanford* (SLAC, Stanford, Calif.).

Weidenmüller, H.A., 1960, Phys. Rev. Letters 4, 299.

Weidenmüller, H.A., 1962, Phys. Rev. 112, 1375.

Weidenmüller, H.A., 1965, Nucl. Phys. 69, 113.

Weinberg, S., 1958, Phys. Rev. 112, 1375.

Weinberg, S., 1966, Phys. Rev. Letters 17, 617.

Weinberg, S., 1967, Phys. Rev. Letters 18, 188.

Weinberg, S., 1971, Phys. Rev. Letters 27, 1688.

Weinberg, S., and S.B. Treiman, 1959, Phys. Rev. 116, 465.

Weisberger, W.I., 1965, Phys. Rev. Letters 14, 1047.

Weisberger, W.I., 1966, Phys. Rev. 143, 1302.

Weissenberg, A.O., 1967, *Muons* (North-Holland Publ. Comp., Amsterdam).

Weisskopf, M.C., J.P. Carrico, H. Gould, E. Lipworth and T.S. Stein, 1968, Phys. Rev. Letters 21, 1645.

Weitkamp, W.G., D.W. Storm, D.C. Shreve, W.J. Braithwaite and D. Bondansky, 1968, Phys. Rev. 165, 1233.

Weizsäcker, von C.F., 1935, Z. Phys. 96, 431.

Wentzel, G., 1953, Phys. Rev. 91, 1573.

Werntz, C., 1960, Nucl. Phys. 16, 59.

Wessell, W.R., and P. Phillipson, 1964, Phys. Rev. Letters 13, 23.

Weyl, H., 1929, Z. Phys. 56, 330.

Wheeler, J.H., 1949, Rev. Mod. Phys. 21, 133.

Wheeler, J.H., and J. Tiomno, 1949, Rev. Mod. Phys. 21, 153.

Wigner, E.P., 1957, *Proc. Robert A. Welch Found. Conf. on Chemical Research, Houston, Texas,* Ed. W.O. Milligan (The Robert A. Welch Foundation, Houston, Texas) Vol. 1, p. 88.

Wigner, E., 1932, Nachr. Adad. Wiss. Göttingen, II Math. - Phys. Kl. 31, 546.

Wilkinson, D.H., 1956a, Phil. Mag. 1, 379.

Wilkinson, D.H., 1956b, Phil. Mag. 1, 1032.

Wilkinson, D.H., 1958a, Phys. Rev. 109, 1603.

Wilkinson, D.H., 1958b, *Proc. Rehovoth Conf. on Nuclear Structure*, Ed. H.J. Lipkin (North-Holland Publ. Comp.).

Wilkinson, D.H., 1960, *Nuclear Spectroscopy*, Part B, Ed. F. Ajzenberg – Selove (Academic Press Inc.) p. 852.

Wilkinson, D.H., 1964, Phys. Rev. Letters 13, 571.
Wilkinson, D.H., 1966, *Isobaric Spin in Nuclear Physics,* Eds. J.D. Fox and D. Robson (Academic Press, New York–London) p. 30.
Wilkinson, D.H., 1970, Phys. Letters 31B, 447.
Wilkinson, D.H., 1971a, Phys. Rev. Letters 27, 1018.
Wilkinson, D.H., 1971b, Proc. Roy. Soc. (Edin) A70, 307.
Wilkinson, D.H., 1971c, Nucl. Phys. A178, 65.
Wilkinson, D.H., 1972, Nucl. Phys. A179, 289.
Wilkinson, D.H., and D.E. Alburger, 1970, Phys. Rev. Letters 24, 1134.
Wilkinson, D.H., and D.E. Alburger, 1971, Phys. Rev. Letters 26, 1127.
Wilkinson, D.H., and W.D. Hay, 1966, Phys. Letters 21, 80.
Wilkinson, D.H., and B.E.F. Macefield, 1970, Nucl. Phys. A158, 110.
Wilkinson, D.H., and M.E. Mafethe, 1966, Nucl. Phys. 85, 97.
Wilkinson, D.H., D.E. Alburger, D.R. Goosman, K.W. Jones, E.K. Warburton, G.T. Garvey and R.L. Williams, 1971, Nucl. Phys. A166, 661.
Williams, A., 1964, Nucl. Phys. 52, 324.
Williams, A., 1968, Nucl. Phys. A117, 238.
Wolfenstein, L., 1958, Nuovo Cimento 7, 706.
Wolfenstein, L., 1962, *Proc. Intern. Conf. on High Energy Physics, CERN, Geneva*, p. 821.
Wolfenstein, L., 1964, Phys. Rev. Letters 12, 562.
Wolfenstein, L., and J. Ashkin, 1952, Phys. Rev. 85, 947.
Wolfenstein, L., and E.M. Henley, 1971, Phys. Letters 36B, 28.
Wong, C.W., 1964, Nucl. Phys. 56, 213.
Wong, C.W., 1965, Nucl. Phys. 71, 385.
Wong, C.W., 1970, Nucl. Phys. A151, 323.
Wu, C.S., and S.A. Moszkowski, 1966, *Beta-decay* (Interscience Publishers).
Wu, C.S., E. Ambler, R.W. Hayward, D.D. Hoppes and R.P. Hudson, 1957, Phys. Rev. 105, 1413.
Wycech, S., 1969, Nucl. Phys. B14, 133.

Yalçin, C., 1971, Nucl. Phys. A169, 201.
Yalçin, C., 1972, Nucl. Phys. A183, 203.
Yalçin, C., and N. Birsen, 1970, Phys. Letters 33B, 203.
Yalçin, C., and C.T. Yap, 1970, Nucl. Phys. A153, 424.
Yamaguchi, Y., 1954, Phys. Rev. 95, 1628.
Yano, A.F., 1964, Phys. Rev. Letters 12, 110.
Yano, A.F., F.B. Yano and H.P.C. Rood, 1971, Phys. Letters 37B, 189.
Yap, C.T., 1967, Nucl. Phys. A100, 619.
Yap, C.T., 1968, Nucl. Phys. B5, 369.
Yap, C.T., and C.S. Tee, 1971a, Nucl. Phys. A165, 497.
Yap, C.T., and C.S. Tee, 1971b, Nucl. Phys. A169, 609.
Yip, P.C.-Y., Y. Nogami and C.K. Ross, 1971, Nucl. Phys. A176, 505.
Yovonovich, M.L., and V.S. Evseev, 1963, Phys. Letters 6, 333.
Yukawa, H., 1935, Proc. Phys. Math. Soc. (Japan) 17, 48.

Zachariasen, F., and G. Zweig, 1965, Phys. Rev. Letters 14, 794.
Zel'dovich, Y.B., and S.S. Gershtein, 1961, Sov. Phys. - Uspekki 3, 593.
Zelvinskii, V.G., 1967, Sov. J. Nucl. Phys. 4, 733.
Zweig, G., 1964, CERN reports 8182/TH, 401 and 8419/TH, 412, unpublished.

INDEX